T0337030

Organic and Nanocomposite Optical Materials

MATERIALS RESEARCH SOCIETY
SYMPOSIUM PROCEEDINGS VOLUME 846

Organic and Nanocomposite Optical Materials

Symposium held November 28–December 3, 2004, Boston, Massachusetts, U.S.A.

EDITORS:

Alexander Cartwright
State University of New York
Buffalo, New York, U.S.A.

Thomas M. Cooper
Air Force Research Laboratory
Dayton, Ohio, U.S.A.

Shashi P. Karna
Army Research Laboratory
Aberdeen, Maryland, U.S.A.

Hachiro Nakanishi
Tohoku University
Sendai, Japan

Materials Research Society
Warrendale, Pennsylvania

CAMBRIDGE
UNIVERSITY PRESS

University Printing House, Cambridge CB2 8BS, United Kingdom

One Liberty Plaza, 20th Floor, New York, NY 10006, USA

477 Williamstown Road, Port Melbourne, VIC 3207, Australia

314-321, 3rd Floor, Plot 3, Splendor Forum, Jasola District Centre, New Delhi - 110025, India

79 Anson Road, #06-04/06, Singapore 079906

Cambridge University Press is part of the University of Cambridge.

It furthers the University's mission by disseminating knowledge in the pursuit of education, learning and research at the highest international levels of excellence.

www.cambridge.org
Information on this title: www.cambridge.org/9781558997943

Materials Research Society
506 Keystone Drive, Warrendale, PA 15086
http://www.mrs.org

First published 2005
First paperback edition 2013

Single article reprints from this publication are available through University Microfilms Inc., 300 North Zeeb Road, Ann Arbor, MI 48106

CODEN: MRSPDH

A catalogue record for this publication is available from the British Library

ISBN 978-1-558-99794-3 Hardback
ISBN 978-1-107-40903-3 Paperback

CONTENTS

Preface xiii

Materials Research Society Symposium Proceedings xiv

NONLINEAR OPTICAL
PROPERTIES MATERIALS

Spectral Profiles of Two-Photon Absorption: Coherent Versus Two-Step Two-Photon Absorption 3
S. Polyutov, I. Minkov, F. Gel'mukhanov, K. Kamada, A. Baev, and H. Ågren

Theoretical Study on Open-Shell Nonlinear Optical Systems 13
Masayoshi Nakano, Ryohei Kishi, Nozomi Nakagawa, Tomoshige Nitta, Takashi Kubo, Kazuhiro Nakasuji, Kenji Kamada, Koji Ohta, Benoît Champagne, Edith Botek, Satoru Yamada, and Kizashi Yamaguchi

NONLINEAR OPTICAL MATERIALS

Optical Responses of Conjugated Polymers by TDDFT in Real-Space and Real-Time Approach 27
Nobuhiko Akino and Yasunari Zempo

Writing Laser Power Dependence of Second Harmonic Generation in an Optically Poled Azo-Dye/Polymer Film 33
C.H. Wang, Oliver Y.-H. Tai, and Yuxiao Wang

Strong Enhancement of the Two Photon Absorption Cross Section of Porphyrin J-Aggregates in Water 39
Elisabetta Collini, Camilla Ferrante, and Renato Bozio

ORGANOMETALLIC OPTICAL
MATERIALS I

Structure Property Characterization of Nonlinear Optical Materials 47
Joy E. Rogers, Jonathan E. Slagle, Daniel G. McLean, Benjamin C. Hall, Thomas M. Cooper, and Paul A. Fleitz

Invited Paper

Platinum-Functionalized Chiral Molecular Squares as Light-Emitting Materials 53
Lin Zhang, Yu-Hua Niu, Alex K.-Y. Jen, and Wenbin Lin

Optical and Electronic Properties of Metal-Containing Poly-ynes and Their Organic Precursors 59
Marek Jura, Olivia F. Koentjoro, Paul R. Raithby, Emma L. Sharp, and Paul J. Wilson

ORGANOMETALLIC OPTICAL MATERIALS II

Novel Iridium Complexes With Polymer Side-Chains 67
Elisabeth Holder, Veronica Marin, Emine Tekin, Dmitry Kozodaev, Michael A.R. Meier, Bas G.G. Lohmeijer, and Ulrich S. Schubert

Phosphorescence Quantum Efficiency and Intermolecular Interaction of Iridium(III) Complexes in Co-Deposited Films With Organic Semiconducting Hosts 73
Yuichiro Kawamura, Kenichi Goushi, Jason Brooks, Julie J. Brown, Hiroyuki Sasabe, and Chihaya Adachi

PLASMONICS

Percolation-Enhanced Supercontinuum and Second-Harmonic Generation From Metal Nanoshells 81
Charles Rohde, Keisuke Hasegawa, Aiqing Chen, and Miriam Deutsch

Linear and Nonlinear Transmission of Surface Plasmon Polaritons in an Optical Nanowire 87
N.C. Panoiu and R.M. Osgood Jr.

Coherent Oscillations of Breathing Modes in Metal Nanoshells 93
Arman S. Kirakosyan and Tigran V. Shahbazyan

Surface Plasmon Excitations and Emission Light Due to Molecular Luminescence on Metal Thin Film 99
Futao Kaneko, Susumu Toyoshima, Yasuo Ohdaira, Kazunari Shinbo, and Keizo Kato

ELECTRO-OPTIC AND ELECTRONIC MATERIALS

Functionalized Guanidines for Electro-Optic Materials 107
Nicholas Buker, Kimberly A. Firestone, Marnie Haller, Lafe Purvis, David Lao, Robert Snoeberger, Alex K.-Y. Jen, and Larry R. Dalton

Highly Ordered Pseudo-Discotic Chromophore Systems for Electro-Optic Materials and Devices 115
Nishant Bhatambrekar, Scott Hammond, Jessica Sinness, Olivier Clot, Harry Rommel, Antao Chen, Bruce Robinson, Alex K.-Y. Jen, and Larry Dalton

Synthesis of Dendridic NLO Chromophores for the Improvement of Order in Electro-Optics 121
Jessica Sinness, Olivier Clot, Scott R. Hammond, Nishant Bhatambrekar, Harrison L. Rommel, Bruce Robinson, Alex K.-Y. Jen, and Larry Dalton

Optical Switching Devices Using Redox Polymer Nanosheet 127
Jun Matsui, Kenichi Abe, Masaya Mitsuishi, Atsushi Aoki, and Tokuji Miyashita

Size-Dependence of the Linear and Nonlinear Optical Properties of GaN Nanoclusters 133
Andrew C. Pineda and Shashi P. Karna

Current-Voltage Characteristics in Organic Semiconductor Crystals: Space Charge vs. Contact-Limited Carrier Transport 141
J. Reynaert, V.I. Arkhipov, J. Genoe, G. Borghs, and P. Heremans

Photoinduced Memory Effects Based on Poly(3-hexylthiophene) and Al Interface 151
Keiichi Kaneto, Mitsuru Ujimoto, and Wataru Takashima

ORGANIC AND HYBRID LIGHT EMITTING DEVICES

Large Area Microcontact Printing Presses for Plastic Electronics 159
Hee Hyun Lee, Etienne Menard, Nancy G. Tassi, John A. Rogers, and Graciela B. Blanchet

Invited Paper

New Light Emitting Polymers and High Energy Hosts for Triplet Emission 165
Chris S.K. Mak, Scott E. Watkins, Charlotte K. Williams, Nicholas R. Evans, Khai Leok Chan, Sung Yong Cho, Andrew B. Holmes, Clare E. Boothby, Richard H. Friend, Anna Hayer, and Anna Köhler

Highly Luminescent Composite Films From Core-Shell Oxide Nanocrystals 171
Valérie Buissette, Mélanie Moreau, Thierry Gacoin, Thierry Le Mercier, and Jean-Pierre Boilot

Influence of Halides on the Luminescence of Oxide/Anthracene/Polymer Nanocomposites 177
Dorothée V. Szabó, Heike Reuter, Sabine Schlabach, Christoph Lellig, and Dieter Vollath

Ambipolar Injection in a Submicron Channel Light-Emitting Tetracene Transistor With Distinct Source and Drain Contacts 183
J. Reynaert, D. Cheyns, D. Janssen, V.I. Arkhipov, G. Borghs, J. Genoe, and P. Heremans

NANOCOMPOSITE OPTICAL MATERIALS

Self-Assembled Oligonucleotide Semiconductor Conjugated to GaN Nanostructures for Biophotonic Applications 193
A. Neogi, J. Li, A. Sarkar, P.B. Neogi, B. Gorman, T. Golding, and H. Morkoc

Supernatant Controlled Synthesis of Monodispersed Zinc Sulfide Spheres and Multimers 201
Yanning Song and Chekesha M. Liddell

Organic/Inorganic Hybrid Silicate Materials for Optical Applications; Highly Fluorinated Hybrid Glasses Doped With (Erbium-Ions/CdSe Nanoparticles) for Laser Amplifier Material 207
Kyung M. Choi and John A. Rogers

Microscopic Theory of Surface-Enhanced Raman Scattering in Noble-Metal Nanoparticles 213
Vitaliy N. Pustovit and Tigran V. Shahbazyan

Photoelectrochemical Behaviors of Pt/TiO_2 Nanocomposite Thin Films Electrodes Prepared by PLD/Sputtering Combined System 223
Takeshi Sasaki, William T. Nichols, Jong-Won Yoon, and Naoto Koshizaki

POSTER SESSION: ORGANOMETALLIC OPTICAL MATERIALS

Luminescence Properties of Eu^{3+}:Y_2O_3 and Eu^{3+}:Lu_2O_3 Nanoparticles, Ceramics and Thin Films 231
Kai Zhang, D. Hunter, S. Mohanty, J.B. Dadson, Y. Barnakov, and A.K. Pradhan

Layered Double Hydroxides as a Matrix for Luminescent Rare Earth Complexes 237
Natalia G. Zhuravleva, Andrei A. Eliseev, Alexey V. Lukashin, Ulrich Kynast, and Yuri D. Tretyakov

Novel and Efficient Electroluminescent Organo-Iridium Phosphorescent OLED Materials 243
Heh-Lung Huang, Kou-Hui Shen, Miao-Cai Jhu, Mei-Rurng Tseng, and Jia-Ming Liu

POSTER SESSION: LINEAR AND NONLINEAR OPTICAL PROPERTIES OF ORGANIC AND NANOCOMPOSITE MATERIALS

Fabrication of Micro-Optical Devices by Holographic Interference of High Photosensitive Inorganic-Organic Hybrid Materials (Photo-HYBIRMER) 251
Dong Jun Kang, Jin-Ki Kim, and Byeong-Soo Bae

Fabrication of Polydiacetylene Nanocrystals Deposited With Silver Nanoparticles for a Nonlinear Optical Material 257
Tsunenobu Onodera, Hidetoshi Oikawa, Hitoshi Kasai, Hachiro Nakanishi, and Takashi Sekiguchi

Size-Effect on Fluorescence Spectrum of Perylene Nanocrystal Studied by Single-Particle Microspectroscopy Coupled With Atomic Force Microscope Observation 263
Hideki Matsune, Tsuyoshi Asahi, Hiroshi Masuhara, Hitoshi Kasai, and Hachiro Nakanishi

Synthesis and Spectroellipsometric Characterization of Y_2O_3-Stabilized ZrO_2-Au Nanocomposite Films for Smart Sensor Applications 269
George Sirinakis, Richard Sun, Rezina Siddique, Harry Efstathiadis, Michael A. Carpenter, and Alain E. Kaloyeros

Synthesis of Hydrophilic Two-Photon Absorptive Fullerene-diphenylaminoflourene Dyads for Molecular Self-Assembly in Water 275
Sarika Verma, Tanya Hauck, Prashant A. Padmawar, Taizoon Canteenwala, Long Y. Chiang, and Kenneth P.H. Pritzker

Effects of Deposition Angle on the Optical Properties of Helically Structured Films 281
Jason B. Sorge, Andy C. van Popta, Jeremy C. Sit, and Michael J. Brett

POSTER SESSION: ORGANIC AND HYBRID LIGHT EMITTING DEVICES: ELECTRICAL AND OPTICAL PROPERTIES

Fast-Switching, High-Contrast Electrochromic Thin Films Prepared Using Layer-by-Layer Assembly of Charged Species 289
Jaime C. Grunlan

A Novel Light-Emitting Mixed-Ligand Iridium(III) Complex With a Polymeric Terpyridine-PEG Macroligand: Synthesis and Characterization 295
Elisabeth Holder, Veronica Marin, Michael A.R. Meier, Dmitry Kozodaev, and Ulrich S. Schubert

Degradation of $Ru(bpy)_3^{2+}$-Based OLEDs 301
Velda Goldberg, Michael Kaplan, Leonard Soltzberg, Joseph Genevich, Rebecca Berry, Alma Bukhari, Sherina Chan, Megan Damour, Leigh Frigugliettí, Erica Gunn, Karen Ho, Ashley Johnson, Yin Yin Lin, Alisabet Lowenthal, Seiyam Suth, Regina To, Regina Yopak, Jason D. Slinker, George G. Malliaras, Samuel Flores-Torres, and Hector D. Abruña

An Efficient Top-Emitting Electroluminescent Device on Metal-Laminated Plastic Substrate 307
L.W. Tan, X.T. Hao, K.S. Ong, Y.Q. Li, and F.R. Zhu

ORGANIC PHOTONIC BANDGAP STRUCTURES

Complex, 3D Photonic Crystals Fabricated by Atomic Layer Deposition 315
J.S. King, D. Gaillot, T. Yamashita, C. Neff, E. Graugnard, and C.J. Summers

Prototyping of Three-Dimensional Photonic Crystal Structures Using Electron-Beam Lithography 321
G. Subramania and J.M. Rivera

Two-Dimensional Magneto-Photonic Crystal Circulators 327
Zheng Wang and Shanhui Fan

Optical Properties of Polystyrene Opals Infiltrated With Cyanine Dyes in the Form of J-Aggregates 333
F. Marabelli, D. Comoretto, D. Bajoni, M. Galli, and L. Fornasari

Author Index 339

Subject Index 343

🎖 *Ribbon Award Winner*

🏆 *Trophy Award Winner*

PREFACE

Symposium DD, "Organic and Nanocomposite Optical Materials," was held November 28–December 3 at the 2004 MRS Fall Meeting in Boston, Massachusetts. The 121 papers were presented in twelve sessions including three poster sessions.

The presentations during this six-day symposium began with a tutorial on organic nanophotonics presented by Alexander Cartwright and Paras Prasad. The symposium showcased the merging of traditional organic and inorganic optical materials science into new disciplines. Materials synthesis ranged from chromophore development to photonic bandgap structures. The materials classes discussed included nonlinear optical materials, organometallics and nanocomposites. The technological applications reflected the broad range of materials, including light-emitting devices, plasmonics, nonlinear optics, electro-optics and photonic bandgap structures. We would like to thank all of the participants for their contributions and hope the papers presented here stimulate further discussion and research.

Alexander Cartwright
Thomas M. Cooper
Shashi P. Karna
Hachiro Nakanishi

April 2005

MATERIALS RESEARCH SOCIETY SYMPOSIUM PROCEEDINGS

Volume 807— Scientific Basis for Nuclear Waste Management XXVII, V.M. Oversby, L.O. Werme, 2004, ISBN: 1-55899-752-0
Volume 808— Amorphous and Nanocrystalline Silicon Science and Technology—2004, R. Biswas, G. Ganguly, E. Schiff, R. Carius, M. Kondo, 2004, ISBN: 1-55899-758-X
Volume 809— High-Mobility Group-IV Materials and Devices, M. Caymax, E. Kasper, S. Zaima, K. Rim, P.F.P. Fichtner, 2004, ISBN: 1-55899-759-8
Volume 810— Silicon Front-End Junction Formation—Physics and Technology, P. Pichler, A. Claverie, R. Lindsay, M. Orlowski, W. Windl, 2004, ISBN: 1-55899-760-1
Volume 811— Integration of Advanced Micro- and Nanoelectronic Devices—Critical Issues and Solutions, J. Morais, D. Kumar, M. Houssa, R.K. Singh, D. Landheer, R. Ramesh, R. Wallace, S. Guha, H. Koinuma, 2004, ISBN: 1-55899-761-X
Volume 812— Materials, Technology and Reliability for Advanced Interconnects and Low-k Dielectrics—2004, R. Carter, C. Hau-Riege, G. Kloster, T-M. Lu, S. Schulz, 2004, ISBN: 1-55899-762-8
Volume 813— Hydrogen in Semiconductors, N.H. Nickel, M.D. McCluskey, S. Zhang, 2004, ISBN: 1-55899-763-6
Volume 814— Flexible Electronics 2004—Materials and Device Technology, B.R. Chalamala, B.E. Gnade, N. Fruehauf, J. Jang, 2004, ISBN: 1-55899-764-4
Volume 815— Silicon Carbide 2004—Materials, Processing and Devices, M. Dudley, P. Gouma, P.G. Neudeck, T. Kimoto, S.E. Saddow, 2004, ISBN: 1-55899-765-2
Volume 816— Advances in Chemical-Mechanical Polishing, D. Boning, J.W. Bartha, G. Shinn, I. Vos, A. Philipossian, 2004, ISBN: 1-55899-766-0
Volume 817— New Materials for Microphotonics, J.H. Shin, M. Brongersma, F. Priolo, C. Buchal, 2004, ISBN: 1-55899-767-9
Volume 818— Nanoparticles and Nanowire Building Blocks—Synthesis, Processing, Characterization and Theory, O. Glembocki, C. Hunt, C. Murray, G. Galli, 2004, ISBN: 1-55899-768-7
Volume 819— Interfacial Engineering for Optimized Properties III, C.A. Schuh, M. Kumar, V. Randle, C.B. Carter, 2004, ISBN: 1-55899-769-5
Volume 820— Nanoengineered Assemblies and Advanced Micro/Nanosystems, J.T. Borenstein, P. Grodzinski, L.P. Lee, J. Liu, Z. Wang, D. McIlroy, L. Merhari, J.B. Pendry, D.P. Taylor, 2004, ISBN: 1-55899-770-9
Volume 821— Nanoscale Materials and Modeling—Relations Among Processing, Microstructure and Mechanical Properties, P.M. Anderson, T. Foecke, A. Misra, R.E. Rudd, 2004, ISBN: 1-55899-771-7
Volume 822— Nanostructured Materials in Alternative Energy Devices, E.R. Leite, J-M. Tarascon, Y-M. Chiang, E.M. Kelder, 2004, ISBN: 1-55899-772-5
Volume 823— Biological and Bioinspired Materials and Devices, J. Aizenberg, C. Orme, W.J. Landis, R. Wang, 2004, ISBN: 1-55899-773-3
Volume 824— Scientific Basis for Nuclear Waste Management XXVIII, J.M. Hanchar, S. Stroes-Gascoyne, L. Browning, 2004, ISBN: 1-55899-774-1
Volume 825E—Semiconductor Spintronics, B. Beschoten, S. Datta, J. Kikkawa, J. Nitta, T. Schäpers, 2004, ISBN: 1-55899-753-9
Volume 826E—Proteins as Materials, V.P. Conticello, A. Chilkoti, E. Atkins, D.G. Lynn, 2004, ISBN: 1-55899-754-7
Volume 827E—Educating Tomorrow's Materials Scientists and Engineers, K.C. Chen, M.L. Falk, T.R. Finlayson, W.E. Jones Jr., L.J. Martinez-Miranda, 2004, ISBN: 1-55899-755-5
Volume 828— Semiconductor Materials for Sensing, S. Seal, M-I. Baraton, N. Murayama, C. Parrish, 2005, ISBN: 1-55899-776-8
Volume 829— Progress in Compound Semiconductor Materials IV—Electronic and Optoelectronic Applications, G.J. Brown, M.O. Manasreh, C. Gmachl, R.M. Biefeld, K. Unterrainer, 2005, ISBN: 1-55899-777-6
Volume 830— Materials and Processes for Nonvolatile Memories, A. Claverie, D. Tsoukalas, T-J. King, J. Slaughter, 2005, ISBN: 1-55899-778-4
Volume 831— GaN, AlN, InN and Their Alloys, C. Wetzel, B. Gil, M. Kuzuhara, M. Manfra, 2005, ISBN: 1-55899-779-2
Volume 832— Group-IV Semiconductor Nanostructures, L. Tsybeskov, D.J. Lockwood, C. Delerue, M. Ichikawa, 2005, ISBN: 1-55899-780-6

MATERIALS RESEARCH SOCIETY SYMPOSIUM PROCEEDINGS

Volume 833— Materials, Integration and Packaging Issues for High-Frequency Devices II, Y.S. Cho, D. Shiffler, C.A. Randall, H.A.C. Tilmans, T. Tsurumi, 2005, ISBN: 1-55899-781-4
Volume 834— Magneto-Optical Materials for Photonics and Recording, K. Ando, W. Challener, R. Gambino, M. Levy, 2005, ISBN: 1-55899-782-2
Volume 835— Solid-State Ionics—2004, P. Knauth, C. Masquelier, E. Traversa, E.D. Wachsman, 2005, ISBN: 1-55899-783-0
Volume 836— Materials for Photovoltaics, R. Gaudiana, D. Friedman, M. Durstock, A. Rockett, 2005, ISBN: 1-55899-784-9
Volume 837— Materials for Hydrogen Storage—2004, T. Vogt, R. Stumpf, M. Heben, I. Robertson, 2005, ISBN: 1-55899-785-7
Volume 838E—Scanning-Probe and Other Novel Microscopies of Local Phenomena in Nanostructured Materials, S.V. Kalinin, B. Goldberg, L.M. Eng, B.D. Huey, 2005, ISBN: 1-55899-786-5
Volume 839— Electron Microscopy of Molecular and Atom-Scale Mechanical Behavior, Chemistry and Structure, D. Martin, D.A. Muller, E. Stach, P. Midgley, 2005, ISBN: 1-55899-787-3
Volume 840— Neutron and X-Ray Scattering as Probes of Multiscale Phenomena, S.R. Bhatia, P.G. Khalifah, D. Pochan, P. Radaelli, 2005, ISBN: 1-55899-788-1
Volume 841— Fundamentals of Nanoindentation and Nanotribology III, D.F. Bahr, Y-T. Cheng, N. Huber, A.B. Mann, K.J. Wahl, 2005, ISBN: 1-55899-789-X
Volume 842— Integrative and Interdisciplinary Aspects of Intermetallics, M.J. Mills, H. Clemens, C-L. Fu, H. Inui, 2005, ISBN: 1-55899-790-3
Volume 843— Surface Engineering 2004—Fundamentals and Applications, J.E. Krzanowski, S.N. Basu, J. Patscheider, Y. Gogotsi, 2005, ISBN: 1-55899-791-1
Volume 844— Mechanical Properties of Bioinspired and Biological Materials, C. Viney, K. Katti, F-J. Ulm, C. Hellmich, 2005, ISBN: 1-55899-792-X
Volume 845— Nanoscale Materials Science in Biology and Medicine, C.T. Laurencin, E. Botchwey, 2005, ISBN: 1-55899-793-8
Volume 846— Organic and Nanocomposite Optical Materials, A. Cartwright, T.M. Cooper, S. Karna, H. Nakanishi, 2005, ISBN: 1-55899-794-6
Volume 847— Organic/Inorganic Hybrid Materials—2004, C. Sanchez, U. Schubert, R.M. Laine, Y. Chujo, 2005, ISBN: 1-55899-795-4
Volume 848— Solid-State Chemistry of Inorganic Materials V, J. Li, M. Jansen, N. Brese, M. Kanatzidis, 2005, ISBN: 1-55899-796-2
Volume 849— Kinetics-Driven Nanopatterning on Surfaces, E. Wang, E. Chason, H. Huang, G.H. Gilmer, 2005, ISBN: 1-55899-797-0
Volume 850— Ultrafast Lasers for Materials Science, M.J. Kelley, E.W. Kreutz, M. Li, A. Pique, 2005, ISBN: 1-55899-798-9
Volume 851— Materials for Space Applications, M. Chipara, D.L. Edwards, S. Phillips, R. Benson, 2005, ISBN: 1-55899-799-7
Volume 852— Materials Issues in Art and Archaeology VII, P. Vandiver, J. Mass, A. Murray, 2005, ISBN: 1-55899-800-4
Volume 853E—Fabrication and New Applications of Nanomagnetic Structures, J-P. Wang, P.J. Ryan, K. Nielsch, Z. Cheng, 2005, ISBN: 1-55899-805-5
Volume 854E—Stability of Thin Films and Nanostructures, R.P. Vinci, R. Schwaiger, A. Karim, V. Shenoy, 2005, ISBN: 1-55899-806-3
Volume 855E—Mechanically Active Materials, K.J. Van Vliet, R.D. James, P.T. Mather, W.C. Crone, 2005, ISBN: 1-55899-807-1
Volume 856E—Multicomponent Polymer Systems—Phase Behavior, Dynamics and Applications, K.I. Winey, M. Dadmun, C. Leibig, R. Oliver, 2005, ISBN: 1-55899-808-X
Volume 858E—Functional Carbon Nanotubes, D.L. Carroll, B. Weisman, S. Roth, A. Rubio, 2005, ISBN: 1-55899-810-1
Volume 859E—Modeling of Morphological Evolution at Surfaces and Interfaces, J. Evans, C. Orme, M. Asta, Z. Zhang, 2005, ISBN: 1-55899-811-X
Volume 860E—Materials Issues in Solid Freeforming, S. Jayasinghe, L. Settineri, A.R. Bhatti, B-Y. Tay, 2005, ISBN: 1-55899-812-8
Volume 861E—Communicating Materials Science—Education for the 21st Century, S. Baker, F. Goodchild, W. Crone, S. Rosevear, 2005, ISBN: 1-55899-813-6

Prior Materials Research Society Symposium Proceedings available by contacting Materials Research Society

Nonlinear Optical Properties Materials

Mater. Res. Soc. Symp. Proc. Vol. 846 © 2005 Materials Research Society

Spectral profiles of two-photon absorption: Coherent versus two-step two-photon absorption

S. Polyutov[1], I. Minkov[1], F. Gel'mukhanov[1], K. Kamada[2], A. Baev[3], and H. Ågren[1]
[1]Theoretical Chemistry, Roslagstullsbacken 15, Royal Institute of Technology, S-106 91 Stockholm, Sweden
[2] Photonics Research Institute, National Institute of Advanced Industrial Science and Technology, AIST Kansai Center, Osaka, Japan
[3]Institute for Lasers, Photonics and Biophotonics, The State University of New York at Buffalo, Buffalo, New York 14260-3000

Abstract

We present a theory of two-photon absorption in solutions which addresses the formation of spectral shapes taking account of the vibrational degrees of freedom. The theory is used to rationalize observed differences between spectral shapes of one- and two-photon absorption. We elaborate on two underlying causes, one trivial and one non-trivial, behind these differences. The first refers simply to the fact that the set of excited electronic states constituting the spectra will have different relative cross sections for one- and two- photon absorption. The second reason is that the two-step and coherent two-photon absorption processes are competing, making the one- and two-photon spectral bands different even considering a single final state. The theory is applied to the N-101 molecule [di-phenyl-amino-nitro-stilbene] which was recently studied experimentally in the paper [T.-C. Lin, G.S. He, P.N. Prasad, and L.-S. Tan, J. Mater. Chem., **14**, 982, 2004.]

1 Introduction

Understanding the formation of two-photon spectra is essential in order to tailor structure-property relations for two-photon materials and so to improve their use in technical applications. One salient feature of two-photon spectra is that they in general are very different from their one-photon counterparts. Two-photon absorption of polyatomic molecules is strongly influenced by vibrational interaction, providing a general broadening of the spectra with fine structure. As for other types of electronic transitions this merely reflects the fact that the potential surface change going from the ground to the final state of the optical excitation. In conventional theory of photoexcitation using the sudden approximation the spectral profile is defined by the Franck-Condon factors between the ground and the excited state. This means that the two-photon profile has to copy the profile of one-photon absorption as the consequence of the Franck-Condon principle is exactly the same for the two processes. However, experiments, e.g. [1, 2], indicate a strong violation of this statement. The aim of our article is to explain the reason for this phenomenon. Our approach is based essentially on the dynamical theory developed in our previous articles [3, 4, 5, 6, 7]. Our explanation is based on two notions: First of all, considering the full spectrum, one-photon as well as coherent TPA absorption are formed by resonance transitions to a set of excited excited electronic states, each with its particular cross sections for the two processes. Secondly, the conventional coherent TPA absorption (one-step absorption) is accompanied by two-step absorption which has the same order of magnitude as one-step TPA and which in general both broaden and distort the profile of the particular band. A third reason, which in general is less important and indeed not operating for the here studied case, is that the spectral shapes of one- and two-photon absorption can be different also due to the vibronic coupling between the states.

2 Formulation of the theory

We consider the nonlinear interaction of the optical field with a many-level molecule (Fig.1).

This optical field, $\mathbf{E}$, induces a nonlinear polarization, $\mathbf{P}$:

$$\mathbf{E} = \frac{\boldsymbol{\mathcal{E}}}{2} e^{-\imath\omega + \imath kz} + c.c., \qquad \mathbf{P} = \boldsymbol{\mathcal{P}} e^{-\imath\omega + \imath kz} + c.c., \qquad k = \frac{\omega}{c} \tag{1}$$

An extraction of the fast variables in the Maxwell equations results in the following wave equation for the light intensity $I = c\varepsilon_0 |\mathcal{E}|^2/2$

$$\left(\frac{\partial}{\partial z} + \frac{1}{c}\frac{\partial}{\partial t}\right) I = -kc\Im m(\boldsymbol{\mathcal{E}}^* \cdot \boldsymbol{\mathcal{P}}), \quad \mathbf{P} = Sp(\mathbf{d}\rho). \tag{2}$$

The SI system of units is used here. To find the polarization of the system shown in Fig.1 we need the transition dipole moments, $\mathbf{d}_{ij}$, and the density

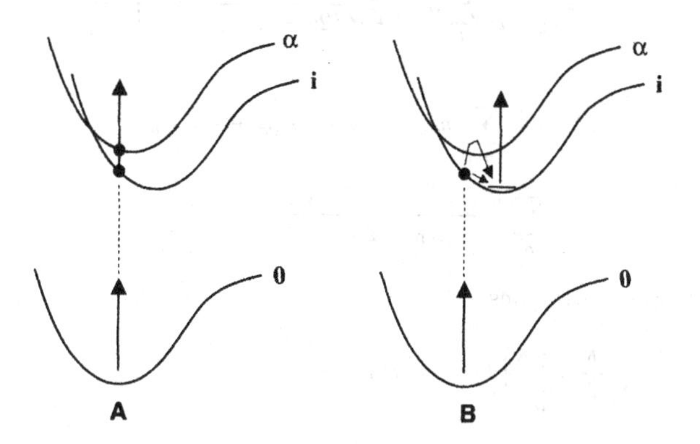

Figure 1: Scheme of transitions

matrix ρ_{ij}, of the absorber:

$$\left(\frac{\partial}{\partial t}+\Gamma_{ii}\right)\rho_{ii}=\sum_{\alpha}\tilde{\Gamma}_{\alpha\alpha}\rho_{\alpha\alpha}+\frac{\imath}{\hbar}[\rho,V]_{ii},\qquad \left(\frac{\partial}{\partial t}+\Gamma_{\alpha\alpha}\right)\rho_{\alpha\alpha}=\frac{\imath}{\hbar}[\rho,V]_{\alpha\alpha},$$

$$\left(\frac{\partial}{\partial t}+\Gamma_{ij}(\omega)\right)\rho_{ij}=\frac{\imath}{\hbar}[\rho,V]_{ij},\qquad \mathrm{Tr}\rho=N,\qquad V=-\mathbf{E}\cdot\mathbf{d}, \tag{3}$$

where the concentration of the absorbing molecules is denoted by N. To be specific we consider randomly oriented molecules and long pulses.

The final equation for the intensity is the following

$$\left(\frac{\partial}{\partial z}+\frac{1}{c}\frac{\partial}{\partial t}\right)I(z,t)=-N(\sigma I(z,t)+\sigma^{(2)}I^2(z,t)),\qquad \sigma^{(2)}=\sigma_1^{(2)}+\sigma_2^{(2)}. \tag{4}$$

The first term at the right-hand side of eq.(4) describes linear off-resonant photoabsorption with a cross section

$$\sigma=\frac{k}{3\hbar\varepsilon_0}\sum_i\frac{d_{i0}^2\Gamma_{i0}\langle 0_0|\nu_i\rangle^2}{(\omega-\omega_{i\nu_i,00})^2+\Gamma_{i0}^2} \tag{5}$$

The effective TPA cross section[3, 4] is the sum of one-step (coherent)

$$\sigma_1^{(2)} = \frac{k}{c\varepsilon_0^2\hbar^3}\sum_i \langle S_{i0}\rangle \sum_{\nu_i} \frac{\Gamma_{i0}(\omega)\langle 0_0|\nu_i\rangle^2}{(2\omega - \omega_{i\nu_i,00_0})^2 + \Gamma_{i0}^2(\omega)},$$

$$\langle S_{i0}\rangle = \frac{1}{15}\sum_{nm}(s_{nn}s_{mm} + s_{nm}s_{nm} + s_{nm}s_{mn}),$$

$$s_{mn} = \sum_\alpha \frac{d_{i\alpha}^{(m)} d_{\alpha 0}^{(n)}}{\omega - \omega_{\alpha 0}} + \frac{d_{i0}^{(m)} \Delta d_{ii}^{(n)}}{2(\omega - \omega_{i0})}, \tag{6}$$

and two-step contributions

$$\sigma_2^{(2)} = \frac{k}{c\varepsilon_0^2\hbar^3}\sum_{\alpha\nu_\alpha}\Big[d_{i0}^2 d_{\alpha i}^2 \Lambda_\alpha^{i0} P_{i0}(\omega)\frac{R_{ii}}{\Gamma_{ii}} + \\ + \sum_\beta d_{\beta 0}^2 d_{i\alpha}^2 \Lambda_{i\alpha}^{\beta 0} P_{\beta 0}(\omega)\frac{R_{\beta\beta}}{\Gamma_{\beta\beta}}\frac{\Gamma^{\beta i}}{\Gamma_{ii}}\Big]\frac{\Gamma_{\alpha i}(\omega)\langle 0_i|\nu_\alpha\rangle^2}{(\omega - \omega_{\alpha\nu_\alpha,i0_i})^2 + \Gamma_{\alpha i}^2(\omega)} \tag{7}$$

The first term at the right-hand side of eq.(7) describes the two-step absorption $0 \to i \to \alpha$, while the second two-step contribution corresponds to the one-photon absorption $0 \to \beta$ than non-radiative decay $\beta \to i$ with the rate $\Gamma^{\beta i}$ and one photon absorption $i \to \alpha$. We assume that the pulse is long and the molecule has time to relax to the lowest vibrational level of the electronic level i. The case of short pulse differs qualitatively as it is illustrated in Fig.1. Here

$$R_{jj} = \frac{\Gamma_{jj}}{I(z,t)} e^{-\Gamma_{jj}t} \int_{-\infty}^{t} e^{\Gamma_{jj}\tau} I(z,\tau)d\tau. \tag{8}$$

One can show that $R_{jj} \approx 1$ for long pulse. We introduced the probability of non-resonant absorption $P_{i0}(\omega)$ and the anisotropy factor $\Lambda_{i\alpha}^{\beta 0}$

$$P_{i0}(\omega) = \Gamma_{i0}(\omega)\Big[\frac{1}{(\omega - \omega_{i0}^V)^2 + \Gamma_{i0}^2(\omega)} + \frac{1}{(\omega + \omega_{i0}^V)^2}\Big], \quad \Lambda_{i\alpha}^{\beta 0} = \frac{1 + 2\cos^2(\mathbf{d}_{\beta 0}, \mathbf{d}_{i\alpha})}{15} \tag{9}$$

We use the following notations: $\Delta\mathbf{d}_{ii} = \mathbf{d}_{ii} - \mathbf{d}_{00}$; $\mathbf{d}_{i\alpha}$ is the transition dipole moment between electronic states i and α; ω_{i0}^V is the frequency of vertical transition between ground and excited states; $\omega_{\alpha\nu_\alpha,i\nu_i}$ is the frequency of electron-vibrational transition $|i\nu_i\rangle \to |\alpha\nu_\alpha\rangle$; $\langle\nu_i|\nu_\alpha\rangle$ is the Franck-Condon (FC) amplitude between vibrational states ν_i and ν_α of electronic states i and α.

2.1 Non-resonant absorption and hierarchy of relaxation times

Three characteristic time scales govern the dynamics of the studied system: The pulse duration, τ_p, the lifetime of the excited state, Γ_{ii}^{-1}, and the decay time, $\Gamma_{ij}^{-1}(\omega)$, of the coherence ρ_{ij} $(i \neq j)$. The decay of the coherence consists of two qualitatively different contributions;

$$\Gamma_{ij}(\omega) = \frac{1}{2}(\Gamma_{ii} + \Gamma_{jj}) + \gamma_{ij}(\omega). \tag{10}$$

The first contribution is related to the decay rates of populations of the i and j states and does in general depend on the density of the medium. The so-called dephasing rate, $\gamma_{ij}(\omega)$, is proportional to the total concentration, N_{tot}, of molecules in a solution. Near the resonance $\gamma_{ij}(\omega_{res}) \simeq \bar{v}\sigma_{ij}N_{tot}$, where σ_{ij} is the cross section of the dephasing collisions of the absorbing molecule with the solvent particles; $\bar{v}$ is a thermal velocity. It is striking that the expression for $\gamma_{ij}(\omega_{res})$, being strictly valid for gas phase, gives the correct order of magnitude even for liquids [8]: $\Gamma_{ij}(\omega_{res}) \sim \gamma_{ij} \sim 0.1-0.01$ eV ($1/\gamma_{ij}(\omega_{res}) \sim 7-70$ fsec). When the detuning from resonant frequency is larger than the inverse duration of collision $1/\tau_{coll}$ the collisional broadening is small

$$\gamma_{ij}(\omega) \ll \gamma_{ij}(\omega_{res}), \qquad |\omega - \omega_{res}|\tau_{coll} \gg 1 \tag{11}$$

The decay rate, Γ_{ii}, of the population, ρ_{ii}, is mainly determined by the intramolecular radiative, Γ_{ii}^r, and non-radiative (internal conversion), Γ_{ii}^{nr}, transitions: $\Gamma_{ii} = \Gamma_{ii}^r + \Gamma_{ii}^{nr}$. Optical experiments[9] with organic molecules show that the lowest excited electronic state i has the longest lifetime $\Gamma_{ii}^{-1} \sim$ 1 nsec. The picture differs qualitatively for the higher electronic states, α, which have short lifetimes in the order of $\Gamma_{\alpha\alpha}^{-1} \lesssim 1$ psec owing to cascades of non-radiative, internal, conversions to the lowest excited state. This allows to estimate the relative dephasing rate in the off-resonant region, $\omega \sim \omega_{\alpha 0}$, as

$$\frac{\Gamma_{\alpha 0}(\omega)}{\Gamma_{ii}} \gtrsim 10^3 \tag{12}$$

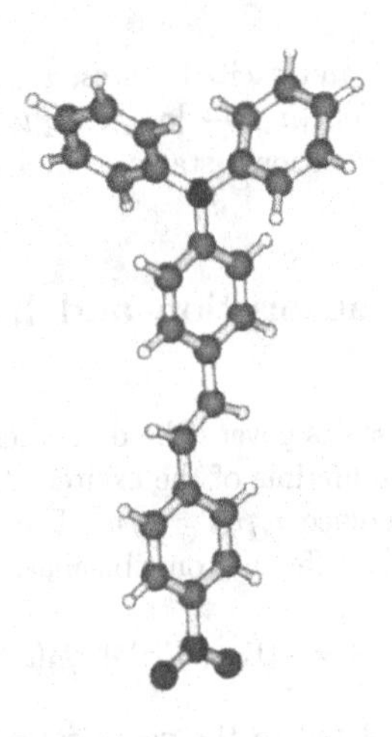

Figure 2: N-101 molecule.

The experimental data[10] show that the one-photon absorption drops down fast with decrease of the frequency and that ratio of probability of non-resonant and resonant one-photon absorptions is $P^{NR}_{\alpha0}(\omega_{\alpha0}/2)/P^{R}_{\alpha0}(\omega_{\alpha0}) = 4\Gamma_{\alpha0}(\omega_{\alpha0}/2)\Gamma_{\alpha0}(\omega_{\alpha0})/\omega^2_{\alpha0} \lesssim 10^{-4}$. This gives the same estimate as eq.(12).

3 Results of simulations of the N-101 molecule

We here present results of preliminary simulations of the two-photon absorption of the N-101 molecule (Fig.2) studied recently in experiment [1].

Our simulations are based on the response theory outlined in ref.[11]. In the simulations we take into account the two first excited electronic states: $\omega_{10} = 2.52$ eV $\omega_{20} = 3.54$ eV (see Fig.1).

The many dimensional FC amplitudes are simulated using the harmonic approximation. The gradients of the excited state potentials are calculated making use of a code for analytical excited state gradients implemented in the DALTON suite of programs.

First of all we simulated the spectra of one-photon and coherent two-

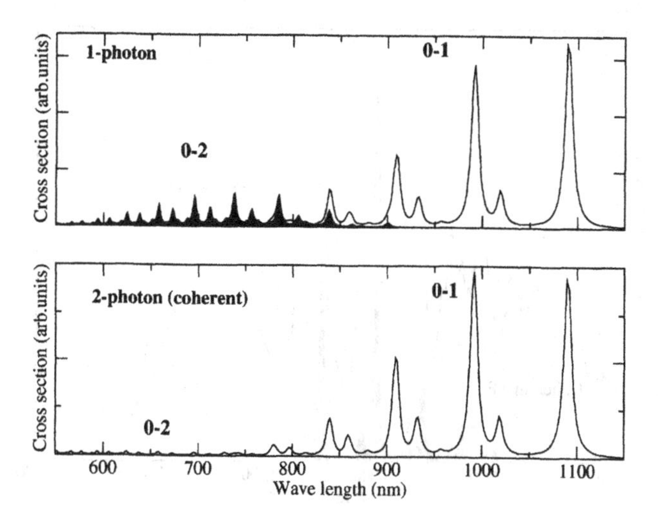

Figure 3: Spectra of one-photon and of coherent two-photon absorption of the N-101 molecule.

photon absorption, Fig.3. To compare these profiles we doubled the wave length in the spectrum of the one photon absorption. One can see that the one-photon spectrum consists of two vibrational bands related to the transitions to the first $(0 \to 1)$ and the second $(0 \to 2)$ electronic states. This is in qualitative agreement with experiment [1], although the intensity of the shortwave band is weaker than observed.

One reason for this is the contributions from upper electronic states which are close lying to state nr. 2. The peak intensity of the $0 \to 2$ transition $\propto (d_{20}\langle\nu_2|0_0\rangle)^2)$ is smaller than the intensity of the $0 \to 1$ excitation $(\propto (d_{10}\langle\nu_1|0_0\rangle)^2)$. This is because of the transitional dipole moments $d_{10}^2/d_0^2 \approx 2$ and large vibrational broadening of the transition $0 \to 2$. Contrary to the one-photon spectrum the coherent TPA profile is formed mainly due to the TPA transition $0 \to 1$. The reason for this is that that the TPA cross section of the $0 \to 2$ transition $(\propto (d_{20}\Delta d_{22}\langle\nu_2|0_0\rangle)^2)$ is much smaller than the $0 \to 1$ TPA cross section $(\propto (d_{10}\Delta d_{11}\langle\nu_1|0_0\rangle)^2)$.

The other reason of the distinction of the TPA spectrum from the spectrum of one-photon absorption is that two-step contributions modify the TPA

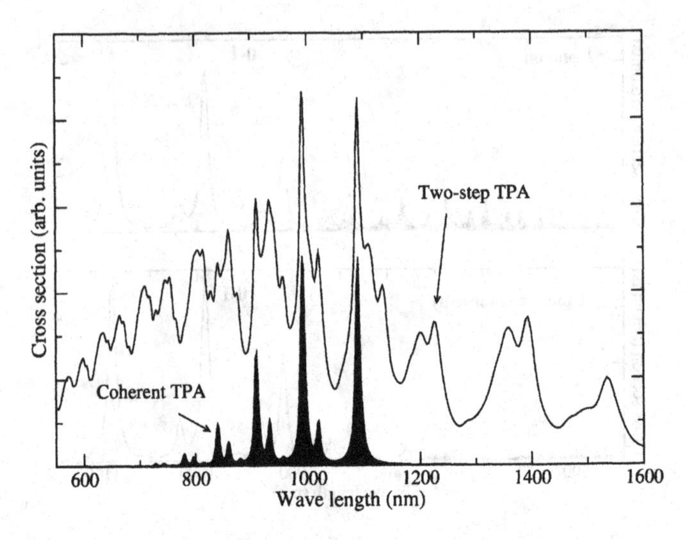

Figure 4: Total profile of two-photon absorption of the N-101 molecule. The shaded region shows the coherent contribution.

profile considerably, see Fig.4. In simulations of the total TPA profile (Fig.4) we assumed according to eq.(12) that $\Gamma_{\alpha 0}(\omega)/\Gamma_{ii} = 10^3$ in the expression for the probability of off-resonant population of the excited electronic state. As one can see from the simulation results the two-step contribution has large vibrational broadening.

4 Summary

We have shown that the spectrum of one-photon absorption of the N-101 molecule consists of two vibrationally broadened bands related to the transitions to the first and the second excited electronic states. The vibrational profile of the second electronic state displays larger vibrational broadening. This as well as the smaller value of the transitional dipole moment explains the smaller intensity of the second band. Our simulations of the coherent two-photon absorption show strong suppression of the shortwave length band comparing with the one-photon absorption. This is because the relative two-photon cross section of the second band is much smaller than the relative intensity of the one-photon absorption. The experimental spectra show the similar behavior except for a larger broadening of the first band. To explain this discrepancy we calculated the two-photon spectrum also taking into account two-step two-photon absorption. Our simulations show that the contribution of the two-step process is comparable with the coherent one-step two-photon absorption and that it results in an extra vibrational broadening of the spectral profile. We take this as yet another indication of the need to consider off-resonant step-wise contributions even when operating at resonant conditions for the coherent multi-photon excitation.

Acknowledgments

This work was supported by the Swedish Research Council (VR).

References

[1] T.-C. Lin, G.S. He, P.N. Prasad, and L.-S. Tan, *J. Mater. Chem.*, **14**, 982 (2004).

[2] G.S. He, T.-C. Lin, J. Dai, and P.N. Prasad *J. Phys. Chem.*, **120**, 5275 (2004).

[3] F. Gel'mukhanov, A. Baev, P. Macak, Y. Luo, and H. Ågren, *J. Opt. Soc. Am.*, **B 19**, 937 (2002).

[4] A. Baev, F. Gel'mukhanov, P. Macak, Y. Luo, and H. Ågren *J. Chem. Phys.* , **117**, 6214 (2002).

[5] A. Baev, F. Gel'mukhanov, O. Rubio-Pons, P. Cronstrand, and H. Ågren, *J. Opt. Soc. Am.*, **B 21**, 384 (2004).

[6] A. Baev, O. Rubio-Pons, F. Gel'mukhanov, and H. Ågren, *J. Phys. Chem.*, **A 108**, 7406 (2004).

[7] A. Baev, V. Kimberg, S. Polyutov, F. Gel'mukhanov, and H. Ågren, *J. Opt. Soc. Am. B.* **00**, 000 (2004).

[8] N. Allard and J. Kielkopf, *Rev. Mod. Phys.* , **54**, 1103 (1982).

[9] N.J. Turro, *Modern Molecular Photochemistry* (Benjamin, New York, 1978).

[10] K. Kamada (to be published).

[11] T. Helgaker and H. J. A. Jensen and P. Jørgensen and J. Olsen and K. Ruud and H. Ågren and A.A. Auer and K.L. Bak and V. Bakken and O. Christiansen and S. Coriani and P. Dahle and E. K. Dalskov and T. Enevoldsen and B. Fernandez and C. Hättig and K. Hald and A. Halkier and H. Heiberg and H. Hettema and D. Jonsson and S. Kirpekar and R. Kobayashi and H. Koch and K. V. Mikkelsen and P. Norman and M. J. Packer and T. B. Pedersen and T. A. Ruden and A. Sanchez and T. Saue and S. P. A. Sauer and B. Schimmelpfenning and K. O. Sylvester-Hvid and P. R. Taylor and O. Vahtras, *DALTON* a molecular electronic structure program – Release 1.2. (2001).

Mater. Res. Soc. Symp. Proc. Vol. 846 © 2005 Materials Research Society DD1.4

Theoretical Study on Open-Shell Nonlinear Optical Systems

Masayoshi Nakano,[1] Ryohei Kishi,[1] Nozomi Nakagawa,[1] Tomoshige Nitta,[1] Takashi Kubo,[2] Kazuhiro Nakasuji,[2] Kenji Kamada,[3] Koji Ohta,[3] and Benoît Champagne,[4] Edith Botek,[4] Satoru Yamada,[2] and Kizashi Yamaguchi[2]

[1] Division of Chemical Engineering, Department of Materials Engineering Science, Graduate School of Engineering Science, Osaka University, Toyonaka, Osaka 560-8531, Japan
[2] Department of Chemistry, Graduate School of Science, Osaka University, Toyonaka, Osaka 560-0043, Japan
[3] Photonics Research Institute, National Institute of Advanced Industrial Science and Technology (AIST), Ikeda, Osaka 563-8577, Japan
[4] Laboratoire de Chimie Théorique Appliquée Facultés Universitaires Notre-Dame de la Paix, rue de Bruxelles, 61, 5000 Namur, Belgium

ABSTRACT

The static second hyperpolarizabilities (γ) of open-shell organic nonlinear optical (NLO) systems composed of singlet diradical molecules are investigated using ab initio molecular orbital (MO) and density functional theory (DFT) methods. It is found that neutral singlet diradical systems with intermediate diradical characters tend to enhance γ as compared to those with small and large diradical characters. This suggests that the diradical character is a novel control parameter of γ for singlet diradical systems.

INTRODUCTION

So far, the investigation of nonlinear optical (NLO) properties has been mostly limited to the closed-shell systems though pioneering studies have been carried out to enhance the field-induced charge fluctuations in charged radical states [1-4] in view of designing multi-functional materials, for example, combining magnetic and optical properties [5-7]. On the basis of the perturbational analysis of virtual excitation processes of γ, the charged radical systems having symmetric resonance structure with invertible polarization (SRIP) [8-10] are expected to have unique negative second hyperpolarizability (γ), which is the microscopic origin of the third-order NLO property, in the off-resonant region and to exhibit large electron-correlation dependences [1,11]. Some crystals composed of these charged radical systems are predicted to present unique third-order NLO properties (negative γ) as well as high electrical conductivity [3,9]. Further, the open-shell molecules and their clusters involving nitronyl nitroxide radicals, which are important in realizing organic ferromagnetic interaction, also belong to the SRIP systems and are predicted to give negative off-resonant γ values.

The characteristic of open-shell NLO systems has also been determined from the viewpoint of the chemical bonding nature [12-14]. In previous studies [12,14], we have predicted a remarkable variation in γ of H_2 model with increasing the bond distance and have suggested a significant enhancement of γ in the intermediate correlation (intermediate bond breaking) regime. This feature is understood by the fact that the intermediate bonding electrons are sensitive to the applied field, leading to large fluctuation of electrons. Such intermediate bond breaking nature in the intermediate correlation regime is, for example, expected to be realized by the increase in the spin multiplicity in open-shell neutral systems [15]. The investigation of the effect of introducing charge into open-shell systems on γ is also important in view of the spin-control of novel open-shell NLO systems since the control of spin state is often achieved by introducing the charges into systems in molecular magnetism [16,17]. Recently, the effects of spin multiplicity on γ of open-shell neutral and/or charged π-conjugated systems have been investigated using simple open-shell model compounds, and the magnitude of πhas turned out to sensitively depend on the spin and charged states [15,18].

In this study, we examine novel open-shell singlet NLO systems, whose NLO properties can be controlled by the diradical character [19-21]. The dependence of πon the diradical character is investigated using diradical model compounds, *p*-quinodimethane (PQM) models with different both-end carbon-carbon (C-C) bond lengths, by highly correlated *ab initio* molecular orbital (MO) and hybrid density functional theory (DFT) method. The πvalue of a real diradical molecule, i.e., 1,4-bis-(imidazole-2-ylidene)-cyclohexa-2,5-diene (BI-2Y), which is expected to possess intermediate diradical character, is also examined. On the basis of the present and previous results, we clarify the features of a new class of open-shell NLO systems.

STRUCTURE-PROPERTY RELATIONSHIPS FOR γBASED ON VIRTUAL EXCITATION PROCESSES

In this section, our structure-property relation for molecular systems is briefly explained [1,8]. The perturbation formula of static πcan be partitioned into three types of contributions (I), (II) and (III): [1]

$$\gamma = \gamma^{(\mathrm{I})} + \gamma^{(\mathrm{II})} + \gamma^{(\mathrm{III})} = \sum_{n=1} \frac{(\mu_{n0})^2(\mu_{nn})^2}{E_{n0}^3} - \sum_{n=1} \frac{(\mu_{n0})^4}{E_{n0}^3} + \sum_{\substack{m,n=1 \\ (m \neq n)}} \frac{(\mu_{n0})^2(\mu_{mn})^2}{E_{n0}^2 E_{m0}}. \quad (1)$$

Here, μ_{n0} is the transition moment between the ground and the *n*th excited states, μ_{mn} is the transition moment between the *m*th and the *n*th excited states, μ_{nn} is the difference of dipole moments between the ground and the *n*th excited states and E_{n0} is the transition energy given by $(E_n - E_0)$. From these equations, apparently, the contributions of types (I) and (III) are positive in sign, whereas the contribution of type (II) is negative. For conventional molecular compounds with large positive γ, there are two characteristic cases: (i) $|\gamma^{(\mathrm{I})}| >> |\gamma^{(\mathrm{II})}| \sim |\gamma^{(\mathrm{III})}|$ $(\gamma > 0)$ and (ii) $|\gamma^{(\mathrm{I})}| = 0$, $|\gamma^{(\mathrm{II})}| < |\gamma^{(\mathrm{III})}|$ $(\gamma > 0)$. In the case (i), the compounds have large asymmetric charge distributions which are responsible for large μ_{nn}, whereas in the case (ii), the compounds are centrosymmetric systems in which the contributions of type (I) disappear. The third case, i.e., (iii) $|\gamma^{(\mathrm{I})}| = 0$, $|\gamma^{(\mathrm{II})}| > |\gamma^{(\mathrm{III})}|$ $(\gamma < 0)$, is interesting because the systems with negative static γ are rare. Such systems are symmetric ($\mu_{nn} = 0$) and exhibit strong dipole transition moment between the ground and the first excited states ($|\mu_{0n}| > |\mu_{nm}|$). This indicates that the symmetric systems with large ground-state polarizability (α_0) tend to exhibit negative γ. As typical

(a) Anion radical state of *s*-indacene

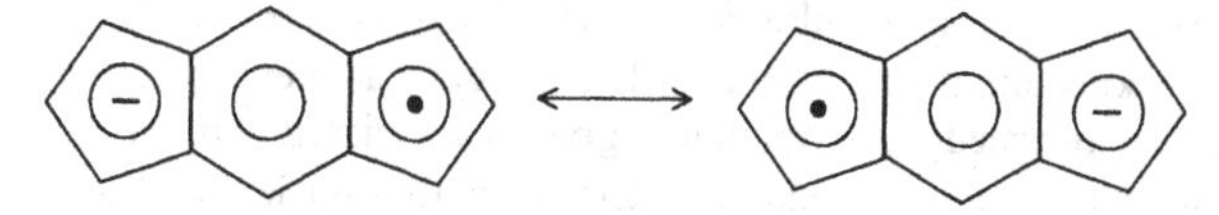

(b) Cation radical state of tetrathiapentalene (TTP)

(c) Nitronyl nitroxide radical

Figure 1. Typical compounds with large SRIP contribution.

compounds belonging to the case (iii), several ionic radical and neutral radical species have been presented, i.e., (a) anion radical state of *s*-indacene [1], (b) cation radical state of tetrathiapentalene (TTP) [9] and (c) nitronyl nitroxide radical [5-7] (see Fig. 1). These compounds tend to possess a large contribution of resonance structures with invertible polarization (SRIP) [2], allowing them to satisfy our criteria for the system to have a negative γ (case (iii)). This can be understood by the fact that the system with a large SRIP exhibits relative increase in the magnitude of the transition moment (μ_{n0}) and relative decrease in the transition energy (E_{n0}) between the ground and the first dipole-allowed excited states, the feature of which leads to a relative enhancement of type (II) (negative contribution). From the previous results for these systems, we have also predicted a sensitive and remarkable electronic structure dependences of the γ. Therefore, the open-shell compounds with large SRIP contributions are expected to be NLO candidates controlled by slight external electromagnetic field and/or chemical perturbation, e.g., charging and intermolecular interactions.

SPIN POLARIZATION EFFECTS ON γ FOR NUTRAL OPEN-SHELL SYSTEMS

Open-shell systems can be classified according to the strength of electron correlation, i.e., weak-, intermediate- and strong (magnetic)- correlation regimes, which can be associated to an equilibrium-, intermediate- and long- bond distance regions of a homogeneous neutral diatomic molecule [13]. We have found the remarkable variation of γ according to increasing the bond distance and have suggested the enhancement of γ in the intermediate correlation regime [12,14,15]. In addition, the amplitude of the electron correlation is expected to change by modulating the spin and/or the charge of a system.

A first parameter to tune the NLO responses is the spin state. From the previous studies on the γ of C_5H_7 radicals, the static longitudinal γ of small-size neutral π-conjugated model is expected to increase with the spin multiplicity. The variation in π value has been related to the degree of bond breaking and the electron correlation regime as mentioned above. It is predicted that in particular, the intermediate correlation regime is associated with the largest π value whereas for the weak and strong correlation regimes, the third-order NLO responses are smaller [12,14,15]. Therefore, this type of open-shell NLO system is expected to have attractive magnetic properties in addition to the largely enhanced NLO properties. It is also predicted that the charged defects introduced in such open-shell systems causes the enhancement of negative contributions (type (II), see Eq. (1)), sometimes leading to negative total π[17].

Another way to optimize the NLO response consists in considering singlet diradical systems. As shown in the variation in (hyper)polarizabilities of H_2 in the dissociation process [12,14], the variations in bond breaking nature, i.e., variations in the degree of electron correlation, is known to be well described by the diradical character [13,19,20], which is an index of the degree of diradical nature and ranges from 0 to 1 as a function of the bond dissociation. In fact, several diradical compounds with various diradical characters have been synthesized and different control schemes of the diradical nature have been proposed [22-25]. In this study, therefore, we investigate the dependence of π on the diradical character using neutral singlet diradical model compounds. Although the singlet diradical systems are useless as molecular magnets, the singlet diradical systems with intermediate and large diradical characters, i.e., belonging to the intermediate and strong correlation regimes, are expected to exhibit π values much larger than those with small diradical characters, i.e., belonging to the weak correlation regime. As a result, the diradical NLO systems are expected to be a candidate for tunable NLO systems controlled by a diradical character.

MODEL COMPOUNDS AND CALCULATION METHOD

Fig. 2(a) shows the structure of the *p*-quinodimethane (PQM) molecule in its singlet ground state. It is described by two resonance forms, *i.e.*, the quinoid and diradical forms. In general, the experimental structures of conjugated diradical systems are well reproduced by the spin-restricted (R) B3LYP and unrestricted (U) B3LYP methods. Indeed, the optimized structure for the present system with D_{2h} symmetry (R_1 = 1.351 Å, R_2 = 1.460 Å and R_3 = 1.346 Å) at the UB3LYP level using 6-311G* basis set is the same as that at the RB3LYP level. The optimized parameters suggest that PQM presents a large degree of quinoid form instead of diradical nature since R_1 and R_3 are like a C-C double bond length and are smaller than R_2 having a C-C single bond nature. In order to increase the degree of diradical nature, we consider several PQM models with bond-length R_1 changing from 1.35 Å to 1.7 Å under the constraint of $R_2 = R_3$ =1.4 Å. We first determine the diradical character from spin-unrestricted Hartree-Fock (UHF) calculations. The diradical character y_i related to the HOMO-*i* and LUMO+*i* is defined by the weight of the doubly-excited configuration in the MC-SCF theory and is formally expressed in the case of the spin-projected UHF (PUHF) theory as [19]

$$y_i = 1 - \frac{2T_i}{1+T_i^2}, \tag{2}$$

where T_i is the orbital overlap between the corresponding orbital pairs. T_i can also be represented by using the occupation numbers (n_j) of UHF natural orbitals (UNOs):

$$T_i = \frac{n_{\mathrm{HOMO}-i} - n_{\mathrm{LUMO}+i}}{2}. \tag{3}$$

Since the PUHF diradical characters amount to 0 % and 100 % for closed-shell and pure diradical states, respectively, y_i represents the diradical character, *i.e.*, the instability of the chemical bond. Table I gives the diradical characters y calculated from Eqs. (2) and (3) using HOMO and LUMO of UNOs for PQM systems with various R_1 values. As expected, it is found that the diradical character y increases when increasing the bond length R_1 starting from the optimized (equilibrium) geometry, which possesses a low y value (0.146), i.e., quinoid like structure. Fig. 2(b) shows an optimized structure of 1,4-bis-(imidazole-2-ylidene)-cyclohexa-2,5-diene (BI-2Y) at the RB3LYP/6-31G** level. Contrary to the PQM, the diradical character y of BI-2Y calculated using the UNO's becomes 0.4227, which indicates that the BI-2Y is an intermediate diradical system.

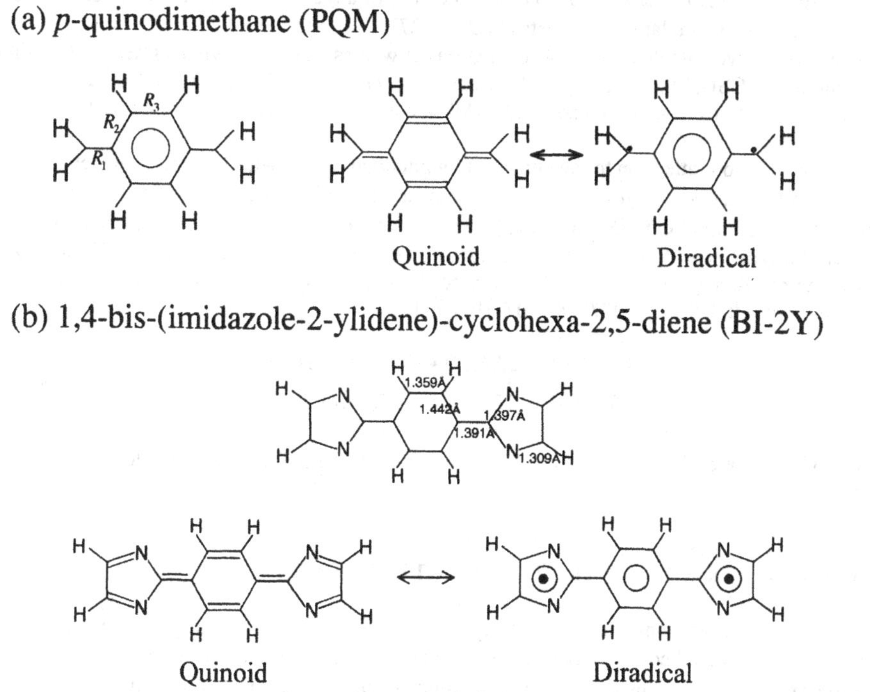

(a) *p*-quinodimethane (PQM)

R_1 R_2 R_3

Quinoid Diradical

(b) 1,4-bis-(imidazole-2-ylidene)-cyclohexa-2,5-diene (BI-2Y)

1.359Å 1.442Å 1.397Å 1.391Å 1.309Å

Quinoid Diradical

Figure 2. Singlet diradical molecules, *p*-quinodimetahne (PQM) model with several parameters R_1, R_2 and R_3 (a) and 1,4-bis-(imidazole-2-ylidene)-cyclohexa-2,5-diene (BI-2Y). The geometry of BI-2Y is optimized at the RB3LYP/6-31G** level.

Table I. Diradical character y for R_1, R_2 and R_3 in the PQM models[1)] shown in Fig. 2(a).

R_1	R_2	R_3	y
1.351	1.460	1.346	0.146
1.350	1.400	1.400	0.257
1.400	1.400	1.400	0.335
1.450	1.400	1.400	0.414
1.500	1.400	1.400	0.491
1.560	1.400	1.400	0.576
1.600	1.400	1.400	0.626
1.700	1.400	1.400	0.731

[1)] The first row corresponds to the equilibrium geometry with D_{2h} symmetry in the singlet ground state optimized at the RB3LYP/6-311G* level

We use the 6-31G*+p basis set with p exponent of 0.0523 on carbon (C) atoms and 0.0582 on nitrogen (N) atoms [25] since several studies have demonstrated that the use of a split-valence or split-valence plus polarization basis set augmented with a set of p and/or d diffuse functions on the second-row atoms enables us to reproduce the γ of large- and medium-size π-conjugated systems calculated with larger basis sets [17,21,26,27]. We apply the RHF, UHF and UHF-coupled-cluster with single and double excitations as well as a perturbative treatment of the triple excitations (UCCSD(T)). Among the DFT schemes, the R and U BHandHLYP exchange-correlation functionals have been adopted. All calculations have been performed using the Gaussian 98 program package [28].

We confine our attention to the dominant longitudinal components of static γ, which is considered to be a good approximation to off-resonant dynamic γ [4,29]. The static γ can be obtained by the finite field (FF) approach [30] which consists in the fourth-order differentiation of energy E with respect to different amplitudes of the applied external electric field. We adopt a 4-point procedure (equivalent to a 7-point procedure for an asymmetric case) using field amplitudes of 0.0, 0.0010, 0.0020, and 0.0030 a.u.:[2]

$$\gamma = \{E(3F) - 12E(2F) + 39E(F) - 56E(0) \\ +39E(-F) - 12E(-2F) + E(-3F)\}/36(F)^4. \tag{4}$$

Here, $E(F)$ indicates the total energy in the presence of the field F applied in the longitudinal direction.

SECOND HYPERPOLARIZABILITY DENSITY ANALYSIS

In this study, we employ the second hyperpolarizability (γ) density analysis [2] in order to characterize the spatial contributions of electrons to γ. The contribution obtained from a pair of positive and negative γ densities gives a local contribution of electrons to the total γ. The static γ value can be expressed by

$$\gamma = -\frac{1}{3!}\int r\rho^{(3)}(r)d^3r, \tag{5}$$

where

$$\rho^{(3)}(\boldsymbol{r}) = \left.\frac{\partial^3 \partial(\boldsymbol{r})}{\partial F \partial F \partial F}\right|_{F=0}. \tag{6}$$

This third-order derivative of the electron density with respect to the applied electric fields, $\rho^{(3)}(\boldsymbol{r})$, is referred to as the γ density. The positive and negative values of γ densities multiplied by F^3 correspond respectively to the field-induced increase and decrease in the charge density (in proportion to F^3), which induce the third-order dipole moment (third-order polarization) in the direction from positive to negative γ densities. Therefore, the γ density map represents the relative phase and magnitude of change in the third-order charge densities between two spatial points with positive and negative values. In order to explain how to analyze the spatial contribution of electrons to γ, let us consider a pair of localized γ densities with positive and negative values. The sign of the contribution to γ is positive when the direction from positive to negative γ density coincides with the positive direction of the coordinate system. The sign becomes negative in the opposite case. Moreover, the magnitude of the contribution to γ associated with this pair of γ densities is proportional to the distance between them.

RELATION BETWEEN γ AND DIRADICAL CHARACTER IN PQM MODELS

Figure 3 shows the results calculated by the RHF, UHF and UBHandHLYP methods as well as the UCCSD(T) result, which is considered to be the most reliable for the present systems. It turns out that the γ value at the UCCSD(T) level increases with increasing y in the weak diradical region ($y = 0.1$-0.5), attains a maximum in the intermediate region ($y \approx 0.5$) and then decreases in the region corresponding to large diradical character ($y > 0.5$). The maximum γ value at the UCCSD(T) level (77500 a.u.) for a y value close to 0.5 is 3.3 times as large as the γ value at the equilibrium geometry (23300 a.u.) ($y = 0.146$). The RHF γ value is negative while its magnitude increases strongly with the diradical character. Such incorrect behavior is predicted to originate from the triplet instability of the RHF solution in the intermediate and strong correlation regimes [13,31-32]. At the UHF level, γ monotonically decreases with increasing diradical character, so that for y smaller than 0.3 they are larger than the UCCSD(T) reference value whereas they are smaller in the intermediate and large diradical character regions. In contrast, the UBHandHLYP method provides a similar qualitative description of the variation in γ to that at the UCCSD(T) level. Nevertheless, the UBHandHLYP γ value overshoots the UCCSD(T) value in the region with small diradical character ($y < 0.3$) while it is slightly smaller in the region with intermediate and large diradical characters ($y > 0.4$). As a result, it is predicted that hybrid DFT methods with exchange-correlation functionals specifically-tuned for reproducing the NLO properties of small- and intermediate-size diradical species could provide satisfactory results in comparison with the more elaborated UCCSD(T) scheme.

The variation in γ for the PQM models with increasing diradical character can be explained from the analogy to the dissociation of H_2. The increase of γ in the intermediate diradical character region is predicted to be caused by the virtual excitation processes (type (III) in Eq. (1)) with zwitterionic contribution between the radicals on both-end carbon atoms, which corresponds to the intermediate dissociation region for the H_2 molecule [12]. In an analogous

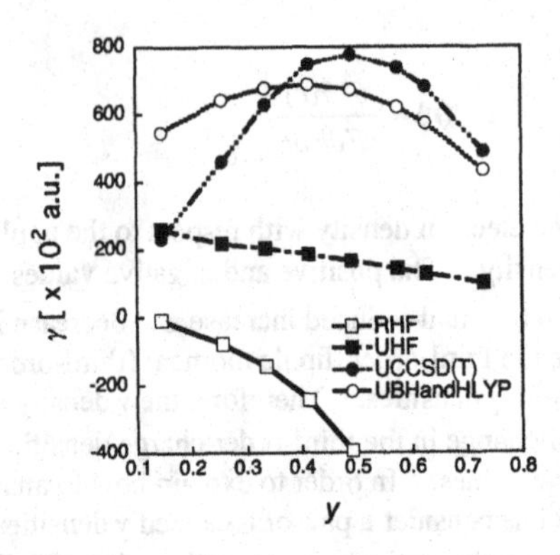

Figure 3. γ value versus diradical character y for PQM models calculated by the RHF, UHF, UCCSD(T) and UBHandHLYP methods.

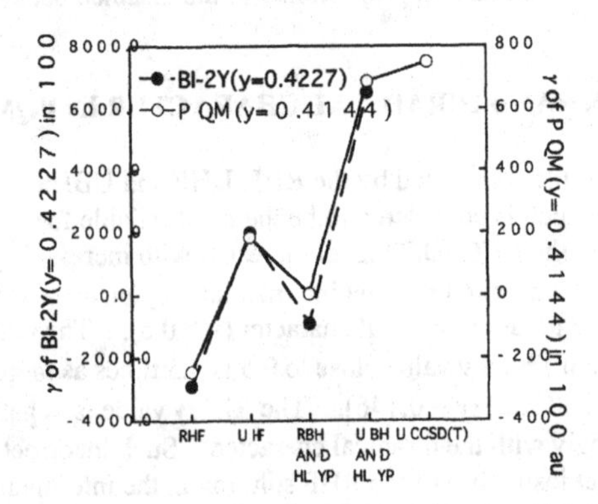

Figure 4. Variation in γ values of BI-2Y and PQM (y = 0.4144) for calculation methods (RHF, UHF, RBHandHLYP, UBHandHLYP and UCCSD(T)).

way, the intermediately spatial polarized wave-functions for α and β electrons, i.e., spin polarization, on both-end carbon sites in the PQM with intermediate diradical character are predicted to contribute to the enhancement of β through the virtual excitation processes involving the zwitterionic nature as compared to the case with small diradical character (stable bond nature), giving a relatively small polarization. On the other hand, in the region with large radical character, the localized spins on both-end sites exhibit less charge polarization to the applied external electric field due to its strong correlation nature, so that the β value decreases again.

SPATIAL CONTRIBUTIONS OF ALPHA AND BETA ELECTRONS TO γ IN BI-2Y

As an example of real diradical systems, we consider BI-2Y shown in Fig. 2(b), which is predicted to have intermediate diradical character y (= 0.4227). Comparisons of γ values at the RHF, UHF, RBHandHLYP and UBHandHLYP levels between PQM model with a similar diradical character (y = 0.4144) and BI-2Y are shown in Fig. 4. Relative variations in γ for the calculation methods are similar to each other though the magnitudes of γ values for BI-2Y are one order of magnitude larger than for PQM model. The γ values at the RHF and RBHandHLYP levels are turned out to be negative in sign, which indicate that the |type (II)| contribution overcomes the |type (III)| contribution. In contrast, the spin-unrestricted methods are shown to correct the sign of γ though the magnitude of γ at the UHF level is still much less than that of the most reliable result obtained using the UCCSD(T) method. Judging from the similarity of the γ values of the PQM model at the UBHandHLYP and UCCSD(T) levels (Fig. 4), the γ value for BI-2Y at the UBHandHLYP level is expected to reproduce that at the UCCSD(T) level. The γ value for BI-2Y at the UBHandHLYP/6-31G*+p level (6534 x 10^2 a.u.) is found to be significantly larger than that for diphenylacetylene (750 x 10^2 a.u.) at the MP2/6-31G*+d level, which is a closed-shell molecule with a similar size π-conjugation length. This result supports our prediction that the open-shell systems with intermediate diradical characters tend to enhance the πvalue compared to the singlet closed-shell systems (y = 0) with a similar π-conjugation length.

(a) UBHandHLYP/6-31G*+p

Total γ = 6458 x 10^2 a.u.

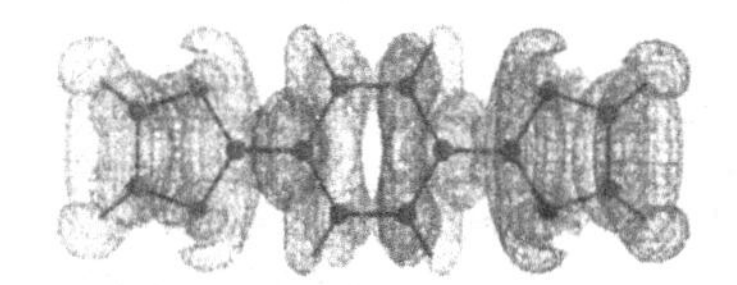

(b) RBHandHLYP/6-31G*+p

Total β= -880 x 10^2 au

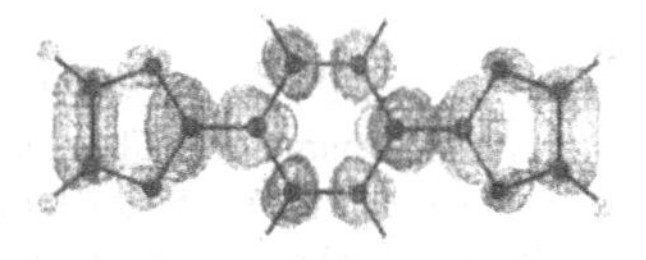

(c) β and β electron contribution at the UBHandHLYP level

α electron contribution (3229 x 10^2 a.u.)

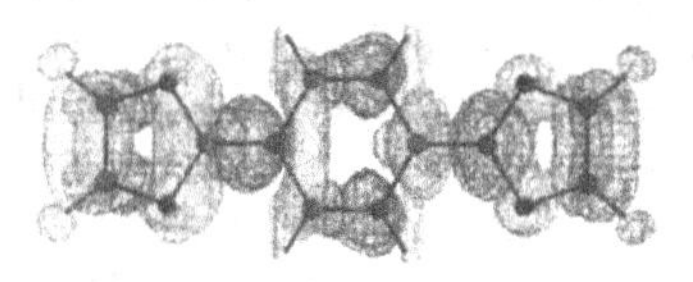

β electron contribution (3229 x 10^2 a.u.)

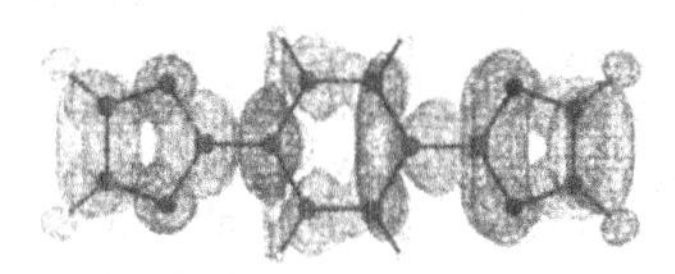

Figure 5. πdensity distributions and πvalues of BI-2Y at the UBHandHLYP (a) and RBHandHLYP (b) levels as well as the α and β electron contributions at the UBHandHLYP level (c). Yellow and blue meshes show the positive and negative βdensity distributions, respectively.

Figure 5 shows the γ density distributions for BI-2Y at the UBHandHLYP (a) and RBHandHLYP (b) levels. The contributions of α and β electrons at the UBHandHLYP level are also shown in Fig. 5(c). The β values integrated by β density calculated using the box dimensions ($-10 \leq x \leq 10$ Å, $-6 \leq y \leq 6$ Å and $-6 \leq z \leq 6$ Å) are found to be in agreement with the FF results within an error of 1-2 %. As expected, the π-electron contributions are dominant at both levels. For the results at the UBHandHLYP level, the positive and negative contributions are significantly cancelled with each other in the central benzene ring, whereas the primary contributions with positive sign come from both-end imidazole rings, leading to the positive total π The shape of the π electron π density distribution, which is asymmetric about the longitudinal axis, turns out to be inverted with respect to its π analog. The π electron π density distributions are larger on the left-hand imidazole ring, while the π electron ones are larger on the right-hand ring. Such features are reflected by the broken symmetry of the unrestricted spin wave-function. For the result at the RBHandHLYP level, the dominant contribution comes from the both-end imidazole ring due to the cancellation between positive and negative densities in the central benzene ring, similarly to the case of UBHandHLYP. However, the amplitude of total π density distribution at the RBHandHLYP level is found to be significantly reduced as compared to that at the UBHandHLYP level in the whole region, and also the sign of density at the RBHandHLYP level is inverted with respect to that at the UBHandHLYP, leading to negative total π with small amplitude. Such difference is predicted to be also observed between the RHF and UHF based results. Judging from the similarity between the π values evaluated at the UBHandHLYP level and UCCSD(T) level, the unrestricted spin based methods are predicted to provide a good starting point for estimating NLO properties of diradical systems with intermediate or strong diradical character.

SUMMARY

We have investigated the π values for open-shell systems and have elucidated their relation with respect to diradical character. The neutral systems with intermediate diradical character exhibit larger π value than those with small or large diradical characters. The enhancement of π in the intermediate diradical character region is predicted to be caused by the virtual excitation processes (type (III)) involving the zwitterionic nature between both-end radical sites. This model is substantiated by the intermediate spin multiplicity state of neutral π-conjugated systems, e.g., quartet C_5H_7, and neutral singlet diradical molecules, e.g., *p*-quinodimethane and BI-2Y. On the other hand, from our previous studies, the introduction of charged defects turns out to cause the relative enhancement of negative contribution (type (II)), so that charged open-shell symmetric systems satisfying SRIP condition, e.g., anion radical pentalene and cation radical pentafulvalene, tend to show negative total π value. As a result, it is predicted that the sign and amplitudes of π values of open-shell NLO systems can be controlled by tuning the spin multiplicity, diradical character and charged states. Furthermore, these systems are expected to be candidates for multi-functional materials, e.g., nonlinear optical property combined with magnetic property and/or electric conductivity, in which these properties could be mutually controllable.

ACKNOWLEDGMENTS

This work was supported by Grant-in-Aid for Scientific Research (No. 14340184) from Japan Society for the Promotion of Science (JSPS). E.B. thanks the Interuniversity Attraction Pole on "Supramolecular Chemistry and Supramolecular Catalysis" (IUAP N°. P5-03) for her postdoctoral grant. B.C. thanks the Belgian National Fund for Scientific Research for his Senior Research Associate position.

REFERENCES

1. M. Nakano and K. Yamaguchi, *Chem. Phys. Lett.* **206**, 285-292 (1993).
2. M.Nakano, I. Shigemoto, S. Yamada and K. Yamaguchi, *J. Chem. Phys.* **103**, 4175-4191 (1995).
3. S. Di Bella, I. Fragalà, I. Ledoux, T. J. Marks, *J. Am. Chem. Soc.*, **117**, 9481 (1995).
4. S. P. Karna, *J. Chem. Phys.* **104**, 6590, (1996) *erratum* 1996, *105*, 6091.
5. M Nakano, S. Yamada and K. Yamaguchi, Synthetic Metals **71**, 1691-1692 (1995).
6. S. Yamada, M. Nakano, I. Shigemoto, S. Kiribayashi and K. Yamaguchi, *Chem.Phys. Lett.* **267**, 438-444 (1997).
7. M. Nakano, S. Yamada and K. Yamaguchi, *Bull. Chem. Soc. Jpn.* **71**, 845-850 (1998).
8. M. Nakano, S. Kiribayashi, S. Yamada, I. Shigemoto, K. Yamaguchi, *Chem. Phys. Lett.* **262**, 66-73 (1996).
9. M. Nakano, S. Yamada and K. Yamaguchi, *J. Phys. Chem. A* **103**, 3103-3109 (1999).
10. S. Yamada, M. Nakano, M. Takahata, R. Kishi, T. Nitta, and K. Yamaguchi, *J. Phys. Chem. A* **108**, 4151- 4155 (2004).
11. M. Nakano and K. Yamaguchi, "Analysis of nonlinear optical processes for molecular systems", *Trends in Chemical Physiscs* Vol. 5 (Research trends, 1997) pp.87-237.
12. M. Nakano, H. Nagao, K. Yamaguchi, *Phys. Rev. A* **55**, 1503 (1997).
13. S. Yamanaka, M. Okumura, M. Nakano, K. Yamaguchi, *J. Mol. Structure* **310**, 205 (1994).
14. M. Nakano, S. Yamada K. Yamaguchi, JCMSE, in press.
15. M. Nakano, T. Nitta, K. Yamaguchi, B. Champagne, E. Botek, *J. Phys. Chem. A*, **108**, 4105 (2004).
16. M. Nakano, B. Champagne, E. Botek, R. Kishi, T. Nitta, K. Yamaguchi, Proceedings of CIMTEC, Synthetic Metals., in press.
17. B. Champagne, E. Botek, M. Nakano, T. Nitta, and K. Yamaguchi, J. Chem. Phys., submitted.
18. K. Yamaguchi, T. Kawakami, D. Yamaki, Y. Yoshioka, "*Theory of Molecular Magnetism*", "*Molecular Magnetism*", K. Ito, M. Kinoshita, Eds., (Kodansha, and Gordon and Breach, 2000), pp.9-48.
19. K. Yamaguchi, "*Self-Consistent Field Theory and Applications*", R. Carbo and M. Klobukowski, Eds. (Elsevier: Amsterdam, 1990), p. 727.
20. K. Yamaguchi, M. Okumura, K. Takada, S. Yamanaka, *Int. J. Quantum Chem. Symp.* **27**, 501 (1993).
21. M. Nakano, R. Kishi, T. Nitta, T. Kubo, K. Nakasuji, K. Kamada, K. Ohta, B. Champagne, E. Botek, and K. Yamaguchi, J. Phys. Chem. A, submitted.
22. D. Scheschkewitz, H. Amii, H. Gomitzka, W. W. Schoeller, D. Bourissou, G. Bertrand, *Angew. Chem. Int. Ed.* **43**, 585 (2004)

23. D. R. McMasters, J. Wirz, *J. Am. Chem. Soc.* **123**, 238 (2001).
24. H. Sugiyama, S. Ito, M. Yoshifuji, *Angew. Chem. Int. Ed.* **42**, 3802 (2003).
25. E. Niecke, A. Fuchs, F. Baumeister, M. Nieger, W. W. Schoeller, *Angew. Chem. Int. Ed. Engl.*, **34**, 555 (1995)
26. G. J. B. Hurst, M. Dupuis, E. Clementi, *J. Chem. Phys.* **89**, 385 (1988).
27. B. Champagne, B. Kirtman, in Chap. 2, "*Handbook of Advanced Electronic and Photonic Materials and Devices*", Vol. 9, "*Nonlinear Optical Materials*", H. S. Nalwa, Ed. (Academic Press, New York, 2001), pp.63.
28. M.J. Frisch, G.W. Trucks, H.B. Schlegel, G.E. Scuseria, M.A. Robb, J.R. Cheeseman, V.G. Zakrzewski, J.A. Montgomery, R.E. Stratmann, J.C. Burant, S. Dapprich, J.M. Millam, A.D. Daniels, K.N. Kudin, M.C. Strain, O. Farkas, J. Tomasi, V. Barone, M. Cossi, R. Cammi, B. Mennucci, C. Pomelli, C. Adamo, S. Clifford, J. Ochterski, G.A. Petersson, P.Y. Ayala, Q. Cui, K. Morokuma, D.K. Malick, A.D. Rabuck, K. Raghavachari, J.B. Foresman, J. Cioslowski, J.V. Ortiz, B.B. Stefanov, G. Liu, A.P. Liashenko, A.P. Piskorz, I. Komaromi, R. Gomperts, R.L. Martin, D.J. Fox, T. Keith, M.A. Al-Laham, C.Y. Peng, A. Nanayakkara, C. Gonzalez, M. Challacombe, P.M.W. Gill, B.G. Johnson, W. Chen, M.W. Wong, J.L. Andres, M. Head-Gordon, E.S. Replogle, and J.A. Pople, GAUSSIAN 98, Revision A11, Gaussian Inc., Pittsburgh, PA, 1998.
29. E. Botek, B. Champagne, *Chem. Phys. Lett.* **387**, 130 (2004).
30. H.D. Cohen and C.C.J. Roothaan, *J. Chem. Phys.* **43**, S34 (1965).
31. M. Okumura, S. Yamanaka, W. Mori, K. Yamaguchi, *J. Mol. Structure*, **310**, 177 (1994).
32. D. J. Thouless, "*The Quantum Mechanics of Many-Body Systems*" (Academic Press, New York 1961).
33. M.-A. Ozaki, *J. Math. Phys.*, **26**, 1521 (1985).

Nonlinear Optical Materials

Mater. Res. Soc. Symp. Proc. Vol. 846 © 2005 Materials Research Society DD2.3

Optical Responses of Conjugated Polymers by TDDFT in Real-Space and Real-Time Approach

Nobuhiko Akino and Yasunari Zempo
Tsukuba Research Laboratory
Sumitomo Chemical Co., Ltd.
6 Kitahara, Tsukuba, 300-3294
JAPAN

ABSTRACT

The time dependent density functional theory (TDDFT) has applied to study the optical responses of the conjugated polymers such as poly(*p*-phenylenevinylene) and poly(9,9-dialkyl-fluorene). In our study, the real-space grid representation is used for the electron wavefunctions in contrast to a conventional basis set on each atom. In the calculations of the optical responses, the real-time approach is employed, where we follow the linear responses of the systems under externally applied perturbations in the real time. Since a real polymer is too large to handle, we have calculated the oligomers with different length and observed the spectrum peak is redshifted as the length of oligomer increases. The property of the polymer is extrapolated as the infinitely long oligomer. The estimated polymer spectra agree with the experiments reasonably well.

INTRODUCTION

Polymer light emitting diodes (PLEDs) have been of interest for displays and other lighting applications. The low cost and the ease of processing are the advantages of the conjugated polymers over the inorganic materials and the low molecules as the former can be deposited by spin-coating over large area. Moreover, the color tuning and efficiency are considered to be controlled by the manipulation of the molecular structures. These make the conjugated polymers good candidates in the LEDs and other applications.

The simplest PLED consists of the polymer layer sandwiched between the cathode and the anode. The radiative recombination of the injected electrons and holes in the polymer layer results in the emission of light. The emission color is determined by the nature of polymer and the device optimization requires us to understand the fundamental physics of the charge injection, transport, and recombination. Thus, it is essential to study from both the material design and the device optimization to achieve better LEDs.

In this study, we have focus on the spectra of the polymers, which is one of the most important properties in real applications as it determines the emission frequency. To study the spectra of conjugated polymers, we have used the time dependent density functional theory (TDDFT), which has been widely used and recognized as a powerful tool in studies of the electron excitations, the dielectric properties, and the optical properties. However, since a real polymer is too large to handle, we have performed the calculations of the oligmers with different length and attempted to extrapolate the properties in the polymer.

In the next section we describe the theory with our numerical details. Then, the results of its applications to poly(*p*-phenylenevinylene) and poly(9,9-dialkyl-fluorene) are given. In the last section, we summarize our results with discussion.

THEORY AND NUMERICAL DETAILS

The fundamental equation of time dependent density functional theory (TDDFT) is given by

$$\left\{-\frac{\hbar^2}{2m}\vec{\nabla}^2+\sum_a V_{ion}(\vec{r}-\vec{R}_a)+e^2\int d\vec{r}'\frac{n(\vec{r}',t)}{|\vec{r}-\vec{r}'|}+V_{xc}[n(\vec{r})]+V_{ext}(\vec{r},t)\right\}\psi_i(\vec{r},t)=i\hbar\frac{\partial}{\partial t}\psi_i(\vec{r},t) \quad (1)$$

where $n(\vec{r},t)=\sum_i|\psi_i(\vec{r},t)|^2$ is the electron density[1]. The V_{xc} is the exchange-correlation potential and the usual local density approximation (LDA) is used in our study. The Troullier-Martins pseudopotentials in the separable form are also used so that the only valence electrons are considered[2]. In order to calculate the optical responses[3], first we have computed the optimized electron density for a given structure by the time independent DFT[4,5] where the real-space grid representation[6] is used for the electron wavefunctions in contrast to a conventional basis set on each atoms. Then, we applied an external field to the system as a perturbation and followed the linear responses of the system in the real time. The electron wave functions are perturbed by a phase factor,

$$\phi_i(t=0)=e^{ikx}\phi_i^{(0)} \quad (2)$$

The wave functions are then evolve by the time dependent equation of motion eq.(1). For the time evolution of wave function, we have used the Taylor expansion of the time evolution operator up-to 4th order.

For the optical response, the desired physical observable is the dynamic polarizability $\alpha(\omega)$, which is equal to the Fourier transform of the time dependent dipole moment $\mu(t)$,

$$\alpha(\omega)=\frac{1}{k}\int dt\, e^{-i\omega t}\mu(t) \quad (3)$$

where $\mu(t)$ is given by

$$\mu(t)=\frac{e^2}{\hbar}\int dt\, e^{i\omega t}zn(\vec{r},t) \quad (4)$$

Another useful quantity is the strength function $S(\omega)$, whose integral is the total oscillator strength and is given by

$$S(\omega)=\frac{2m\omega}{\pi\hbar e^2}\mathrm{Im}\,\alpha(\omega) \quad (5)$$

In our calculations, the uniform spatial grids are used to represent the electron wave functions. The important parameters are the grid spacing Δx, and the total number of grid points Nx. The grid spacing Δx is determined so that the Kohn-Sham eigenvalues converge within 0.1 eV, and this accuracy is achieved with $\Delta x = 0.3$ Å in our systems. The volume, $Nx(\Delta x)^3$, is determined by setting the distance between the edge and any atoms to be at least 4 Å.

In the real time calculations, the time step Δt and the total number of time step T have to be considered as additional important parameters. The time step Δt has to be short enough so that the single-particle Hamiltonian can be treated as static. One may expect the dependence $\Delta t \sim (\Delta x)^2$. We have found that $\Delta t = 0.001\hbar/eV$ is needed for carbon structures with a grid size of $\Delta x = 0.3$ Å. The total length of time evolution, T, is related to the effective energy resolution in the strength function. As our preliminary test, $T = 10\hbar/eV$ gives the energy resolution of $\hbar/T = 0.1eV$. An example of the effect of the total length of time evolution is shown in Figure 1.

where the molecule C_{60} is considered. The solid and dashed lines correspond to the case of T=5000 and 2000 respectively. The energy resolution is clearly improved in the longer time length where one can clearly see the low erngy absorption peaks. The calculated low energy peaks are ~3.5, ~4.4, and ~5.6 eV, which agree well with the experimentally observed peaks at ~3.8, 4.8, and 5.8 eV. However, since the total time length of T=5000 in this example is not long enough, the results may need some caution.

One of the advantages of the real-time approach is that the CPU time scales with the number of particles N and the spatial dimension D as $ND \sim N^2$, which becomes advantage especially when we need to handle a large system. Another advantage of the real-time calculation is that one can obtain spectra in whole energy region which can be seen in Figure 1.

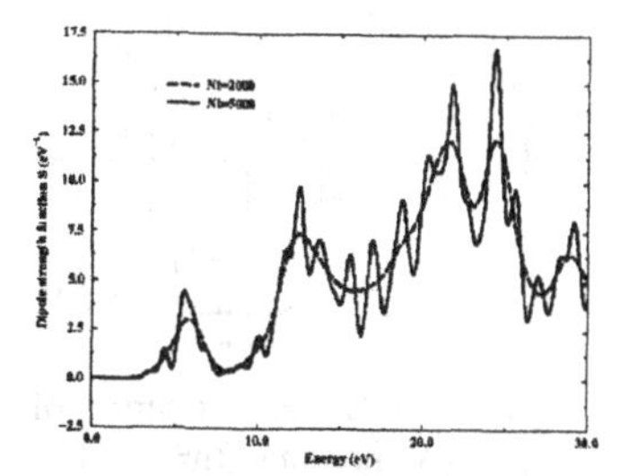

Figure 1. Strength function of the molecule C_{60}. The solid and dashed lines correspond to T=5000 and 2000, respectively.

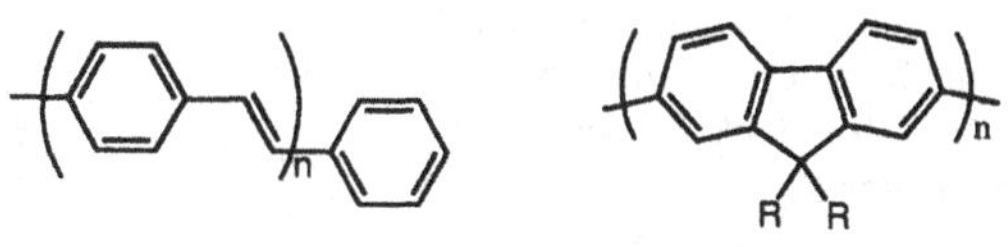

Figure 2. Structures of oligo(*p*-phenylenevinylene) PPVn [left]and oligo(9,9-dialkyl-fluorene) [right]where n is the number of repeating unit.

RESULTS

Poly(*p*-phenylenevinylene) : PPV

Oligo- and poly(*p*-phenylenevinylene) and its derivatives have been intensively studied both experimentally and theoretically as their potential applicability to the PLEDs. The structures of oligo-PPV*n* used in our calculations are shown in Figure 2, where *n* represents the number of repeating units. Since it is impossible to study the optical response of the polymer itself, we have considered the oligo-PPV*n* to extrapolate the property of the polymer. This is important because the understanding of the relation between the length *n* and the optical property is considered as one of the essential information to design new molecules, thus polymers. The calculated absorption spectra for n=1 to 4 are shown in Figure 3. One can clearly see that the energy of the peak decreases as the number *n* increases. To estimate the polymer property, we have plotted the energy at peak as an inversely proportional function with *n* as shown in Figure 4 and extrapolated the value at $n = \infty$ as a polymer property. The estimated energy at peak is ~1.93eV, which is smaller than the experimentally observed energy of ~2.2eV[7]. This inconsistency is due to the problem of DFT which underestimates the band gap.

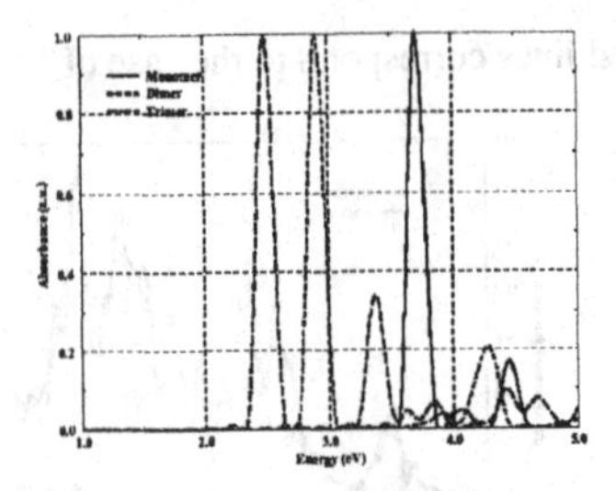

Figure 3. Optical properties of the PPV monomer (n=1, solid line), dimmer (n=2, dashed line), and trimer (n=3, long dashed

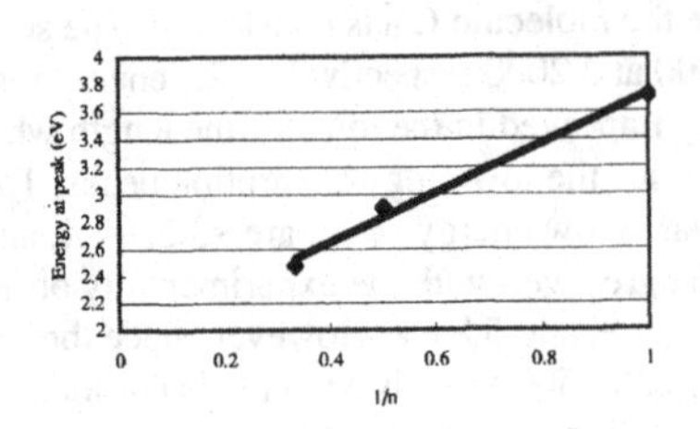

Figure 4. The relation between the peaks and the inverse of the number of repeating units.

Poly(9,9-dialkyl-fluorene)

This material is also well known and has been well studied as a candidate for the blue emitting material in PLED[8]. The structure of oligo-(9,9-dialkyl-fluorene) is shown in Figure 2. For this material, we have performed the calculations with n=2,3 and 4, and the obtained spectra are shown in Figure 5. The peak wavelength are estimated at ~370, 412, and 440 *nm* for n = 2, 3, and 4, respectively. These values should be compared with the experimatally observed peaks at ~329, 350, and 362 *nm*. The systematic overestimation of peak wavelength is due to the inherent problem in DFT as mention above. The redshift of peak wavelength as a function of n is observed, which give us the insight to the n-dependence of the band gap in fluorene.

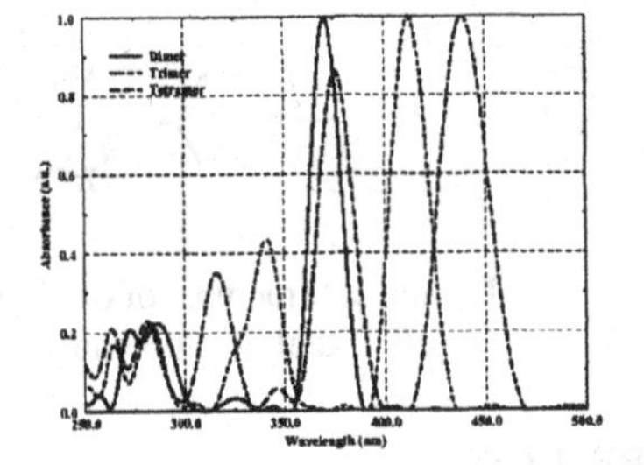

Figure 5. Optical properties of the fluorene dimer (n=2, solid line), trimer (n=3, dashed line), and tetramer (n=4, long dashed line).

CONCLUSIONS AND DISCUSSIONS

We have applied the TDDFT to study of the optical responses of conjugated polymers. In our calculations, the real-space and real-time approach is used. In the real-space calculation, the optimaized electron density for a given structure is calculated with the uniform spatial grid representation. To study the optical response, we have applied the external field and followed the time evolution of the dipole moment as a function of time from which the dynamic polarizability and the strength functions can be calculated.

This method has been applied to study the optical responses of oligo-(p-phenylenevinylene) and oligo-(9,9-dialkyl-fluorene) . In both cases, the peak wavelength are redshifted as the number of units, n, increases. The linear relation between the peak wavelength and the $1/n$ is observed in both materials. Then, the responses of the polymer are estimated by extrapolation at $n = \infty$.

As the number of units, n, in this study is small compared to the real polymer, it is interesting to study the oligmers with larger n to see whether or not the linear relation continues.

ACKNOWLEGMENTS

We acknowledge Prof. Yabana for providing a code as our base code and also for useful discussions and encouragements. Part of calculations in this study has been performed at the system of National Research Grid Initiative (NAREGI) in Japan.

REFERENCES

1. E. Runge and E. K. U. Gross, Phys. Rev. Lett. **52**, 997 (1984)
2. N. Troullier and J. L. Martins, Phys. Rev. **B43**, 1993 (1991)
3. K. Yabana and G. F. Bertisch, Phys. Rev. **B54**, 4484 (1996)
4. P. Hohenberg and W. Kohn, Phys. Rev. **136**, B864 (1964)
5. W. Kohn and L. J. Sham, Phys. Rev. **140**, A1133 (1965)
6. J. Chelikowsky, N. Troullier, K. Wu, and Y. Saad, Phys. Rev. 50, 11355 (1994)
7. J. Cornil, D. Beljonne, Z. Shuai, T. W. Hagler, I. Campbell, D.D.C. Bradley, J. L. Bredas, C. W. Spangler, and K. Muellen, Chem. Phys. Lett. **247**, 425 (1995)
8. A. W. Grice, D. D. C. Bradley, M. T. Bernius, M. Inbasekaran, E. P. Woo, W. W. Wu, Appl. Phys. Lett. **75** (1990)3270

Mater. Res. Soc. Symp. Proc. Vol. 846 © 2005 Materials Research Society DD2.4

Writing Laser Power Dependence of Second Harmonic Generation in an Optically Poled Azo-dye/polymer Film

C. H. Wang, Oliver Y.-H. Tai, and Yuxiao Wang
Center of Organic Materials for Advanced Technology and Department of Physics
National Sun Yatsen University, Kaohsiung, Taiwan 80424

Abstract

The time and power dependence of the optical poling process of an azo chromophore in the PMMA matrix at room temperature is investigated. A model previously proposed is used to account for the writing process of the optical poling of the chromophore/PMMA system. Theoretical predictions of the growth rate and the plateau SHG intensity are found to be in good agreement with the experimental result.

Introduction

Second harmonic generation (SHG) induced in optically poled chromophores in the polymer film has attracted much attention in recent years, due to the potential application in electro-optic modulation and frequency doubling devices [1-4] An optically isotropic (or centrosymmetric) medium does not exhibit a second order nonlinear optical process. To induce the second order optical nonlinearity in a centrosymmetric medium, local centrosymmetry needs to be broken. This can be accomplished by applying an external electric field on the sample. Local centrosymmetric symmetry can also be broken by a simultaneous pumping of the medium with a fundamental field E_{ω} and a second harmonic field $E_{2\omega}$, known as optical poling[1]. If both fundamental and second harmonic beams propagate co-linearly, the second order optical susceptibility $\chi^{(2)}$ is proportional to $E_{\omega}(z)^2E_{2\omega}(z)^*\exp(i\Delta kz)$. Here z is the propagation coordinate and $\Delta k(=k_{2\omega}-2k_{\omega})$ is the wave vector mismatch. In contrast to electric filed poling, no electrodes are needed to create a macroscopic second order optical nonlinearity, and undesirable effects such as charge injection and inhomogeneous space charge distribution associated with electric poling can be avoided.

Previous studies of the optical poling effect mainly deals with the reading process, especially with the decay of the SHG intensity [4-10]. The power dependence of the writing process has however not been adequately studied; the understanding of the dependence of these parameters is important for devise consideration. We report here results on the time and laser power dependence of the writing process of a nonlinear optical polymer film consisting of an azo-chromophore in polymethylmethacrylate (PMMA).

Experiment

The azo-chromophore used in the experiment is n-[((4'-[(N, N-diethyl) amino-4-azobenzene] oxyl) carbonyl)-1-propyne (NACP), which is synthesized using a scheme similar to Tang, et al.[11] The molecular structure of the chromophore is given by:

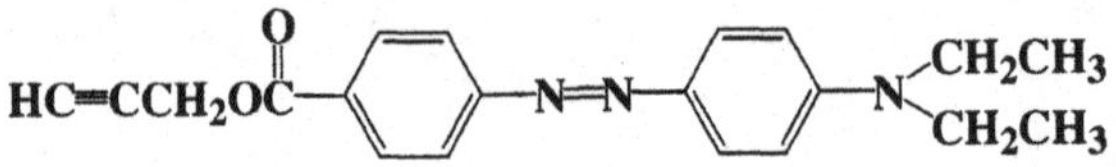

$$HC{\equiv}CCH_2OC(=O)-C_6H_4-N{=}N-C_6H_4-N(CH_2CH_3)_2$$

In THF (tetrahydrofuran) solution, the chromophore has a strong structure-less absorption band in the region of 360-550 nm, characteristic of the azo group. PMMA used has a weight average molecular weight $M_w \approx 12,000$. The NACP/PMMA film used in the experiment is made first by dissolving NACP (10 wt. %) and PMMA (90 wt. %) in chloroform, and then spin-coating the solution with appropriate viscosity on a glass substrate. Residual solvent is removed by placing the film on the glass substrate in a vacuum oven at 40^0 C overnight. Optical experiment is carried out at room temperatures.

A mode-locked Nd: YAG laser (Spectra Physics Pro-190) delivering 10 ns pulses at 1064 nm with a 50 Hz-repetition rate provides the fundamental laser beam. A BBO crystal is inserted in the fundamental beam to frequency double part of the beam to yield the second harmonic radiation at the wavelength of 532nm. A polarizer is inserted to ensure parallel polarization of the two beams. For the detection, the writing process is interrupted at a scheduled interval by inserting a green (532 nm) blocking filter, hence leaving only the fundamental beam to induce SHG in the sample. The induced SHG radiation is detected by a photomultiplier tube (PMT). For the SHG detection, a shutter is placed in front of the PMT, which is synchronized to become open upon the insertion of the green blocking filter.

Results and Discussion

We measure the SHG intensity induced in the NACP/PMMA film as a function of time. The SHG signal grows gradually and reaches a plateau in about 100 minutes. As the writing process is stopped by blocking the second harmonic pumping component, the SHG signal starts to decay with a very long decay time, exceeding 10 hours. The decay portion can be fit to two exponentials with rather different relaxation times.[12]

The gradual growth and long decay process suggests that the trans-cis transformation resulting from the optical field-electron interaction. The dynamical process involved in the present optical poling experiment is associated with the change of the orientational distribution of azo chromophores in the trans-state oriented to yield a non-centrosymmetric environment, which is required for the occurrence of second harmonic generation (SHG). In a recent work, we have proposed a phenomenological equation describing the orientation dynamics of the chromophore involved in optical poling.[12] The model is built upon focusing on the population of azo-chromophores that is optically poled. A kinetic equation is written for the poled population, which is then solved and the solution is related to the second harmonic intensity, given by[12]

$$I_{SHG}(t) = \eta I_{\omega}^{2} N_{1}^{2}(t) = I_{o}[1 - \exp(-\gamma t)]^{2} \tag{1}$$

The equation allows one to describe the time and power dependence of the second order optical susceptibility induced by optical pooling. The duration for the SHG intensity to reach the plateau is determined by the effective rate constant γ; parameter I_0 represents the plateau of the SHG intensity. Here η is a parameter that depends on the optical design employed in the experiment and is independent of the laser intensity. $I_0 = \eta I_{\omega}^{2} K^{2} / (K + \tau^{-1})^{2}$ and $\gamma = (K + 1/\tau)$, where $K = \phi I_{\omega}^{2} I_{2\omega}$. Equations (1) together with I_0, and γ provides one with a description of the laser power dependence of the writing process.

We carry out measurements of the time dependence of the writing process using different fundamental laser intensities. Shown in Fig. 1 is the growth portion of the SHG intensity data using different fundamental laser intensities at 22, 24, 27, 31 and 36 mW.

The data cover the time dependence at the beginning of the writing process to time when the plateau intensity is reached. By fitting the observed intensity curves to Eq. (1), we have extracted the dependence of γ and I_0 on I_ω (the intensity of the fundamental laser beam).

One notes that as the writing laser power increases, both I_0 and γ increase with increasing the writing laser power. From fitted results, we verify the predicted laser power dependence of I_0 and γ. The effective growth rate γ is the sum of two terms: the population relaxation rate τ^{-1}, and the quantity K, which is equal to $\phi I_\omega^2 I_{2\omega}$. Since $I_{2\omega}$ used in the present optical poling experiment is obtained by doubling the fundamental laser beam with a BBO crystal, the quantity K depends effectively on the fourth power of the fundamental laser intensity; i.e., $K = \phi \varepsilon I_\omega^4$, where ε is the doubling efficiency of the BBO crystal. We fit the power dependence of γ to the ($\alpha + \beta I_\omega^4$)

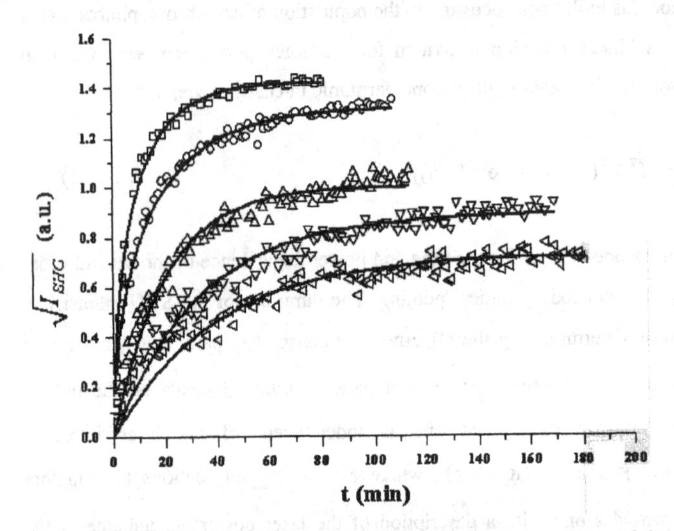

Fig.1: The growth portions of the SHG intensity curves induced by using different writing fundamental laser intensities in a NACP/PMMA film: Dots are experiment data; the solid line is the theoretical fit using Eq. (1). Different curves correspond to different fundamental laser intensities at 22mW (slanted triangles), 24mW (inverted triangles), 27mW (upright triangles), 31mW (circles), and 36mW (squares) used in the experiment.

function, obtaining $\alpha = 0.006$ min^{-1} and $\beta = 8.51x10^{-8}$ min^{-1} mW^{-4} as the best fit, as shown by the solid curve connecting the squares in Fig. 2. The fit shows that population decay rate ($1/\tau$) is 0.006 min^{-1}; hence at the laser power level used in the experiment, the laser power dependence term βI_ω^4 dominates the γ contribution and the contribution to the effective growth rate γ associated with the population decay rate is insignificant. In passing, we have also fitted γ to $(\alpha I_\omega^2 + 2D)$, suggested by ref. 9; the fit gives a negative D and moreover αI_ω^2 does not give good fit to the power dependence of γ. Here D is the rotational diffusion coefficient.

Using the fact that the population decay rate makes an unimportant contribution to effective growth rate γ, one obtains the plateau parameter I_0 to be proportional to I_ω^2. Because the factor K^2 that appears both in the nominator and denominator cancels each other out, the plateau intensity has a quadratic reading power dependence.

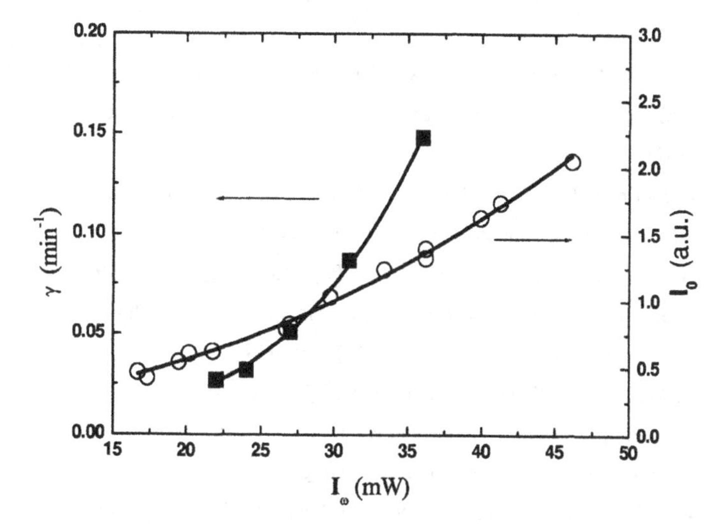

Fig.2: Left vertical axis shows the dependence of the effective growth rate γ on the intensity of the fundamental laser beam. (Solid squares are experimental data and curve is the fit to Eq. (2)). Right vertical axis shows the dependence of the plateau SHG intensity I_o plotted versus the reading laser intensity. Experimental data are empty circles; the solid curve is the theoretical fit Eq. (3), which gives essentially the quadratic power dependence.

In summary, we have studied the writing process in the all optical process for a azo chromophore in PMMA film at room temperature. We have investigated the writing laser power dependence for the growth of the SHG intensity in an all-optical poling process. We have used a previously proposed model to account for the experimental result associated with the writing process. We have shown that the effective growth rate of the induced SHG intensity is proportional to the 4th power of the fundamental laser intensity, and the plateau intensity proportional to I_{ω}^2.

Acknowledgments

This work is supported by a grant from the National Science Council of Taiwan and by a Ministry of Education Program for Promoting Academic Excellent of Universities under the grant number 91-E-FA08-1-4.

References

1. F. Charra, F. Devaux, J. Nunzi, and P. Raimond, Phys. Rev. Lett., 68, 2440 (1992)
2. F. Charra, F. Kajzar, J. M. Nunzi, P. Ramond, and E. Idiart, Opt. Lett. 18, 941 (1993).
3. C. Fiorini, F. Charra, and J. Nunzi, J. Opt. Am. B, 11, 2347 (1994).
4. J. Si, and J. Qiu, Appl. Phys. Lett., 77, 3887 (2000).
5. A. Etile, C. Fiorini, F. Charra, J. Nunzi, Phys. Rev. A, 56, 3888 (1997).
6. K. Kitaoka, J. Si, and T. Mitsuyu, Appl. Phys. Lett., 75, 157 (1999).
7. S. Brasselet and J. Zyss, J. Opt. Am. B, 15, 257 (1998).
8. G. Xu, X. Liu, J. Si, and P. Ye, Opt. Lett., 25, 329 (2000).
9. C. Fiorini, F. Charra, J. Nunzi, and P. Raimond, J. Opt. Am. B, *14*, 1984 (1997).
10. C. Fiorini, and J. Nunzi, Chem. Phys. Lett. 286, 415 (1998).
11. B. Z. Tang, X. X. Kong, X. H. Wan, H. Peng, W. Y. Lam, Macromolecules, 31, 2419 (1998)
12. Writing Laser Power Dependence of Second Harmonic Generation in a New Optically Poled Azo-dye/polymer Film, Yuxiao Wang, Oliver Y.-H. Tai ,and C. H. Wang, (to be submitted)

Mater. Res. Soc. Symp. Proc. Vol. 846 © 2005 Materials Research Society DD2.7

Strong Enhancement of the Two Photon Absorption Cross Section of Porphyrin J-Aggregates in water

Elisabetta Collini[1], Camilla Ferrante[1], Renato Bozio[1]
Chemistry Department, University of Padova
Via Marzolo 1, I-35131 Padova, Italy
[1]Consorzio INSTM, Via Benedetto Varchi 59, I-50132 Firenze, Italy

ABSTRACT

Strong enhancement of the two photon absorption (TPA) cross section at 812 nm is observed for tetrakis(4-sulphonatophenyl)porphyrin diacid (H_4TPPS^{2-}) when a J-type aggregate is formed in water, in comparison to the one observed for the H_4TPPS^{2-} monomer in mixture of water, DMSO and urea. Open aperture Z-scan experiments, performed with ultra-short laser pulses, are employed to measure the TPA absorption cross section. The observed enhancement is discussed in terms of possible electronic cooperative effects in the aggregate.

INTRODUCTION

The study of the TPA properties of porphyrins has been the focus of many recent works, because of the possibility to exploit them as active molecules in many different areas of applied research [1]. One of the most intriguing uses of these dyes can be foreseen in the biological and medical field. Because of their relatively high TPA cross section, and their capability to penetrate in biological tissues, they can be exploited in TPA induced emission in the field of bio-imaging [2]. Furthermore, thanks to the capability to produce singlet oxygen that some metal-porphyrins exhibit upon photoexcitation, they have been already employed as active agents in photodynamic therapy [3]. The possibility, intrinsic in the TPA process, to selectively excite only the active molecules placed in the focal volume of the two-photon pumping beam, would be an important achievement for photodynamic therapy.

Porphyrins substituted with ionic groups are capable to promote aggregation of many monomer units when dissolved in a highly polar solvent, like water. J-aggregation is clearly observed for the diacid form H_4TPPS^{2-} even at low concentrations. The photophysical properties of these aggregates have been thoroughly investigated by steady state absorption and emission spectroscopy, resonance Raman spectroscopy and transient absorption experiments with ultrafast laser pulses [4].

Recently a short communication on a covalently bound porphyrin dimer reports a very strong cooperative enhancement of the TPA cross section when this system is pumped at 790 nm [5].

The formation of a J-type aggregate, giving rise to delocalized excitation when light of suitable wavelength is absorbed, should increase the non linear absorption properties of the system, thanks to cooperative effects. Theoretical studies predict an enhancement of the nonlinear optical response in molecular aggregates [6]. Aim of this work is to find an experimental evidence of such behavior.

EXPERIMENTAL DETAILS

Aqueous solutions of H_4TPPS^{2-} (Porphyrin Products, Inc.) are prepared with bidistilled water acidified with HCl (37%) until pH~1.0 is reached. The analytic concentrations of the four solutions examined are: $2.5x10^{-4}$ M, $5.0x10^{-4}$ M, $7.5x10^{-4}$ M and $1.0x10^{-3}$ M. The lowest value is imposed by the detection limits of the open aperture Z-scan technique.

In order to compare the TPA cross section of J-aggregates with the one of the diacid monomer, it is necessary to reach a diacid monomer concentration of $1x10^{-2}$ M, to get a good signal to noise ratio in the Z-scan experiments. Since such a concentration can not be reached in acidified water without aggregate formation, we used a 1:1 mixture of bidistilled water / dimethylsulfoxide (DMSO) (Sigma-Aldrich, Uvasol) to which urea ~1g/ml was added (water/DMSO/urea mixture).

Measurements of the TPA cross section (σ_{TPA}) at 812 nm are performed by the open aperture Z-scan technique [7]. This technique records the sample transmittance, when it is moved along the propagation direction (z-axis) of a focused beam. If nonlinear absorption occurs in the sample, a dip will appear in the Z-scan trace at the focal position. Under suitable conditions, described in details by Sheik-Bahae et al. [7], it is possible, from a fit of the dip, to extrapolate the σ_{TPA} of the sample, usually expressed in Göppert-Mayer (1GM = 10^{-50} $cm^4 \cdot s \cdot photons^{-1} \cdot molecule^{-1}$). The laser source employed is an amplified Ti:sapphire system delivering 150 fs long pulses centered at 812 nm and at 100 Hz repetition rate. Typical pulse energies, after suitable attenuation, are in the range 0.1-0.4 μJ. The beam waist at the focus is 20 ± 5 μm.

RESULTS AND DISCUSSION

The presence or absence of H_4TPPS^{2-} aggregates in all the water solutions investigated with the Z-scan technique has been ascertained through linear absorption spectra. It is well known that aggregates give rise to new bands in the absorption spectra which are attributed to J or H type excitons, depending on the geometry and shape of the aggregate itself [8]. Figure 1 shows the absorption spectra of H_4TPPS^{2-}, normalized with respect to the analytical dye concentration and the optical path length, for three different solutions. The solid line spectrum corresponds to the $2.5x10^{-4}$ M solution of the dye in water at pH ~ 1, where aggregates are present; the dashed line spectrum pertains to the $1x10^{-5}$ M solution in water at pH ~ 1, where only monomer is present;

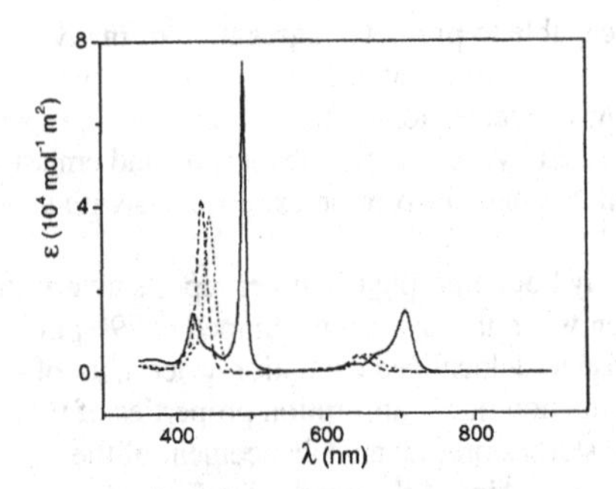

Figure 1. Absorption spectra of H_4TPPS^{2-} in acidified water at c = $2.5x10^{-4}$ M (full line), H_4TPPS^{2-} in acidified water at c = $1.0x10^{-5}$ M (dashed line), H_4TPPS^{2-} in water/DMSO/urea mixture at c = $1.0x10^{-2}$ M (dotted line).

the dotted line spectrum belongs to the $1x10^{-2}$ solution of the dye in water/DMSO/urea mixture.

As it can be readily seen from Figure 1, the dash- and dotted-line spectra, although there is a 10 nm shift for the peaks positions, have a very similar shape and intensity distribution, suggesting that only H_4TPPS^{2-} monomer is present. Another evidence that no aggregation occurs in the water/DMSO/urea mixture is gained by checking that, upon dilution, the bands do not change shape or shift in wavelength.

The spectral shape of the $2.5x10^{-4}$ M solution of H_4TPPS^{2-} in water (full line) differs noticeably form the previous two. In particular there are two peaks in the B band region, at 489 nm and 422 nm, which are ascribed to aggregate formation. The 489 nm peak, red shifted with respect to the monomer B-band, is ascribed to a J-type exciton, while the 422 nm peak, blue shifted with respect to the monomer B-band, is ascribed to an H-type exciton. These two exciton bands arise from the two degenerate transitions contributing to the B-band of the monomer [4].

All the H_4TPPS^{2-} water solutions, in the range of analytic concentration investigated by the Z-scan experiments, show the presence of aggregates. If one considers that the spectra of all four solutions are indistinguishable when normalized with respect to the analytic concentration and optical path length, then the amount of monomer still present should be negligible, within the experimental error. Another information gained form these spectra is that the coherence length of the J-type exciton, that influences the width of the J-aggregate band, does not vary significantly in this concentration range.

Figure 2 shows the Z-scan spectra of the $1x10^{-2}$ M solution of H_4TPPS^{2-} in the water/DMSO/urea mixture (full circles) and of the $1x10^{-3}$ M solution of H_4TPPS^{2-} in water (open squares), measured in the same experimental conditions. The full lines are the result of a fit [7]. This picture clearly shows that the dip in the Z-scan trace of the solution where aggregate is present is almost 3 times deeper than the one observed for the monomer solution, although the analytic concentration of the latter is higher. The fit of the experimental data allows determining values of $\sigma_{TPA} = 30 \pm 15$ GM for H_4TPPS^{2-} in the water/DMSO/urea mixture, where only monomer is present, and $\sigma_{TPA} = 1000 \pm 200$ GM for H_4TPPS^{2-} in water, where only aggregate is present.

The strong enhancement observed can be attributed to the following effects: a) the relative position and shape of the two-photon excited resonance for the aggregate may change with respect to that of the monomer; b) two photon resonance with two-exciton or bi-exciton states, influencing only the TPA cross section of the aggregate;

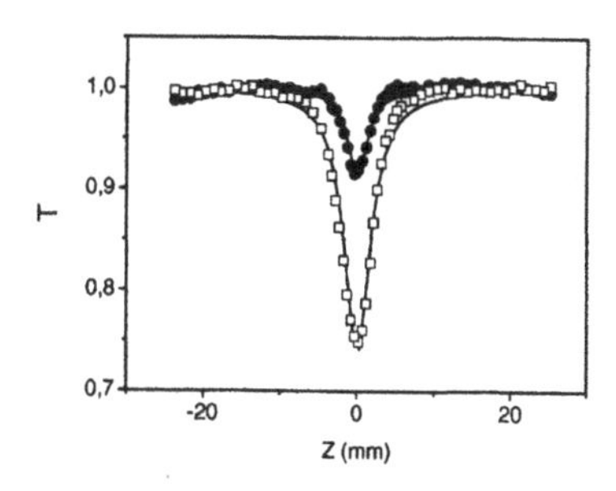

Figure 2 Z-scan traces of H_4TPPS^{2-} in the water/DMSO/urea mixture $C=1x10^{-2}$ M (full circles) and in acidic water $C=1x10^{-3}$ M (open squares), measured in the same experimental conditions.

c) one photon pre-resonance with the Q-bands, which can be more effective in the aggregate than in the monomer.

Contributions from excited triplet state formation can be excluded, since negligible population of these states should form and affect the Z-scan experiment, when 150 fs long pulses at 100 Hz repetition rate are employed.

The first two hypotheses are connected to cooperative effects, which should enhance σ_{TPA} in the aggregate. The influence of the one photon pre-resonance with the Q-bands, can be roughly estimated by means of the following approximate expression for σ_{TPA}, often used to describe TPA spectra [5, 9]:

$$\sigma_{TPA} \propto \frac{|\mu_{10}|^2}{(\omega_{10}-\omega_L)^2} \cdot \frac{|\mu_{12}|^2 \Gamma_{20}}{(\omega_{20}-2\omega_L)^2+\Gamma_{20}^2} \tag{1}$$

where the subscripts 0, 1, 2 refer to the ground state, a one-photon allowed state (in the present work the Q-band) and a two-photon allowed state, respectively; μ_{ij}, ω_{ij}, Γ_{ij} are the dipole moment, the frequency and the HWHM relative to transition between states i and j, respectively; ω_L is the frequency of the incident laser beam. This equation can be used when $|\omega_{10}-\omega_L| >> \Gamma_{10}$ and $|\omega_{10}-\omega_L| >> |\omega_{20}-2\omega_L|$ [9].

The first term on the r.h.s. of Equation 1 describes the one photon pre-resonance and increases monotonously as ω_L approaches ω_{10}, the Q-band transition frequency. All the parameters in it can be evaluated from the linear absorption spectrum of the samples. The ratio of the value calculated for the aggregate over that for the monomer is equal to 6. The pre-resonance effect, therefore accounts only for one fifth of the whole enhancement observed. It is clear that the remaining enhancement is caused by the cooperative interactions, i.e. the aforementioned a) and b) hypotheses.

In order to fully characterize the properties of σ_{TPA} of H_4TPPS^{2-} in its aggregate form it is important also to observe if it varies with the analytic concentration of the water solutions, or with the intensity of the applied laser pulses.

Figure 3 portrays the behavior of σ_{TPA}, measured at 812 nm by open aperture Z-scan experiments under the same experimental conditions, for four solutions with different analytic concentrations. The data show that, within the experimental error, σ_{TPA} is independent of concentration.

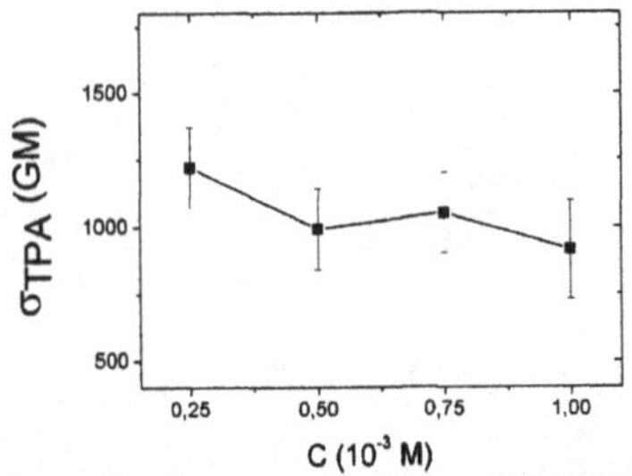

Figure 3 TPA cross sections of H_4TPPS^{2-} aggregates in water as a function of analytic concentration of the solutions. The error bars are estimated from repeated measurements.

Such independence of the cross section σ_{TPA} is in good agreement with the same behavior observed for the linear absorption coefficient of these solutions.
As mentioned above, the molar extinction coefficients of the four solutions are equal in the whole spectral range.

These experimental findings can be rationalized by considering that the coherence length of the excitons, characterizing the linear and non-linear absorption properties of the four solutions investigated, do not vary in this range of concentrations at room temperature. The latter statement does not mean that the effective dimension of the aggregates, i.e. the number of monomers forming it, is invariant in this concentration range. The linear and non-linear absorption properties of an aggregate are solely influenced by the delocalization length of the effective excitation; this dimension is called exciton coherence length and can be smaller than the real length of the aggregate.

Another important factor affecting σ_{TPA} is the intensity of the applied laser pulses. Figure 4 portrays the behaviour of σ_{TPA} as a function of the intensity of the applied laser pulses (calculated with respect to the beam waist at the focal position, and the experimentally measured pulse duration [10]) for the diacid monomer in the water/DMSO/urea mixture (open and full triangles) and for the $1x10^{-3}$ M solution in acidified water, containing the porphyrin aggregate (open and full circles). Open and full symbols refer to two sets of independent measurements. While the value of σ_{TPA} of the monomer seems not to be affected by the intensity of the applied laser pulses, although the high experimental error can mask an intensity dependence, the σ_{TPA} of the aggregate shows a marked increase, as the intensity is increased, well above the experimental error. The behaviour is also reproducible.

This experimental finding points out that the non-linear absorption behaviour in the aggregate is not a pure two-photon absorption process, but is affected by higher order processes, i.e., by terms of the nonlinear susceptibility which are higher than the third order term commonly employed to describe the instantaneous two-photon absorption process. We are currently investigating in more details the origin of this effect.

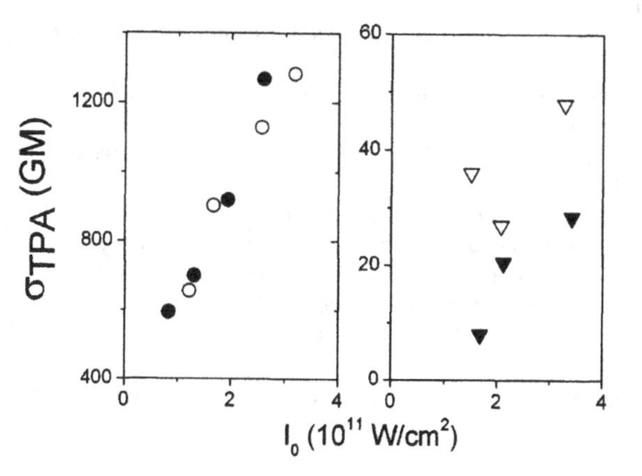

Figure 4 TPA cross section of H_4TPPS^{2-} monomer (triangles) and aggregate (circles) from Z-scan experiments as a function of the applied beam intensity. Open and closed symbols are referred to two sets of independent measurements.

CONCLUSIONS

In this paper a 30-fold enhancement of σ_{TPA} of H_4TPPS^{2-} at 812 nm is observed when aggregation occurs. This enhancement is attributed to cooperative effects, which have been theoretically predicted for molecular aggregates, as well as to a pre-resonance effect with the Q-bands. The latter contribution accounts only for one fifth of the observed enhancement. Furthermore while the σ_{TPA} for H_4TPPS^{2-} monomer seems to be independent from the input pump intensity, although the experimental error can mask such behaviour, the σ_{TPA} in the aggregate increases as the input pump intensity increases.

ACKNOWLEDGMENTS

The authors thank Mr. Gabriele Marcolongo for its technical help. This work was supported by FIRB2001 grant from MIUR (n. RBNE01P4JF) and by the University of Padova.

REFERENCES

1. A. Karotki, M. Drobizhev, M. Kruk, C. Spangler, E. Nickel, N. Mamardashvili, and A. Rebane, *J. Opt. Soc. Am. B* **20**, 321, (2003).
2. W.R. Zipfel, R.M. Williams, and W.W. Webb, *Nature Biotechnology* **21**, 1369, (2003).
3. T.J. Dougherty, *Photochem. Photobiol.* **38**, 337, (1983); M. Yamashita, T. Tamono, S. Kobayashi, K. Torizuka, K. Aizawa, and T. Sato, *Photochem. Photobiol.* **47**, 189, (1988).
4. O. Ohno, Y. Kaizu, and H. Kobayashi, *J. Chem. Phys.* **99**, 4128, (1993); D.M. Chen, T. He, D.F. Cong, Y.H. Zhang, and F.C. Liu, *J. Phys. Chem. A*, **105**, 398, (2001); H. Kano, and T. Kobayashi , *J. Chem. Phys.*, **116**, 184, (2001).
5. A. Karotki, M. Drobizhev, Y. Dzenis, P.N. Taylor, H.L. Anderson, and A. Rebane, *Phys. Chem. Chem. Phys.* **6**, 6, (2004).
6. F.C. Spano, and J. Knoester, *Advances in Magnetic and Optical Resonance*, ed. W.S. Warren (Academic Press, 1994) p 117.
7. M. Sheik-Bahae, A.A. Said, T.H. Wei, D.J. Hagan, and E.W. Van Stryland, *IEEE J. of Quantum Electronics* **26**, 760, (1990).
8. H. Khun, and C. Kuhn, "Chromophore Coupling Effects", *J-Aggregates*, ed. T. Kobayashi (World Scientific, 1996) pp. 1-40.
9. B. Dick, R.M. Hochstrasser, and H.P. Trommsdorff, "Resonant Molecular Optics", *Nonlinear Properties of Organic Molecules and Crystals*, ed. D.S. Chemla and J. Zyss (Academic Press, 1987) pp. 172-178; M. Albota, D. Beljonne, J.L. Brédas, J.E. Ehrlich, J.Y. Fu, A.A. Heikal, S.E. Hess, T. Kogej, M.D. Levin, S.R. Murder, D. McCord-Maughon, J.W. Perry, H. Rockel, M. Rumi, G. Subramanian, W.W. Webb, X.L. Wu, and C. Xu, *Science* **281**, 1653, (1998).
10. Pulse duration is measured through an autocorrelation experiment in a doubling BBO crystal.

Organometallic Optical Materials I

Mater. Res. Soc. Symp. Proc. Vol. 846 © 2005 Materials Research Society DD3.4

Structure Property Characterization of Nonlinear Optical Materials

Joy E. Rogers[a,b], Jonathan E. Slagle[a,c], Daniel G. McLean[a,d], Benjamin C. Hall[a,e], Thomas M. Cooper[a], and Paul A. Fleitz[a]

[a] Materials and Manufacturing Directorate, Air Force Research Laboratory, 3005 Hobson Way Bldg 651, Wright Patterson Air Force Base, Ohio 45433 E-mail: Joy.Rogers@wpafb.af.mil

[b] UES, Inc. Dayton, OH 45432

[c] AT&T Government Solutions, Dayton, OH 45324

[d] Science Applications International Corporation, Dayton, OH 45434

[e] Universal Technology Corporation, Dayton, OH 45431

ABSTRACT

This research is comprised of understanding the linear photophysical properties of various dyes to better understand the more complicated nonlinear optical properties. Determining structure property relationships of a series of structurally closely related chromophores is the key in understanding the drivers for the various photophysical properties. In this paper we survey the effect of physically changing the Pt poly-yne structure on the S_0-S_1 and T_1-T_n absorption properties for each of the chromophores. A series of structurally modified platinum poly-ynes have been studied using experimental methods including UV/Vis absorption and nanosecond laser flash photolysis. We found that with extension of the ligand length both the ground and triplet excited state absorption shift to lower energies. Comparing the absorption properties of the ligands and butadiynes with the platinum containing versions reveal that the S_1 and T_n exciton is localized on one portion of the ligand with extension and not conjugated through the whole molecule. Changing the phosphine R group results in little effect to the absorption properties except when the R group is conjugated in the case of phenyl. However, changing the R group results in varied materials properties.

INTRODUCTION

There has been much interest in the development of organometallic compounds for use in materials science, specifically in the field of developing π-conjugated oligomers and polymers with interesting electronic and optical properties.[1,2,3] Having a transition metal present within the polymer changes the redox, optical, and electronic properties and allows for tunability of these properties. Several reviews were recently published focusing on a variety of transition metal-containing complexes.[4,5,6,7,8] This study is focused on our efforts to synthesize platinum(II) containing phenyl-ethynyl oligomers, which similar systems have been reported on numerous times in the literature by us and others.[9,10,11,12,13,14,15] To understand the structure property relationship of these systems, a series of platinum(II)-containing oligomers with varying ligand lengths and various phosphine groups have been synthesized and are shown in Figure 1. Also for comparison a series of chromophores lacking the central metal (ligands and butadiynes)

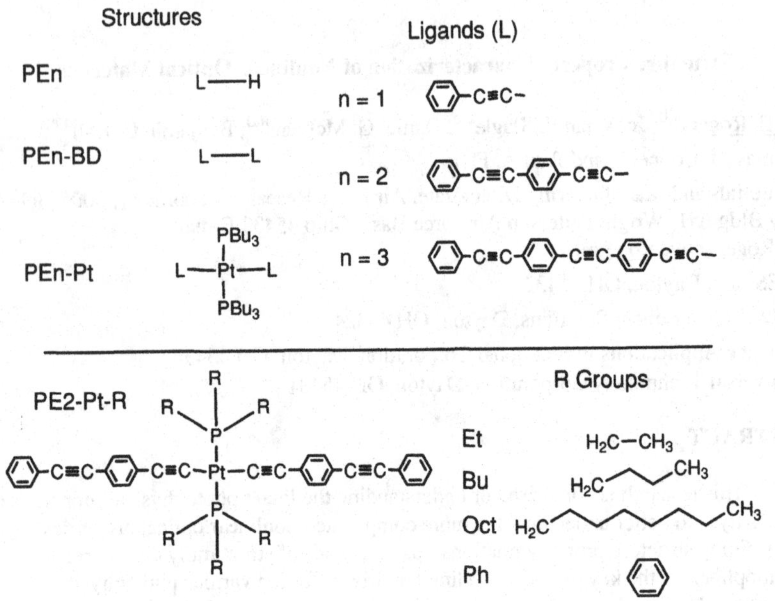

Figure 1. Shown at the top are the structures of the ligands PE1-H, PE2-H, and PE3-H; butadiyne oligomers PE1-BD, PE2-BD, and PE3-BD; and the platinum containing oligomers PE1-Pt, PE2-Pt, and PE3-Pt. Shown below the line are PE2-Pt-Et, PE2-Pt-Bu, PE2-Pt-Oct, and PE2-Pt-Ph.

were either bought (Aldrich) or synthesized and are also shown in Figure 1. In this paper we survey the effect of physically changing the Pt poly-yne structure on the S_0-S_1 and T_1-T_n absorption properties for each of the chromophores.

EXPERIMENTAL DETAILS

Materials shown in Figure 1 were either purchased (Aldrich) or synthesized according to previous methods.[9,11,16] Ground state UV/Vis absorption spectra were measured with a Cary 500 spectrophotometer. Nanosecond transient absorption measurements were carried out using the third and fourth harmonics (355 nm and 266 nm) of a Q-switched Nd:YAG laser (Quantel Brilliant, pulse width ca. 5 ns). Pulse fluences of up to 8 mJ/cm^2 are typically used at the excitation wavelength. A detailed description of the laser flash photolysis has been previously described.[11]

RESULTS AND DISCUSSION

Effect of Ligand Length. Understanding the effect of the ligand length on the overall photophysical properties was our first undertaking to study Pt poly-ynes.[11] In this previously published work we found that extension of the ligand length results in a delocalized S_1 and T_n exciton but to what extent was unknown. To further verify these findings a series of corresponding butadiyne oligomers were synthesized for comparison along with the ligands themselves (Figure 1). Shown in Figure 2 (left) are the ground

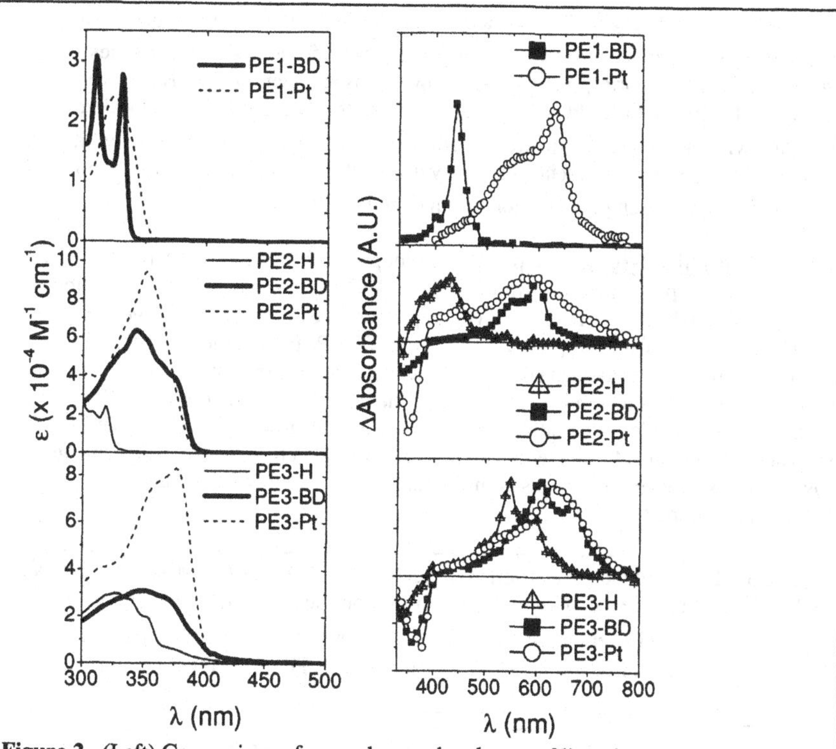

Figure 2. (Left) Comparison of ground state absorbance of ligands (PE2-H and PE3-H), butadiyne oligomers (PE1-BD, PE2-BD, and PE3-BD), and platinum containing oligomers (PE1-Pt, PE2-Pt, and PE3-Pt). All data were obtained at room temperature in benzene. (Right) T_1-T_n absorption spectra observed after nanosecond pulsed 355 nm excitation of PE3-H (8.5 μM), PE2-BD (3.4 μM), PE3-BD (6.0 μM), PE2-Pt (4.5 μM), and PE3-Pt (5.2 μM). PE1-BD (16.0 μM) and PE2-H (14.0 μM) were excited at 266 nm in THF because of small ground state absorption at 355 nm. For PE1-Pt (312 μM) the transient was obtained using picosecond pump-probe absorption exciting at 355 nm. All spectra were normalized at the peak maximum for clarity.

state absorption spectra measured and quantified in benzene. For PE1-H the absorption maximum is at 270 nm in THF. It is interesting to note that the butadiyne spectra overlay well with the corresponding platinum versions and that PE3-H is very similar in energy to the butadiyne and PE3-Pt. This suggests that the S_1 exciton is localized on the PE3 ligand with increasing conjugation and not delocalized through the whole molecule. From this we conclude that extension of the ligand to PE4 (phenyl-ethynyl x 4) would result in ground state absorption of PE4-Pt that is quite similar to PE4-H with a small red-shift.

Also shown in Figure 2 (right) are the T_1-T_n absorption spectra of the various chromophores. Again we find that the butadiyne spectra for PE2 and PE3 species overlay well with the corresponding platinum versions and that the PE3-H spectrum is similar to PE3-BD and PE3-Pt. Thus suggesting that the T_n exciton is also localized on the ligand with increasing conjugation length. For PE1-BD the spectrum is very different from PE1-Pt. In PE1-Pt the metal is strongly involved in the T_1-T_n transition as might be expected due to the strong conjugation through the Pt center.

Effect of Phosphine Group. The phosphine group was changed to show the effect on the photophysical properties. Shown in Figure 3 (left) are the ground state absorption spectra of each in benzene. For the alkyl groups there is essentially no change in the spectra except the molar absorption coefficient for PE2-Pt-Bu is larger. For PE2-Pt-Ph the incorporation of the conjugated phenyl groups results in a red shift of the absorption spectrum. This shows that in general changing the phosphine group does not affect the spectral properties unless a conjugated system is used. We recently reported that PE2-Pt-Oct is actually a liquid at room temperature.[9] Interestingly changing the phosphine group provides a new avenue for changes in materials properties without changing the photophysical properties.

Figure 3. (Left) Ground state absorption spectra of PE2-Pt-Et, PE2-Pt-Bu, PE2-Pt-Oct, and PE2-Pt-Ph in benzene. (Right) T_1-T_n absorption spectra (normalized at the peak maximum) observed after nanosecond pulsed 355 nm excitation of PE2-Pt-Et (3.3 μM), PE2-Pt-Bu (1.8 μM), PE2-Pt-Oct (6.3 μM), and PE2-Pt-Ph (5.1 μM) in benzene.

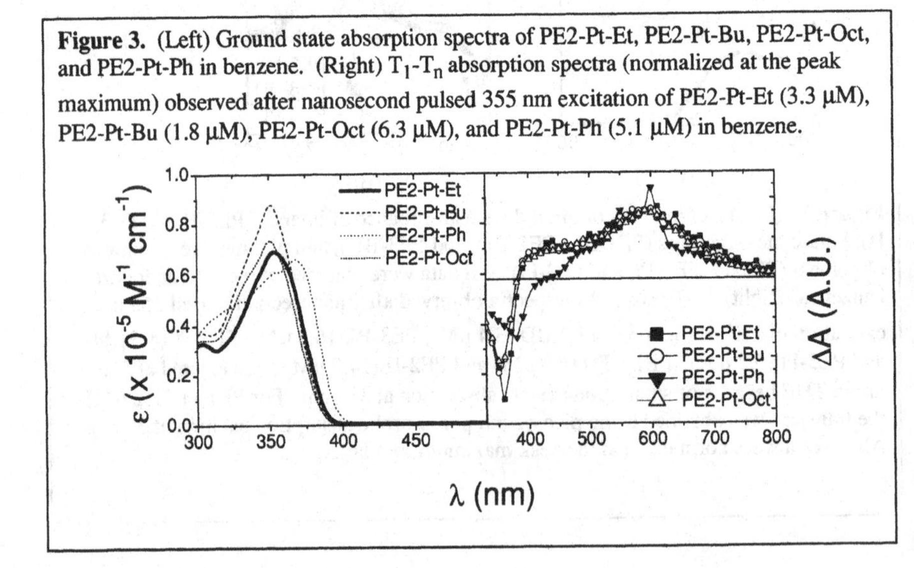

Also shown in Figure 3 (right) are the T_1-T_n absorption spectra obtained in benzene upon 355 nm light absorption. For PE2-Pt-Et, PE2-Pt-Bu, and PE2-Pt-Oct the spectra are very broad and similar with only slight differences in the bleaching region from 350 – 400 nm. Addition of the triphenyl phosphine group results in a more defined T_1-T_n spectrum with small peaks at 550 nm and 600 nm but a similar peak maximum as the others.

CONCLUSIONS

Tailoring the ground state and excited state spectra is accomplished through an understanding of the various measures of modifying the chromophore. We have looked at both changes of the ligand length and the phosphine groups. Through this we found that increasing the ligand length does result in a red shift of both the S_0-S_1 and T_1-T_n absorption spectra but only to a certain point because the exciton becomes more localized on a small part of the chromophore. We found that changing the phosphine group only changes the spectral properties when it is a conjugated group but is important in altering the material properties.

ACKNOWLEDGEMENTS

We are thankful for the support of this work by AFRL/ML Contracts F33615-99-C-5415 for D.G.M., F33615-03-D-5801 for B.C.H., and F33615-03-D-5421 for J.E.S. and J.E.R.

REFERENCES

[1] H.S. Nalwa, *Handbook of Organic Conductive Molecules and Polymers Vols 1-4* (John Wiley & Sons, 1997).
[2] R.A. Skotheim, R.L. Elsenbaumer, J.R. Reynolds, *Handbook of Conducting Polymers, 2nd ed.* (Marcel Dekker, 1998).
[3] A.O. Patil, J. Heeger, F. Wudl, *Chem. Rev.* **88**, 183-200 (1988).
[4] T.M.Cooper, "Transition Metal Acetylide Nanostructures," *Encyclopedia of Nanomaterials and Nanotechnology*, ed. H.S. Nalwa, (American Scientific Publishers, 2003) pp.447.
[5] V. Yam, *Acc. Chem. Res.* **35**, 555, (2002).
[6] V. Yam, K. Lo, K.J. Wong, *Organomet. Chem.* **578**, 3, (1999).
[7] P. Nguyen, P. Gomez-Elipe, I. Manners, *Chem. Rev.* **99**, 1515 (1999).
[8] K.D. Ley, K.S. Schanze, *Coord. Chem. Rev.* **171**, 287, (1998).
[9] T.M. Cooper, B.C. Hall, A.R. Burke, J.E. Rogers, D.G. McLean, J.E. Slagle, P.A. Fleitz, *Chem. Mater.* **161**, 3215, (2004).
[10] L.A. Emmert, W. Choi, J.A. Marshall, J. Yang, L.A. Meyer, J.A. Brozik, *J. Phys. Chem. A* **107**, 11340, (2003).

[11] J.E. Rogers, T.M. Cooper, P.A. Fleitz, D.J. Glass, D.G. McLean, *J. Phys. Chem. A* **106,** 10108, (2002)..

[12] M.I. Bruce, J. Davy, B.C. Hall, Y. Jansen van Galen, B.W. Skelton, A.H. White, A.H. *Appl. Organomet. Chem.* **16**, 559, (2002).

[13] Y. Liu, S. Jiang, K.D. Glusac, D.H. Powell, D.F. Anderson, K.S. Schanze, *J. Am. Chem. Soc.* **124**, 12412, (2002).

[14] A. Kohler, J.S. Wilson, R.H. Friend, M.K. Al-Suti, M.S. Khan, A. Gerhard, H. Bassler, *J. Chem. Phys.* **116**, 9457, (2002).

[15] T.J. McKay, J.A. Bolger, J. Staromlynska, J.R. Davy, *J. Chem. Phys.* **108**, 5537, (1998).

[16] J.E. Rogers, T.M. Cooper, B.C. Hall, D.C. Hufnagle, J.E. Slagle, A.P. Ault, D.G. McLean, P.A. Fleitz, Manuscript in Preparation.

Mater. Res. Soc. Symp. Proc. Vol. 846 © 2005 Materials Research Society DD3.6

Platinum-Functionalized Chiral Molecular Squares as Light-Emitting Materials

Lin Zhang,† Yu-Hua Niu,‡ Alex K.-Y. Jen,‡,* and Wenbin Lin†,*
†Department of Chemistry, CB#3290, University of North Carolina, Chapel Hill, NC 27599, U.S.A.
‡Department of Materials Science and Engineering, Box 352120, University of Washington, Seattle, WA 98195, U.S.A.

ABSTRACT

A family of new chiral metallocycles based on Pt(II) diimine metallocornors and bis(acetylene) bridging ligands have been synthesized, and characterized by a variety of techniques including ^{1}H and ^{13}C NMR, UV-visible, luminescence, infrared, and circular dichroism (CD) spectroscopies, and mass spectrometry. All metallocycles exhibit very strong phosphorescence with quantum yields of 8.3 to 15.7%. Chiral Pt(II)-based molecular squares were used as the light-emitting layer in multiplayer devices, and a maximum brightness of 5470 cd/m^2 with a maximum luminous efficiency of 0.93 cd/A was achieved.

INTRODUCTION

Since Fujita and Stang's seminal work on highly efficient self-assembly of small molecular polygons in the middle of 1990's [1], the synthesis of nanoscopic metallocycles has attracted tremendous interests over the past decade. These metallosupramolecular systems can exhibit interesting functions that are not manifested in the constituent building blocks, including cage-directed stabilization of labile molecules and stereoselective synthesis [2], molecular microporosity [3], and tuning of enantioselectivity of asymmetric catalysts [4]. Metallocycles and metallocages also hold great promise as interesting electrooptical materials owing to their nanoscopic and even mesoscopic nature [5]. The incorporation of metal centers however often introduces non-radiative decay pathways and thus can have deleterious effects in photophysical properties of metallosupramolecular systems. Electroluminescent (EL) devices based on cyclometalated complexes of Ir and Pt have received much attention due to their high quantum efficiencies [6]. Many studies have been reported on the photolumine-scence (PL) of (diimine)Pt(acetylide)$_2$ [7] and the ^{3}MLCT[Pt→π*(diimine)] nature of emissions is now well established [8,9]. We hypothesize that Pt(diimine) moieties can be used as a metallocorner to construct emissive metallocycles by exploiting the ^{3}MLCT-[Pt→π*(diimine)] nature of their emissions. Herein we wish to report the first highly electroluminescent molecular squares.

EXPERIMENTAL DETAILS

Chiral molecular triangle [Pt(4,4'-dtbPy)(L_1)]$_3$ (**1a**) and square [Pt(4,4'-dtbPy)(L_1)]$_4$ (**2a**) were synthesized in 29% and 22% yield via self-assembly between equimolar BINOL-derived 4,4'-bis(alkynyl) linear bridging ligand **L**$_1$-H_2 and Pt(4,4'-dtbPy)Cl_2 in the presence of CuI catalyst and diethylamine at room temperature (Scheme 1). In contrast, a similar reaction between equimolar L_2-H_2 and Pt(4,4'-dtbPy)Cl_2 only afforded molecular square [Pt(4,4'-dtbPy)(L_2)]$_4$ (**2b**) in 17% yield. Molecular squares **2a** and **2b** can also be

Scheme 1. Synthesis of metallocycles **1a**, **2a**, and **2b**

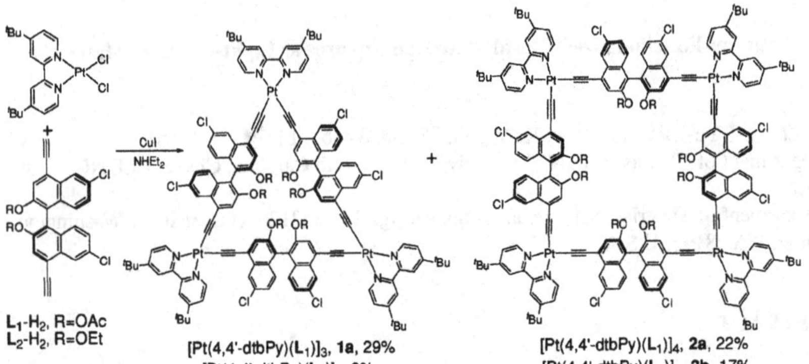

L_1-H_2, R=OAc
L_2-H_2, R=OEt

CuI
$NHEt_2$

[Pt(4,4'-dtbPy)(L_1)]$_3$, **1a**, 29%
[Pt(4,4'-dtbPy)(L_2)]$_3$, 0%

[Pt(4,4'-dtbPy)(L_1)]$_4$, **2a**, 22%
[Pt(4,4'-dtbPy)(L_2)]$_4$, **2b**, 17%

exclusively constructed via stepwise directed-assembly reactions between equimolar Pt(4,4'-dtbPy)Cl_2 and Pt(4,4'-dtbPy)(L_m)$_2$ (**3a**, m=1; **3b**, m=2) in modest yields [10]. All the intermediates and products have been characterized by ^{1}H NMR, ^{13}C{^{1}H} NMR, IR, UV-Vis, fluorescence, and circular dichroism (CD) spectroscopies, MALDI-TOF mass spectrometry, size-exclusion chromatography, and microanalysis.

DISCUSSION

NMR data of all the chiral metallocycles showed a single ligand environment, consistent the formation of cyclic species of D_n symmetry in solution. MALDI-TOF MS data showed the presence of molecular ions due to compounds **1-3** [11]. As expected, the IR spectra showed the disappearance of ν(C≡C-H) stretches at ~3300 cm^{-1} and appearance of strong ν(C≡C) stretches at ~2100 cm^{-1} upon the formation of metallocycles. The formation of **1-3** is further supported by microanalysis results.

In contrast to previously reported metallocycles built from *cis*-Pt(PEt_3)$_2$ metallocorners [12], the formation of **1** and **2** seems to be quite sensitive to the dihedral angle of the two naphthyl units in L_m. This is presumably a result of the rigidity of the Pt(4,4'- dtbPy) metallocorner [13]. Interestingly, although both **1a** and **2a** were isolated from the self--

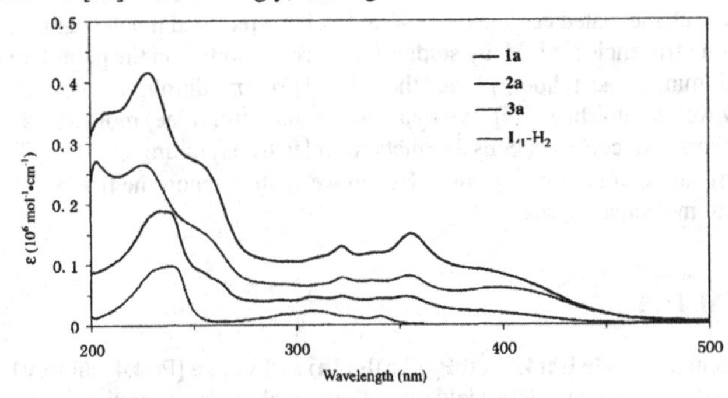

Figure 1. UV-Vis Spectra of L_1-H_2, and **1a**-**3a** in CH_2Cl_2

assembly reaction, **1a** cleanly converts to **2a** in solution with a half-life of 98 hours at 4 mM in CH_2Cl_2 at room temperature.

The electronic spectra of **1a-3a** indicated that the intraligand $\pi \rightarrow \pi^*$ transitions for L_1-H_2 have red-shifted by ~15 nm upon coordination to the Pt(II) centers. Such bathochromic shifts have been well-established in platinum acetylides, owing to the mixing of Pt p-orbitals into the acetylenic $\pi \rightarrow \pi^*$ bands (Figure 1). A new broad band appears at ~400 nm in the Pt(II) complexes and can be assigned to the ^{1}MLCT [Pt $\rightarrow \pi^*$(4,4'-dtbPy)] transition. The CD spectra of these complexes exhibit similar bisignate features to L_1-H_2, assignable to the naphthyl $\pi \rightarrow \pi^*$ transitions. No CD signals were observed in the region that can be attributed to the ^{1}MLCT bands.

1a-3a all exhibit very strong phosphorescence at ~565 nm (Φ_p~8.3–15.7%) in addition to weak fluorescence at ~440 nm in CH_2Cl_2 at room temperature (Table 1). Such efficient singlet-to-triplet intersystem crossing has been previously observed in *cis*-$Pt(PEt_3)_2$–based metallocycles and is undoubtedly assisted by heavy Pt(II) centers [14]. Phosphorescence emanating from the ^{3}MLCT[Pt$\rightarrow \pi^*$(diimine)] transition has been well-established in simple (diimine)$Pt(acetylide)_2$ complexes, and we believe that such triplet excited states can be utilized to construct highly efficient light-emitting devices (LEDs).

Compounds **2a** and **3a** were chosen for device fabrication owing to their high stability and Φ_p's. Photoluminescence studies showed thin films of **2a** and **3a** have significantly red-shifted and broadened emissions at ~730 nm, suggesting severe aggregation of **2a** and **3a** in thin films (Figure 2). Such aggregation is undoubtedly due to the strong intermolecular π-π stacking interactions as a result of the highly aromatic nature of the metallocycles. Weak fluorescence peaks at ~440 nm observed in solutions of **2a** and **3a** have also completely disappeared, probably a result of further facilitation of intersystem crossing in the thin films. Consistent with the aggregation behaviors of **2a** and **3a** in thin films, LED devices with structure ITO/PEDOT:PSS/EL layer/CsF/Al, where PEDOT:PSS denotes poly(ethylene dioxythiophene) doped with poly(styrene sulfonate) and EL layer is **2a** or **3a**, gave very broad emissions at ~730 and showed rather poor performance with highest brightness of < 20 cd/m^2 (Figure 3). Such poor device performance is presumably a result of concentration (aggregation) quenching of triplet emissions.

Table 1. Photoluminescence properties of **1-3**

Complex	λ_{ex} (nm)	$(\lambda_f)_{max}$ (nm)	Φ_f (%)[a]	$(\lambda_p)_{max}$ (nm)	Φ_p (%)[a]
1a	355	436	8.2	564	15.7
2a	356	435	3.2	566	9.4
3a	355	401, 439	3.6	561	8.3
2b	375	443	7.0	584	1.4
3b	369	399, 433	2.7	579	1.8

[a]Quantum yields measured relative to $[Ru(bpy)_3]^{2+}$ in H_2O (Φ=4.2%) in deaerated solution.

To alleviate aggregation quenching, **2a** and **3a** were doped into poly(*N*-vinylcarbazole) (PVK), a well-known hole-transport polymer. The PL spectra of spin-coated films of the blends indicated significant blue-shift of the triplet emissions of **2a** and **3a** (Figure 2). It is also evident from Figure 2 that energy transfer between **2a** and PVK is rather inefficient. Interestingly, the EL spectra of **2a** in a similar device structure showed mostly triplet emission of **2a** at longer wavelength (Figure 3), consistent with a dominant direct charge-trapping mechanism (instead of intermolecular energy transfer) in the EL process.

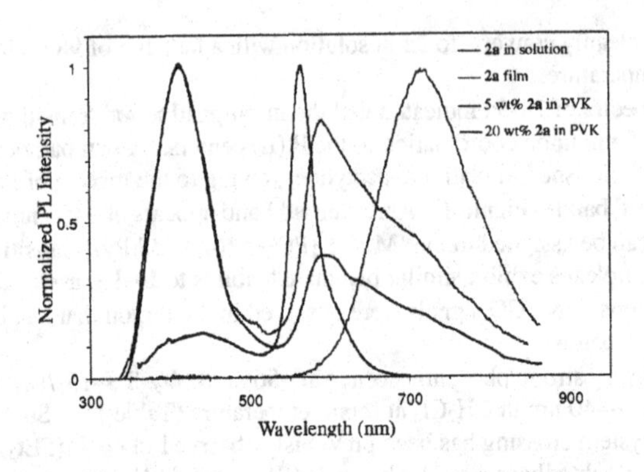

Figure 2. PL of **2a** in solution, thin film, and PVK blends

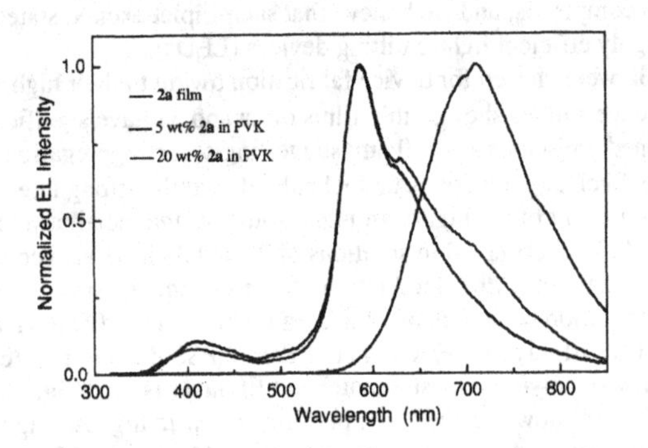

Figure 3. PL of 2a in solution, thin film, and PVK blends

The brightness and efficiency of LEDs based on the blends have been greatly enhanced as shown in Table 1. For the blend with 5 wt% **2a** in PVK, the maximum brightness reaches 5470 cd/m^2 with maximum luminous efficiency of 0.93 cd/A.

Table 2. LED device characteristics for PVK blends of **2a** and **3a**.[a]

EL layer[b]	V_{on}[c]	Q_{max}[d]	LE_{max}[e]	B_{max}[f]	V_{Bmax}[g]
2a	8.5 V	0.46%	0.93 cd/A	5470 cd/m^2	17 V
3a	8.5 V	0.29%	0.65 cd/A	3760 cd/m^2	17.5 V

[a]Device structure: ITO/PEDOT:PSS/EL layer/CsF/Al; [b]Nominal thickness 60 nm; [c]Turn-on voltage corresponding to 1 cd/m^2; [d]Maximum external quantum efficiency; [e]Maximum luminous efficiency; [f]Maximum brightness; [g]Driving voltage corresponding to maximum brightness.

CONCLUSIONS

Chiral molecular triangles and squares containing the Pt(diimine) metallocorners were prepared by a self-assembly process. The chiral molecular square exhibits efficient triplet MLCT emissions and has been used to construct efficient LED devices with high brightness and luminous efficiency. Future efforts will be directed to attaining circularly polarized electroluminescence using chiral metallocycles.

ACKNOWLEDGMENTS

We thank NSF (CHE-0208930) for financial support. W.L. is an A.P. Sloan Fellow, a Beckman Young Investigator, a Cottrell Scholar of Research Corp, and a Camille Dreyfus Teacher-Scholar.

REFERENCES

1. (a) P.J. Stang, B. Olenyuk, *Acc. Chem. Res.* **1997**, *30*, 502. (b) M. Fujita, *Chem. Soc. Rev.* **1998**, *27*, 417-425. (c) B.J. Holliday, C.A. Mirkin, *Angew. Chem. Int. Ed.* **2001**, *40*, 2022-2043. (d)
2. (a) M. Yoshizawa, T. Kusukawa, M. Fujita, S. Sakamoto, K. Yamaguchi, *J. Am. Chem. Soc.* **2001**, *123*, 10454-10459. (b) M. Yoshizawa, Y.Takeyama, T. Kusukawa, M. Fujita, *Angew. Chem. Int. Ed.* **2002**, *41*, 1347-1349.
3. P.H. Dinolfo, J.T. Hupp, *Chem. Mater.* **2001**, *13*, 3113.
4. (a) H. Jiang, A. Hu, W. Lin, *Chem. Commun.* **2003**, 96. (b) H. Jiang, W. Lin, *Org. Lett.* **2004**, *6*, 861-864.
5. H. Jiang, W. Lin, *J. Am. Chem. Soc.* **2004**, *126*, 7426-7427.
6. (a) F.-C. Chen, Y. Yang, M.E. Thompson, J. Kido, J. *Appl. Phys. Lett.* **2002**, *80*, 2308-2310. (b) S. Lamansky, P. Djurovich, D. Murphy, F. Abdel-Razzaq, H.-E. Lee, C. Adachi, P.E. Burrows, S.R. Forrest, M.E. Thompson, *J. Am. Chem. Soc.*;**2001**, *123*, 4304-4312. (c) H. Rudmann, S. Shimada, M.F. Rubner, *J. Am. Chem. Soc.* **2002**, *124*, 4918-4921. (d) J.C. Ostrowski, M.R. Robinson, A.J. Heeger, G.C. Bazan, *Chem. Commun.* **2002**, 784-785. (e) B. Carlson, G.D. Phelan, W. Kaminsky, L. Dalton, X. Jiang, S. Liu, A. K.-Y. Jen, *J. Am. Chem. Soc.* **2002**, *124*, 14162-14172.
7. C.W. Chan, L.K. Cheng, C.M. Che, *Coord. Chem. Rev.* **1994**, *132*, 87-97.
8. (a) J.E. McGarrah, R. Eisenberg, *Inorg. Chem.* **2003**, *42*, 4355-4365. (b) C.E. Whittle, J.A. Weinstein, M.W. George, K.S. Schanze, *Inorg. Chem.* **2001**, *40*, 4053-4062. (c) I.E. Pomestchenko, C.R. Luman, M. Hissler, R. Ziessel, F.N. Castellano, *Inorg. Chem.* **2003**, *42*, 1394-1396. (d) C.J. Adams, S.L. James, X. Liu, P.R. Raithby, L.J. Yellowlees, *J. Chem. Soc., Dalton Trans.* **2000**, 63-67. (e) V. W.-W. Yam, *Acc. Chem. Res.* **2002**, *35*, 555-563.
9. Bis(acetylide)Pt(II) complexes have recently been used as emissive materials in LED devices. See: S.-C. Chan, M.C.W. Chan, Y. Wang, C.-M. Che, K.-K. Cheung, N. Zhu, *Chem. Eur. J.* **2001**, *7*, 4180-4190.
10. **3a** and **3b** were synthesized by treating and Pt(5,5'-dtbPy)$(Cl)_2$ and $\mathbf{L}_m$-H_2 in 1:4 molar ratio in 46% and 62% yield, respectively.

11. The ratio of diffusion coefficients for **1a** and **2a** has been determined to be 1.09 by pulsed gradient NMR spectroscopy, consistent with their trimeric and tetrameric structure.
12. S.J. Lee, A. Hu, W. Lin, *J. Am. Chem. Soc.* **2002**, *124*, 12948-12949.
13. The dihedral angle between the naphthyl groups is governed by the R groups (see Scheme 1) in the 2,2'-positions.
14. S.J. Lee, C.R. Luman, F.N. Castellano, W. Lin, *Chem. Commun.* **2003**, 2124-2125.

Mater. Res. Soc. Symp. Proc. Vol. 846 © 2005 Materials Research Society DD3.7

Optical and Electronic Properties of Metal-containing Poly-ynes and their Organic Precursors

Marek Jura, Olivia F. Koentjoro, Paul R. Raithby, Emma L. Sharp and Paul J. Wilson
Department of Chemistry
University of Bath
Claverton Down
Bath
BA2 7AY
U.K.

ABSTRACT

A series of platinum(II) di-yne complexes with the general formula $[PhPt(PR_3)_2\text{-}C{\equiv}C\text{-}X\text{-}C{\equiv}C\text{-}Pt(PR_3)_2Ph]$ where R = Et, ^{n}Bu and X = a range of extended hetero-aromatic spacer groups, have been prepared by the coupling reaction between two molar equivalents of $[PhPt(PR_3)_2Cl]$ and one molar equivalent of the terminal dialkyne, $HC{\equiv}C\text{-}X\text{-}C{\equiv}CH$, in the presence of CuI catalyst. The complexes have been characterised by IR, NMR and optical spectroscopy. These materials, which are precursors to related platinum(II) poly-yne polymers, have been specifically designed to be incorporated into metal-molecule-metal junctions in nanoelectronic devices.

INTRODUCTION

Over the last two decades the chemistry of conjugated organic polymers has grown into a major research area because of the use of these materials in electronic and photonic applications [1,2]. The synthetic flexibility and ease of processing of these materials as well as the possibility of being able to tailor their properties to achieve a desired function has made them prime candidates for a plethora of applications in materials science. There are reports of their use as laser dyes [3], scintillators [3], light emitting diodes (LEDs) [4], sensors [5], piezoelectric and pyroelectric materials [6] and photoconductors [7]. They are being investigated for use as optical data storage devices [8], and as optical switches and signal processing devices [9], as well as having their nonlinear optical properties utilised [10].

In these organic conjugated polymers, light emission occurs from the singlet excited state (S_1) but emission in LEDs can occur from both the excited singlet (S_1) and triplet (T_1) states. If the technology in this area is to develop further it is essential that the photophysics of the triplet state is well understood. It will then be possible to alter the materials synthetically so that the relative energies of the singlet and triplet states can be manipulated and the emission from the excited triplet state harvested [11]. One successful method of achieving this is to introduce heavy transition metals, such as platinum, into the backbone of the conjugated organic polymers since the presence of the heavy metal produces sufficient spin-orbit coupling to allow intersystem crossing and hence to allow light emission from the triplet excited state to occur [12]. Thus, platinum(II) poly-ynes are good models for the study of the triplet state in organic polymers [13]. In addition, the unusual photophysics of these "rigid-rod" platinum-containing polymers themselves has led to the fabrication of materials that act as highly efficient electroluminescent devices [14] and they have found applications in laser protection equipment [15].

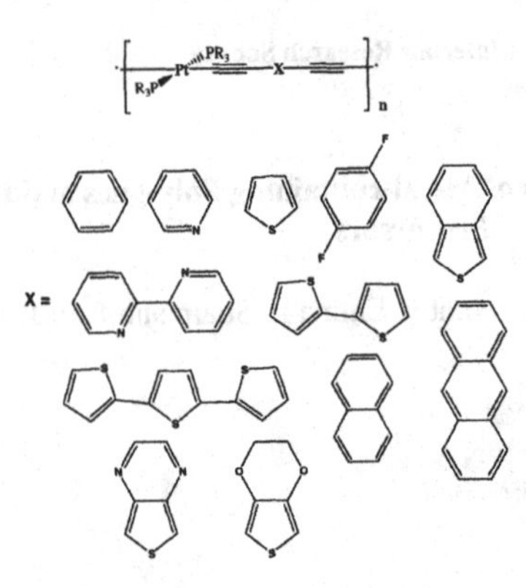

Figure 1. The platinum(II) poly-yne polymers showing a variety of spacer groups, X, that have different donor or acceptor properties. The inclusion of different members of the series can be used to fine tune the properties of the polymers.

We, and others, have investigated a wide range of platinum(II) poly-ynes that have the general formula $[-Pt(PR_3)_2-C{\equiv}C-X-C{\equiv}C-]_\infty$, where R = Et, ^nBu and X = a range of aromatic or hetero-aromatic spacer groups [16], as illustrated in figure 1, which are used to fine tune the electronic properties of the materials. There is a varying degree of donor-acceptor interactions between the metal centres and the conjugated ligands that depends upon the nature of spacer groups.

The results of optical spectroscopic studies on thin films of the full series of platinum(II) poly-ynes show that the absorptions shift to lower energy as X becomes more electron-withdrawing, consistent with a reduction in the optical gap ($S_0 - S_1$). In the series the triplet state, T_1, is also reduced in energy, in such a way that the $S_1 - T_1$ (singlet-triplet) energy gap is independent of X, and that the T_1 state is always 0.7 ± 0.1 eV below the S_1 state [17].

Given that it is now possible to control the size of the band gap in these materials we sought new applications for the platinum-containing "rigid-rod" polymers. In this paper we describe the synthesis and electronic properties of new platinum(II) diyne precursors that have been designed to operate as metal-molecule-metal junctions in nanogaps. These fabricated devices have potential applications in nanoelectronics.

EXPERIMENTAL DETAIL

All reactions unless otherwise stated were performed under a dry argon atmosphere using standard Schlenk or glove-box techniques. Solvents were pre-dried and distilled before use. All chemicals, except where stated otherwise, were obtained from Sigma-Aldrich and used as received. The compounds *trans*-$[PhPt(PEt_3)_2Cl]$ [18] was prepared by a literature method. NMR spectra were recorded on a Bruker WM-250 or AM-400 spectrometer in $CDCl_3$. The ^1H and $^{13}C\{^1H\}$ NMR spectra were referenced to solvent resonances and $^{31}P\{^1H\}$ NMR spectra were referenced to external 85% H_3PO_4. IR spectra were recorded as CH_2Cl_2 solutions, in a NaCl, cell, on a Perkin-Elmer 1710 FT-IR spectrometer. UV/visible absorption spectra, in CH_2Cl_2 solution were recorded on a Perkin-Elmer Lambda-25 spectrometer and emission spectra on a Perkin-Elmer Lambda-35 spectrometer.

Synthesis of the platinum(II) diynes and poly-ynes

The platinum(II) diynes were prepared by the general procedure: To a stirred solution of *trans*-[Pt(PEt_3)$_2$Ph(Cl)] (2 mol equivalents) and the diyne, HC≡C-X-C≡CH, (1 mol equivalent), in $^iPr_2NH$-CH_2Cl_2, under argon, was added CuI (5 mg). The yellow solution was stirred at room temperature overnight after which all volatile components were removed under reduced pressure. The residue was dissolved in ethyl acetate and passed through a short alumina column eluting with hexane-ethyl acetate. Removal of the solvents *in vacuo* afforded the dinuclear platinum complex, [PhPt(PEt_3)$_2$-C≡C-X-C≡C-Pt(PEt_3)$_2$Ph], in good yield upon recrystallisation from ethanol.

DISCUSSION

It is likely that conventional silicon-based microelectronics will reach their limit of miniaturisation within the next decade, when feature lengths shrink below 70 nm. The next generation of electronic devices will require materials that will remain functional on a nanometer scale. Theory predicts that molecule-based integrated circuits could be reduced to a scale many orders of magnitude smaller than silicon-based electronics allow. The idea that a single molecule could be embedded between two electrodes and perform the basic functions of digital electronics was first proposed by Aviram and Ratner in the 1970s [19]. Only recent advances in nanotechnology, however, have made single molecule devices feasible [20]. The key element of these devices is a metal-molecule-metal (M-m-M) junction, as shown in figure 2, where "m" is a long-chain molecule that can be attached through anchoring groups to opposite sides of a fabricated nanogap with dimensions of 2 – 5 nm. Such a molecule, because of the high electron mobility along its path length and because of the easy control of its transport properties through its chemical structure, has all the properties necessary to transport current across nanogaps. Following on from the work of Reed, Tour and others [21, 22] on conjugated organic oligomers and polymers it is valuable to attempt to use the platinum(II) poly-yne systems in metal-molecule-metal junctions as their properties are in some ways superior.

One particular feature of the work on the organic systems is that conductance switching occurs [23] when the conformation of the molecule changes. In the molecule, (figure 3) rotation is possible about the central bond between the two nitro-substituted pyridine rings, and the electronic properties of the molecule change as the angle between the two central rings changes.

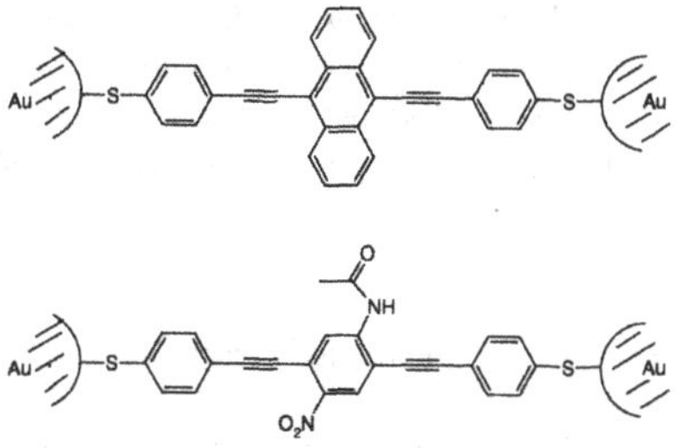

Figure 2. Conjugated organic molecules spanning a gold nanogap. The molecule is tethered to the gold electrodes through thiolate anchors.

This approach has now been applied to the platinum(II) poly-yne materials and a range of molecular precursors with different angular arrangements between the extended conjugated ring systems and their electronic properties determined.

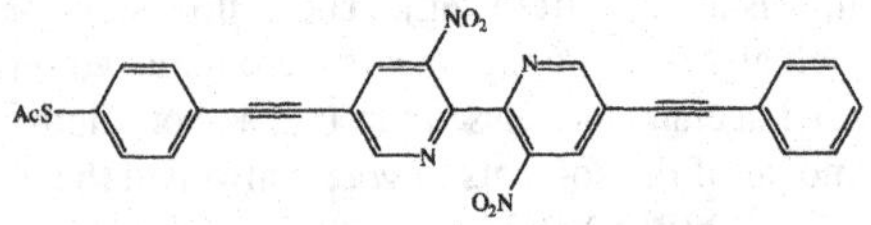

Figure 3. There is rotation about the central bond between the two pyridine rings, and the electronic properties of the molecule alter as the angle between the two rings changes (ref. 23).

The series of platinum(II) diyne complexes prepared to allow the study of the change in electronic properties with conjugation geometry in the central spacer group are illustrated in figure 4.

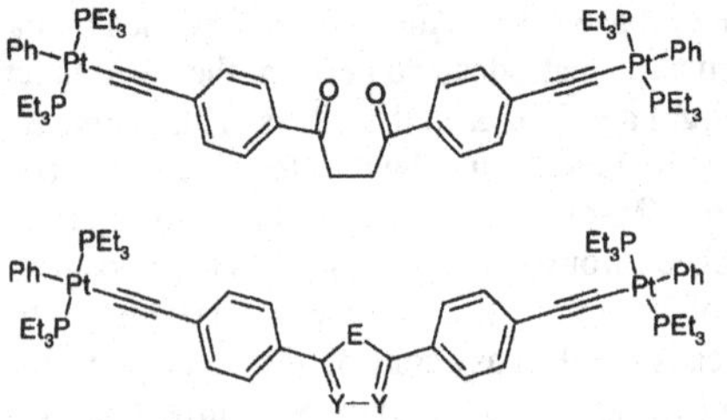

Figure 4. The platinum(II) diyne with the extended central ring system

The solution optical absorption and emission spectra for the four complexes have been measured (Table 1) and the results show that the platinum 1,4–diketone complex **1** with its low level of conjugation absorbs at the lowest wavelength and has an emission maxima at 409 nm. The platinum oxadiazole **2** and platinum thiadiazole **3** complexes exhibit higher wavelengths of maximum absorption, with the thiadiazole species absorbing at the highest wavelength due to lower π^*-overlap of carbon with sulphur compared to oxygen. The platinum thiophene complex **4** exhibits a very similar optical gap to the platinum oxadiazole complex **2** due to inherent internal twisting of the chromophore resulting in loss of conjugation. The corresponding emission spectra follow the same trend as the absorbance spectra, with the sulphur and oxygen containing species emitting in the blue and violet regions of the visible spectrum, respectively.

CONCLUSION

The four new platinum(II) diynes, with extended central heteroaromatic spacer groups, have been synthesised as model precursors for polymeric materials that can be incorporated into metal-molecule-metal junctions. The four complexes have different conformations in the central spacer groups, and these differences are reflected in their absorption and emission spectra indicating that their electronic properties differ. The flexibility of the geometries within these

Table 1 Absorption and emission spectroscopic data for complexes **1-4**

Complex	Absorbance wavelength maxima (nm)	Emission wavelength maxima (nm)
1	349	409
2	366	423
3	385	458
4	384	438

spacer groups implies that they have the potential to change conformation with applied electrical field and act as molecular switches. Work on these and related materials is continuing.

ACKNOWLEDGEMENTS

The authors are grateful to the EPSRC (U.K.), the European Union and the University of Bath for funding to support the project. One of the authors (PJW) is supported by the Royal Society, London. The authors acknowledge the generous loan of platinum metal salts by Johnson Matthey plc. The authors wish to thank Professor Richard Friend, Dr Anna Köhler, Dr Christopher Ford (Cambridge), Dr Muhammad Khan (Sultan Qaboos University, Oman), Dr Sergey Gordeev and Dr Robert Less (Bath) for valuable discussions and support with the project.

REFERENCES

1. J. H. Burroughs, D. D. C. Bradley, A. R. Brown, R. N. Marks, K. Mackay, R. H. Friend, P. L. Burn, and A. B. Holmes, *Nature* **347** 539 (1990).
2. N. Tessler, G. J. Denton, and R. H. Friend, *Nature* **382** 695 (1996).
3. H. Meier, *Angew. Chem. Int. Ed. Engl.* **31** 1399 (1992).
4. A. Kraft, A. C. Grimsdale, and A. B. Holmes, *Angew. Chem. Int. Ed.* **37** 403 (1998); A. Montali, P. Smith, and C. Weder, *Synth. Met.* **97** 123 (1998).
5. T. M. Swager, *Acc. Chem. Res.* **31** 201 (1998); D. T. McQuade, A. E. Pullen, and T. M. Swager, *Chem. Rev.* **100** 2537 (2000).
6. D. K. Das-Gupta in *Introduction to Molecular Electronics,* edited by M. C. Petty, M. R. Bryce and D. Bloor (Edward Arnold, London, 1995) pp. 47-71.
7. H. Bleier in *Organic Materials for Photonics*, edited by G. Zerbi (Elsevier, Amsterdam, 1993) pp. 77-101; R. O. Loutfy. A.-M. Hor, C.-K. Hsiao, G. Baranyi, and P. Kazmaier, *Pure Appl. Chem.* **60** 1047 (1988).
8. L. Feringa. W. F. Jager and B. de Lange, *Tetrahedron* **49** 8267 (1993); G. H. W. Buning in *Organic Materials for Photonics,* edited by G. Zerbi (Elsevier, Amsterdam, 1993) pp.367-397.
9. K. Lösch, *Macromol. Symp.* **100** 65 (1995); J.-M. Lehn. *Supramolecular Chemistry: Concepts and Perspectives* (VCH, Weinheim, 1995) pp. 124-138; P. J. Martin in *Introduction to Molecular Electronics*, edited by M. C. Petty, M. R. Bryce and D. Bloor (Edward Arnold, London, **1995**) pp. 112-141; J.-J. Kim and E.-H. Lee, *Mol. Cryst. Liq. Cryst* **227** 71 (1993); H. Durr, *Angew. Chem. Int. Ed. Engl.* **28** 413 (1989).
10. A. J. Heeger and J. Long Jr., *Optics & Photonics News* **7**(8) 24 (1996); H. S. Nalwa, in *Nonlinear Optics of Organic Molecules and Polymers*, edited by H. S. Nalwa and S.

Miyata (CRC, New York, 1997) pp 611-787; S. R. Marder, B. Kippelen, A. K.-Y. Jen, and N. Peyghambarian, *Nature* **388** 845 (1997).

11. M. A. Baldo, M. E. Thompson, and S. R. Forrest, *Nature* 403 750 (2000); V. Cleave, G. Yahioglu, P. Lebarny, R. H. Friend, and N. Tessler, *Adv. Mater.* **11** 285 (1999).
12. Y. Liu, S. Jiang, K. Glusac, D. H. Powell, D. F. Anderson, and K. S. Schanze, *J. Am. Chem. Soc.* **124** 12412 (2002); V. W-W Yam, *Acc. Chem. Res.* **35** 555 (2002); M. Younus, A. Köhler, S. Cron, N. Chawdhury, M.R.A. Al-Mandhary, M.S. Khan, J. Lewis, N.J. Long, R.H. Friend, and P.R. Raithby, *Angew. Chem. Int. Ed. Engl.* **37** 3036 (1998); N. Chawdhury, A. Köhler, R.H. Friend, W.-Y. Wong, M. Younus, P.R. Raithby, J. Lewis, T.C. Corcoran, M.R. A. Al-Mandhary and M.S. Khan, *J. Chem. Phys.* **110** 4963 (1999).
13. J. S. Wilson, A. S. Doot, A. J. A. B. Seeley, M. S. Khan, A. Köhler, and R. H. Friend, *Nature* **413** 828 (2001); J. S. Wilson, N. Chawdhury, A. Köhler, R. H. Friend, M. R. A. Al-Mandhary, M. S. Khan, M. Younus, and P. R. Raithby, *J. Am. Chem. Soc.*, **123** 9412 (2001); D. Beljonne, H. F. Wittmann, A. Köhler, S. Graham. M. Younus, J. Lewis, P. R. Raithby, M. S. Khan, R. H. Friend, and J. L. Bredas, *J. Chem. Phys.* **105** 3868 (1996).
14. M.A. Baldo, D. F. O' Brien, Y. You, A. Shpustikov, S. Sibley, M. E. Thompson, and S. R. Forrest, *Nature* **395** 151 (1998).
15. J. Staromlynska, T. J. McKay, A. J. Boljer, and J. R. Davy, *J. Opt. Soc. Am. B-Opt. Phys.* **15** 1731 (1998).
16. M. S. Khan, M. R. A. Al-Mandhary, M. K. Al-Suti, A. K. Hisham, P. R. Raithby, B. Ahrens, M. F. Mahon, L. Male, E. A. Marseglia, E. Tedesco, R. H. Friend, A. Köhler, N. Feeder, and S. J. Teat, *J. Chem. Soc., Dalton Trans.* 1358 (2002); M. S. Khan, M. R. A. Al-Mandhary, M. K. Al-Suti, N. Feeder, S. Nahar, A. Köhler, R. H. Friend, P. J. Wilson and P. R. Raithby, *J. Chem. Soc., Dalton Trans.* 2441 (2002); M. S. Khan, M. R. A. Al-Mandhary, M. K. Al-Suti, P. R. Raithby, B. Ahrens, L. Male, R. H. Friend, A. Köhler, and J. S. Wilson, *Dalton Trans.* 65 (2003); M.S. Khan, M.R.A. Al-Mandhary, M.K. Al-Suti, P.R. Raithby, B. Ahrens, M. Mahon, L. Male, C.E. Boothby, and A. Köhler, *Dalton Trans.* 74 (2003); M. S. Khan, M. R. A. Al-Mandhary, M. K. Al-Suti, F. R. Al-Battashi, S. Al-Saadi, B. Ahrens, J. K. Bjernemose, M. F. Mahon, P. R. Raithby, M. Younus, N. Chawdhury, A. Köhler, E. A. Marseglia, E. Tedesco, N. Feeder, and S. J. Teat, *Dalton Trans.* 2377 (2004).
17. J. S. Wilson, A. Köhler, R. H. Friend, M. K. Al-Suti, M. R. A. Al-Mandhary, M. S. Khan and P.R. Raithby, *J. Chem. Phys.* **113** 7627 (2000).
18. K. Siegmann, P. S. Pregosin, and L. M. Venanzi, *Organometallics* **8** 2659 (1988).
19. A. Aviram and M. Ratner, *Chem. Ohys. Lett.* **29** 277 (1974).
20. J. Chen, M. A. Reed, A. M. Rawlett, and J. M. Tour, *Science* **286** 1550 (1999).
21. J. M. Tour, *Acc. Chem. Res.* **33** 791 (2000); J. Chen, W. Wang, M. A. Reed, A. M. Rawlett, D. W. Price, and J. M. Tour, *Appl. Phys. Lett.* **77** 1224 (2000); M. A. Reed, C. Zhou, C. J. Muller, T. P. Burgin, and J. M. Tour, *Science* **278** 252 (1997).
22. D. Feldheim, *Nature* **408** 45 (2000); K. Walzer, E. Marx, N. C. Greenham, R. J. Less, P. R. Raithby, and K. Stokbro, *J. Am. Chem. Soc.* **126** 1229 (2004).
23. Z. J. Donhauser, B. A. Mantooth, K. F. Kelly, L. A. Bumm, J. D. Monnell, J. J. Stapleton, D. W. Price, Jr., A. M. Rawlett, D. L. Allara, J. M. Tour, and P. S. Weiss, *Science* **292** 2303 (2001); A. K. Flatt, S. M. Dirk, J. C. Henderson, D. E. Shen, J. Su, M. A. Reed, and J. M. Tour, *Tetrahedron* **59(43)** 8555 (2003).

Organometallic Optical Materials II

Mater. Res. Soc. Symp. Proc. Vol. 846 © 2005 Materials Research Society

Novel iridium complexes with polymer side-chains

Elisabeth Holder, Veronica Marin, Emine Tekin, Dmitry Kozodaev, Michael A. R. Meier, Bas G. G. Lohmeijer and Ulrich S. Schubert*
Laboratory of Macromolecular Chemistry and Nanoscience,
Eindhoven University of Technology and Dutch Polymer Institute (DPI),
P.O. Box 513, 5600 MB Eindhoven, The Netherlands

ABSTRACT

The focus of the presented research lies on the synthesis of novel charged iridium(III) compounds with potential applications in light-emitting electrochemical cells. The design involves iridium(III)-based materials with polymer side-chains leading to linear light-emitting polymer arrangements.

To study the electro-optical properties of such light-emitting polymers conventional and combinatorial deposition methods are used. Straightforward screening approaches are introduced. The combinatorial efforts engage the processing *via* inkjet printing and the screening of the optical properties using plate reader technologies based on steady state UV-vis and fluorescence. Furthermore, the morphological properties are investigated using optical interferometry and atomic force microscopy (AFM). Overall it can be shown that novel materials can be deposited revealing high-quality thin films, which allow the screening of electro-optical features using combinatorial methods. Some initial local current density studies by AFM have also been performed to characterize the current injection and transport properties of the novel materials.

INTRODUCTION

The synthesis and design of new materials is an important topic in current research due to the growing development of technological relevant applications. Electroluminescent materials have become progressively more interesting due to their potential applications in light-emitting diodes (LEDs) [1,2] and light-emitting electrochemical cells (LECs).[3] Both LEDs and LECs find use in displays and lighting applications but the work performed on LEDs is more mature compared to development performed on LECs. The working principle of the introduced devices is slightly different therefore different classes of iridium compounds were designed.[4] Neutral iridium compounds find applications in LEDs whereas charged compounds find use in LECs. One of the drawbacks in LECs is the rapid degradation on the electrode surface. Another observed feature in both devices is the aggregation of complexes leading to reduce device lifetimes. Some work carried out on ruthenium compounds showed that copolymerization or blending with polymers such as PMMA improved the device performance significantly.[5] Studying the electro-optical properties and the lifetime of potential material compositions is rather time consuming and the structure-property relationships need to be carefully evaluated in order to make a smart design of materials. Therefore a novel processing and screening approach is currently developed. Inkjet printing is being used to deposit test-pixels of the novel electroluminescent materials. Optical screening was carried out in solution and in thin film since the performance can be significantly different. At present the morphology and the local electrical information is characterized and collected by AFM and related techniques.

The current presentation focuses on the design of novel iridium(III) materials with polymer side-chains and their characterization by optical methods. Initial results on the deposition of thin films by inkjet printing as well as their characterization using optical screening tools are shown. In addition, description of the film surfaces by optical interferometry and AFM is presented. Preliminary results on the local current densities are also shown. Parts of this research is already published elsewhere.[4-7]

DISCUSSION

Synthesis and design of linear polymers

The focus of the currently performed research lies on the synthesis of charged iridium(III) complexes immobilized on polymers chains. The utilized polymers consist of linear (a) poly(ethyleneglycol) (PEG), (b) poly(styrene) (PS) and (c) poly(ε-caprolactone) (PCL) side-chains (Figure 1). The structures of the complexes involve bipyridine and terpyridine chelates as well as cyclometalating ligands based on phenylpyridine and derivatives, respectively.

The general approach is schematically introduced as it can be seen in Figure 1. A detailed description of the synthetic procedures can be found elsewhere.[4,6,8]

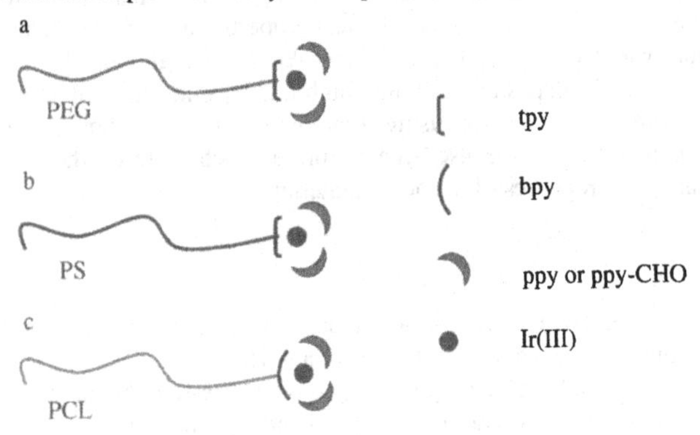

Figure 1. Schematic representation of the investigated iridium(III)-based light-emitting polymers with (a) PEG, (b) PS and (c) PCL side-chains.

UV-vis, emission and cyclic voltammetry properties

Some color-tuning options arise by variation of the cyclometalating ligand set due to the metal-to-ligand based irradiation of iridium(III) complexes. Selected examples of charged iridium(III) compounds (Figure 2, top) are chosen to illustrate the effect in solution and in thin films (Figures 2, bottom and Figure 3). However, in films the effect is minor as can be seen in Figures 2 and 3. The emission features are very similar

(λ_{max} around 550 nm) in the solid state and thus result in yellow emission color for both materials **1** and **2**.

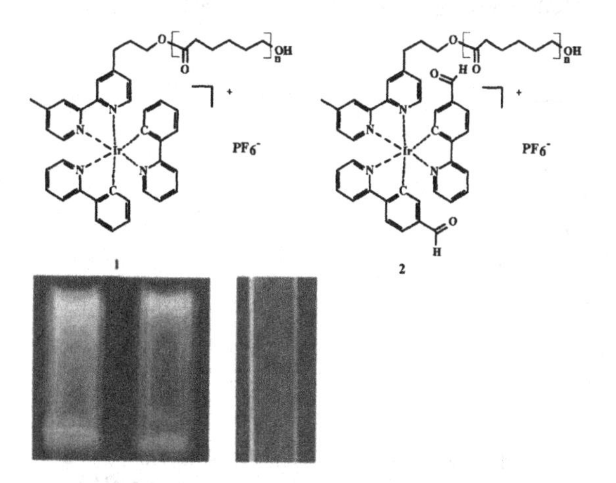

Figure 2. Top: Structures of the investigated PCL-based complexes **1** and **2**. Bottom: Picture of the different emission colors of CH_2Cl_2 solutions of complexes **1** (orange) and **2** (greenish-yellow). A film of **2** on a glass substrate is also displayed to visualize the yellow emission color of the compound in the solid state, respectively.

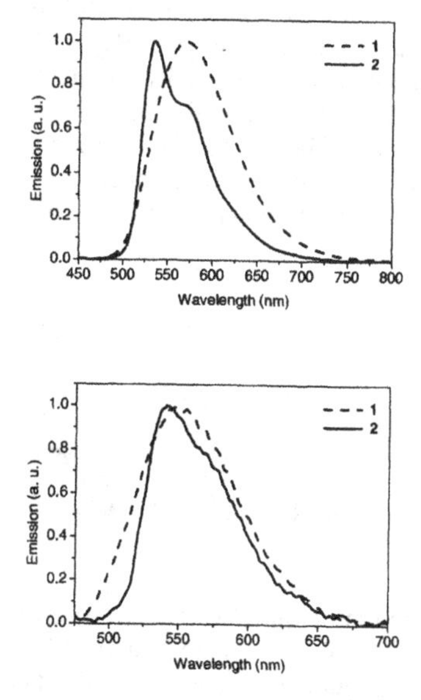

Figure 3. Top: Emission properties of complexes **1** and **2** in CH_2Cl_2 solutions. Bottom: Emission properties of thin films of **1** and **2**, respectively.

Cyclic voltammetry studies are crucial to determine the electrical properties of novel organometallic compounds (characteristic examples of materials **1** and **2** are depicted in Figure 4). Both described materials display reversible oxidation potentials.

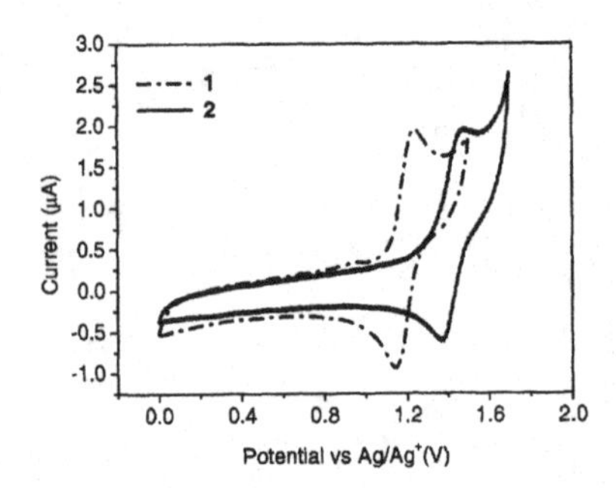

Figure 4. Cyclovoltammetry results of the macroligand complexes **1** (solid line) and **2** (dash-dotted line). All spectra were recorded in CH_2Cl_2 (freshly distilled from CaH_2), containing 0.1 M n-Bu_4PF_6 vs Ag/Ag^+ reference electrode.

Processing by inkjet printing and opto-electrical screening approach

For a fast screening of the thin film properties, small libraries of novel materials such as the iridium(III) polypyridyl polystyrene copolymer **3** have been developed (Figure 5). For this purpose the inkjet printing process was optimized using a photoresist patterned glass substrate to prevent the spreading of the ink (procedure and material from www.microchem.com). Initial investigations of the thin film libraries are shown using a plate reader technology in order to characterize the optical properties of the obtained films in a fast manner (Figure 6). The obtained parameters can be compared to those of compound **2**, due to a similar optical center.

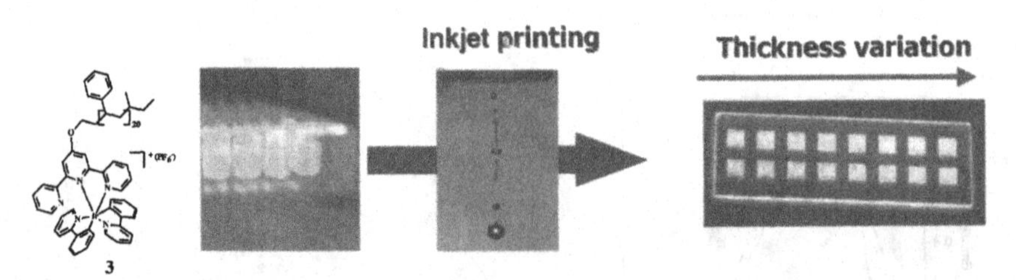

Figure 5. From left to right: Chemical structure of **3**, microtiter plate with the respective inks of **3**, inkjet deposition process, inkjet printed thin film libraries of the iridium(III)-based polymer **3** (structure also schematically shown in Figure 1b. The library is based on film thickness variations of material **3**.

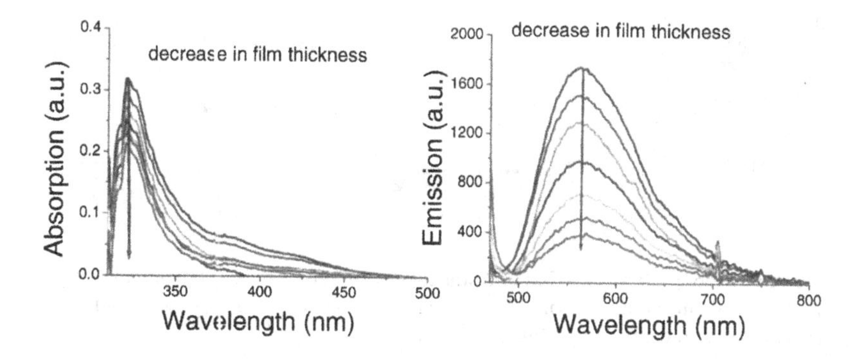

Figure 6. Optical screening of inkjet printed libraries of **3** by a UV-vis/fluorescence plate reader technology: Left: Absorption features of **3** in correlation with the obtained film thickness. Right: Emission features of **3** in correlation with the obtained film thickness. The obtained intensity decreases linearly with the film thickness.

Morphology studies and local current measurements

The film quality was investigated using optical interferometry and AFM (Figure 7 and 8). Optical interferometry is useful to visualize the film quality and thickness on a larger scale. Figure 7 shows a very smooth high-quality thin film of 120 nm thickness. The roughness of inkjet printed films of **3** was determined to be in the order of 1-2%. The local morphological information can also be studied by AFM (Figure 8). The depicted film revealed no defects on the mesoscopic scale. The morphology and quality of the polymer film layer is highly important in order to obtain working devices with large areas. Furthermore, AFM techniques are used in order to determine the local current densities to complement current density studies of larger area thin film devices (Figure 8). The used device setup to detect local currents is displayed in Figure 8 and some initial results of the local current densities are also shown in Figure 8.

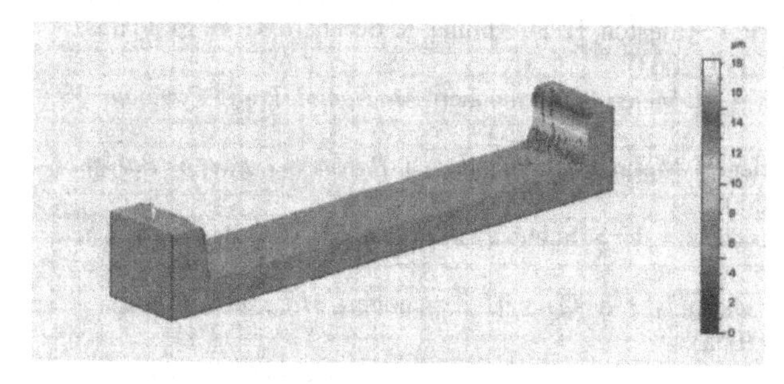

Figure 7. Cross section of a high-quality 120 nm thick film of the iridium(III)-containing polymer **3** investigated by optical interferometry.

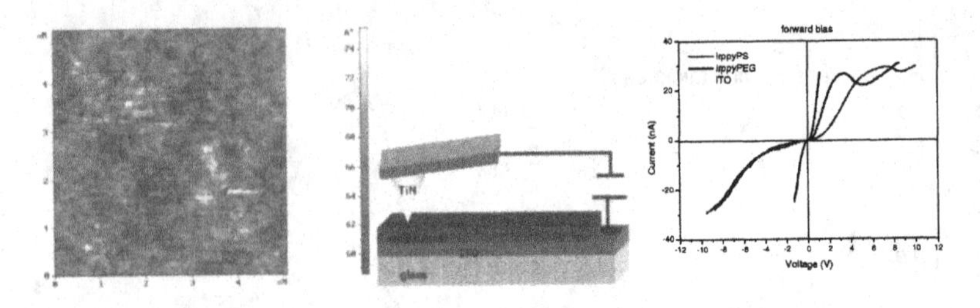

Figure 8. The morphology of a film of **3** was investigated on a mesoscopic level using AFM. The local current density of a PEG-type material *versus* a PS-type material is shown compared to the bias of indium tin oxide.

CONCLUSIONS

The presented research focuses on the design and characterization of novel charged iridium(III) compounds. The introduced materials are based on PEG, PS and PCL polymers. The characterization of the materials is carried out in a conventional fashion. For advanced characterization of thin film libraries optical and electrical screening tools have been introduced. AFM was utilized to characterize film morphologies and local charge injection.

ACKNOWLEDGEMENTS

This work is part of the *Dutch Polymer Institute (DPI)* research program (HTE/CMR and FPS). We thank the *Fonds der Chemischen Industrie* for financial support. We are grateful to Prof. R. Janssen (TU/e) for granting access to the optical set-ups.

REFERENCES

1. A. Köhler, J. S. Wilson, R. H. Friend, *Adv. Mater.* **14**, 701-707 (2002).
2. E. Holder, B. M. W. Langeveld, U. S. Schubert *Adv. Mater.* **17**, (2005), in press.
3. J. Slinker, D. Bernards, P. L. Houston, H. D. Abruna, S. Bernhard, G. G. Malliaras, *Chem. Commun.* 2392-2399 (2003).
4. E. Holder, V. Marin, M. A. R. Meier, U. S. Schubert, *Macromol. Rapid Commun.* **25**, 1491-1496 (2004).
5. E. Holder, M. A. R. Meier, V. Marin, U. S. Schubert, *J. Polym. Sci. Part A: Polym. Chem.* **41**, 3954-3964 (2003).
6. E. Holder, V. Marin, A. Alexeev, U. S. Schubert, *J. Polym. Sci. Part A: Polym. Chem.* **43**, (2005), in press.
7. E. Tekin, E. Holder, V. Marin, B.-J. de Gans, U. S. Schubert, *Macromol. Rapid Commun.* **26**, 293-297 (2005).
8 E. Holder, V. Marin, D. Kozodaev, M. A. R. Meier, B. G.G. Lohmeijer, U. S. Schubert, submitted.

Mater. Res. Soc. Symp. Proc. Vol. 846 © 2005 Materials Research Society DD4.5

Phosphorescence Quantum Efficiency and Intermolecular Interaction of Iridium(III) Complexes in Co-Deposited Films with Organic Semiconducting Hosts

Yuichiro Kawamura[1], Kenichi Goushi[2], Jason Brooks[3], Julie J. Brown[3], Hiroyuki Sasabe[1,2], and *Chihaya Adachi[1,2]
[1]CREST program, Japan Science and Technology Agency (JST), 1-32-12 Higashi, Shibuya, Tokyo 150-0011, Japan
[2]Department of Photonics Materials Science, Chitose Institute of Science & Technology (CIST), 758-65 Bibi, Chitose, Hokkaido 066-8655, Japan
[3]Universal Display Corporation, 375 Phillips Blvd., Ewing, NJ 08618, U.S.A.

ABSTRACT

We accurately measured the absolute photoluminescence (PL) quantum efficiency (η_{PL}) of organic solid-state thin films by using an integrating sphere. We measured the η_{PL} of conventional organic materials used in organic light emitting diodes, such as a tris (8-quinolionolato)aluminum(III) complex (Alq_3), and a phosphorescent 1.5mol%-*fac*-tris(2-phenylpyridyl)iridium(III):4,4'-bis(carbazol-9-yl)-2,2'-biphenyl [$Ir(ppy)_3$:CBP] co-deposited film. Alq_3 and $Ir(ppy)_3$:CBP showed a $\eta_{PL} = 20 \pm 1\%$ and $97 \pm 2\%$, which corresponded well to external electroluminescence efficiency using these materials. We also measured red emitting bis[2-(2'-benzothienyl)pyridinato-*N*,$C^{3'}$] (acetylacetonato) iridium(III) [Btp_2Ir(acac)] with CBP, and the blue complex, bis[(4,6-difluorophenyl)pyridinato -*N*,C^2](picolinato)iridium(III) [FIrpic], with *m*-bis(*N*-carbazolyl)benzene. The maximum η_{PL} values for Btp_2Ir(acac), and FIrpic were 51±1% (at 1.4mol%), and 99±1% (1.2mol%), respectively. These results suggest that the η_{EL} of red phosphorescent OLEDs using Btp_2Ir(acac) as the dopant can be as high as 10%, and blue devices using FIrpic can reach the theoretical limit of 20%.

INTRODUCTION

There has been a great deal of interest in the last decade in developing organic optoelectronic devices. In particular, organic light emitting diodes (OLEDs)[1-2] have achieved nearly 100% electro-photo conversion efficiency[3,4], and ongoing research into light emitting devices such as organic laser diodes[5-8] and organic thin film light-emitting transistors[9] is continuing by many research groups.

Recently, many kinds of new luminescent organic compounds used in these devices have also been extensively developed. Especially, cyclometalated Ir(III) complexes are recognized as promising candidates for phosphorescent dopants because they can emit with high efficiency in room temperature from the triplet metal-to-ligand charge-transfer (^{3}MLCT) state[10,11]. We need to know detailed information about the basic luminescent properties of organic materials to understand the make-up of highly luminescent materials and device architectures.

Absolute photoluminescent (PL) quantum efficiency (η_{PL}) is a principal characteristic of luminescent materials and informs us about photophysical characteristics of organic materials. Although some methods for measuring the η_{PL} in a solution are well established[12], there are few direct methods for obtaining the absolute η_{PL} of organic solid-state thin films[13,14].

We report on a simple and accurate system for measuring the η_{PL} of luminescent organic

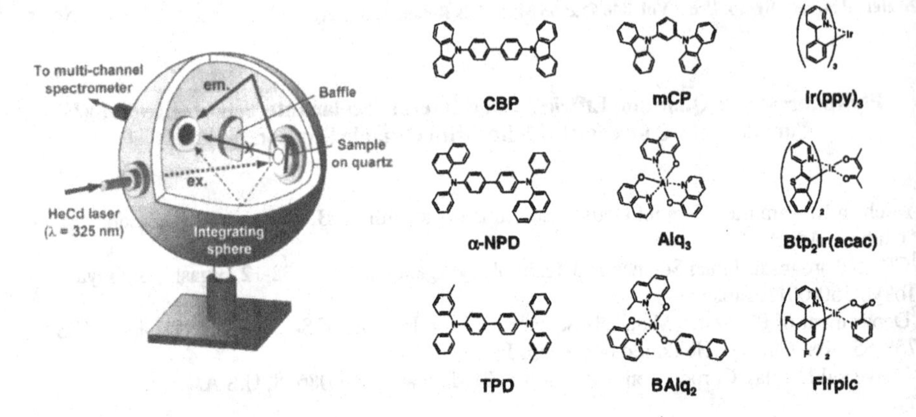

Figure 1. Schematic description of experimental setup using integrating sphere for measuring absolute PL quantum efficiency and chemical structures of the materials.

materials in a solid-state thin film with submicrometer-size thickness using an integrating sphere equipped with a HeCd laser as an excitation source and a multi-channel spectrometer as an optical detector. By applying this method, we determined the absolute η_{PL} of phosphorescent Ir(III) complexes and their dependencies for doping concentrations.

EXPERIMENTAL DETAILS

Figure 1 shows a system for measuring the η_{PL} using an integrating sphere (Labsphere Inc.). [15] The inside of this sphere is coated with a diffusely reflecting material (such as $BaSO_4$) and photons that are generated or reflected in the sphere are channeled to the exit port, maintaining the intensity in proportion to the total light flux. We used a 325-nm cw HeCd laser (Kinmon TK5651) as the excitation source and the laser beam was introduced into the sphere through an optical fiber. The laser was collimated and vertically irradiated the sample attached to a folder on the sphere's wall. A baffle placed between the sample and the exit port prevented the light from directly reaching both the reflected laser and the PL of the sample.

Light intensities, scattered within the sphere, were measured using a multi-channel spectrometer (HAMAMATSU PMA-11). The reflectivity of the sphere was calibrated using a standard light source, ranging from 250 to 800 nm. Measurement was carried out under a dry N_2 gas flow. Here, the η_{PL} is given by

$$\eta_{PL} = \frac{N_{emission}}{N_{absorption}} = \frac{\alpha\int \frac{\lambda}{hc} I_{em}(\lambda)d\lambda}{\alpha\int \frac{\lambda}{hc}[I_{ex}(\lambda) - I'_{ex}(\lambda)]d\lambda}, \quad (1)$$

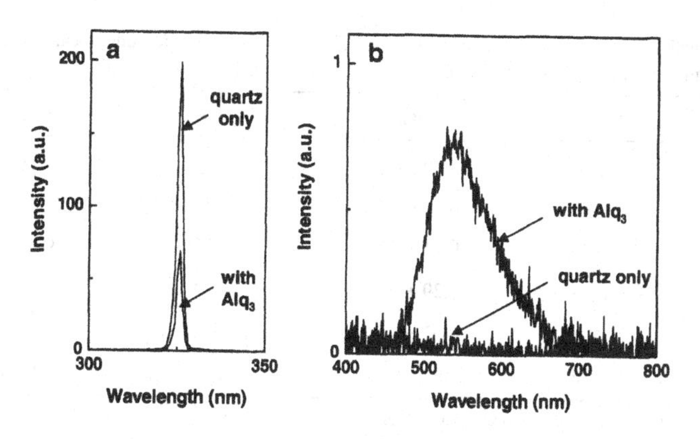

Figure 2. (a) Excitation intensities, and (b) PL spectra of quartz substrate as reference and sample with 200-nm thick Alq_3 on quartz.

where $N_{emission}$ is the emission photon number from the sample, $N_{absorption}$ is the absorbed photon number by the sample, α is the calibration factor for the measurement setup, λ is the wavelength, h is plank's constant, c is the speed of light, $I_{em}(\lambda)$ is the PL intensity from the sample, $I_{ex}(\lambda)$ is the intensity of the excitation laser without the sample, and $I'_{ex}(\lambda)$ is the intensity of the excitation laser with the sample. We needed only two emission spectra of I_{ex} and $I'_{ex} + I_{em}$ for calculating eq. (1) and required no complicated calibrations because α, the calibration factor for the measurement setup, can be canceled in eq. (1). Here, a decrease in the excitation light intensity when the sample was placed inside the sphere could be attributed to absorption by the sample and not to absorption by the transparent quartz substrate. Furthermore, the second contribution of the re-absorption of the reflected excitation light to the PL intensity should be involved because our system can observe the enhancement of the PL intensity from the sample and the decrease of the excitation light intensity by the absorption at the same time even if the re-absorption occurs.

Organic films were fabricated at a thickness of 50-200 nm using conventional thermal vacuum deposition (under 10^{-3} Pa) on pre-cleaned quartz substrates that were sized to 10×16 mm^2 with a 1 mm thickness through a metal mask with an 8 mm diameter opening. Chemical structures of the materials were also shown in Fig. 1.

RESULTS AND DISCUSSIONS

Figure 2 shows the excitation intensities and PL spectra of a quartz substrate as a reference and the sample with a 200-nm-thick tris(8-quinolinolato)aluminum(III) complex (Alq_3) [1] on the quartz. Figure 2 clearly shows that the incidence laser intensity decreased and a broad fluorescence originating from Alq_3 appeared at visible region, caused by absorption of the sample when it was placed in the sphere. From these spectral changes and eq. (1), we confirmed that a neat film of Alq_3, as a standard green emitter, showed an η_{PL} of 20±1%, which agreed well with the external electroluminescent efficiency (η_{EL}). The η_{EL} of OLED is generally given by,

TABLE I Optical characteristics of organic thin films (100 nm). λ_{PL}: luminescence peak wavelength, η_{PL} absolute PL quantum efficiency, τ_{PL} : transient PL lifetime, and k_r: radiative decay rate.

Material	λ_{PL} (nm)	η_{PL}	τ_{PL} ($\times10^{-9}$s)	k_r (s^{-1})
Alq_3	520	0.20 ± 0.01	15	1.3×10^7
$BAlq_2$	492	0.42 ± 0.02	29	1.4×10^7
TPD [(8)]	424	0.41 ± 0.02	0.6	6.8×10^8
α-NPD[(8)]	445	0.29 ± 0.02	3.5	8.0×10^7
CBP [(8)]	393	0.61 ± 0.04	0.7	8.7×10^8
$Ir(ppy)_3$*	512	0.97 ± 0.02	1.3×10^3	7.5×10^5

* $Ir(ppy)_3$ was doped into a CBP host at 1.5 wt% concentration.

$$\eta_{EL} = \eta_{out} \cdot \eta_{e/h} \cdot \eta_{exciton} \cdot \eta_{PL}\,, \qquad (2)$$

where η_{out} is the light out-coupling efficiency, $\eta_{e/h}$ is the charge balance factor, which is the ratio of the electron to the hole, and $\eta_{exciton}$ is the efficiency of the singlet or triplet exciton generation.

Typically, the η_{EL} of Alq_3 based OLED is approximately 1%. By taking into account the singlet exciton generation efficiency ($\eta_{exciton}$) under an electrical excitation of 25%, η_{out} = 20%[16] and $\eta_{e/h}$ = 100%, we obtained an η_{PL} = 20% from eq. (2), which was in good agreement with the absolute PL measurement. We confirmed that the η_{PL} value did not depend on the sample's shape, thickness, and the excitation laser power. These also show that the effect of the re-absorption does not contribute to the measurement.

For further examples, we examined other conventional OLED materials, CBP, *N,N'*-di(*m*-tolyl)-*N,N'*-diphenylbenzidine (TPD), *N,N'*-di(α-naphtyl)-*N,N'*-diphenyl -benzidine (α-NPD)[8] and bis(2-methyl-8-quinolinolato)(4-phenylphenolato) aluminum ($BAlq_2$)[17] with 100-nm-thick films for the η_{PL} measurements, which demonstrated reasonable values when compared with previous reports[14] (Table I).

Furthermore, we estimated the radiative decay rates of these materials. We measured the τ_{PL} values from the transient PL decay using a streak camera. The τ_{PL} was fitted using a single exponential decay as a first approximation. Radiative decay rate constants (k_r) calculated from η_{PL}/τ_{PL} are also listed in Table I. The radiative decay probability from the excited states of the materials is described by k_r. These values, for example, should be useful for material evaluation of laser activities.[8]

Next, we confirmed that the film of a green phosphorescent iridium complex, *fac*-tris(2-phenylpyridyl)iridium(III) [$Ir(ppy)_3$][2], in a 4,4'-bis(carbazol-9-yl)-2,2'- biphenyl (CBP) with a 2 wt% concentration, exhibited an η_{PL} of 97 ± 2%[18]. This result corresponds well to a previous report for a device with $Ir(ppy)_3$ having an η_{EL} of 19%[4]. Figure 3a shows the η_{PL} of the $Ir(ppy)_3$:CBP films as a function of doping percentage. As the concentration of $Ir(ppy)_3$ was increased, significant decreases in η_{PL} were observed resulting in an η_{PL} of less than 3% in the neat film. Additionally, at higher dopant concentrations, the emission peak shifts to a longer wavelength and the full width at half maximum increases. These observations clearly demonstrate the existence of self-quenching interactions between $Ir(ppy)_3$ molecules at increased dopant concentrations.

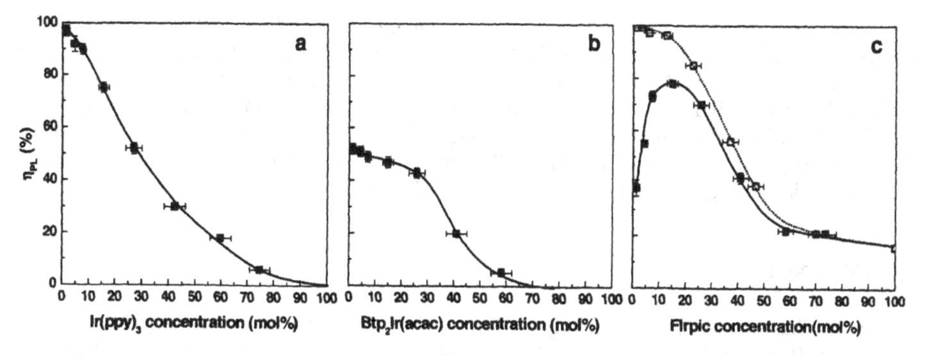

Figure 3. PL quantum efficiency η_{PL} versus dopant concentration in (**a**) Ir(ppy)$_3$:CBP, (**b**) Btp$_2$Ir(acac):CBP, and (**c**) FIrpic:CBP (■) and FIrpic:mCP (□).

The film of bis[2-(2'-benzothienyl)pyridinato-*N*,$C^{3'}$] (acetylacetonato) iridium(III) [Btp$_2$Ir(acac)]:CBP exhibited a maximum η_{PL} of 51±1% at 1.4 mol% (2wt%) (Fig. 3b)[18]. Despite the fact that η_{PL} also decreased with increasing Btp$_2$Ir(acac) concentration, at a dopant concentration of 26mol% (33wt%) the PL intensity was still 90% of its maximum value. This suggests that concentration quenching in the Btp$_2$Ir(acac):CBP film is less effective than observed for the Ir(ppy)$_3$:CBP film. In the case of blue phosphorescent complex, bis[(4,6-difluorophenyl)pyridinato -*N*,C^2] (picolinato)iridium(III) [FIrpic] in CBP, the η_{PL} increased from 38±4% to a maximum of 78±1% going from low to intermediate dopant concentration (1.4–15mol%, corresponding to 2–20wt%) and then decreased with increasing concentration (in Fig. 3c). In contrast to Ir(ppy)$_3$ and Btp$_2$Ir(acac), FIrpic maintained a rather high η_{PL} of 16±1% as a neat film, suggesting that the fluorination on the ppy ligand hinders self-quenching interactions. Since a small amount of emission from the host was observed only at a dopant concentration of 1.4mol% (2wt%), the increase of η_{PL} at low concentrations can not be attributable to poor energy transfer from CBP to FIrpic but rather to a back energy transfer process from the triplet state of FIrpic to the triplet of CBP. Based on the back energy transfer mechanism[19], more than 40% of the excitation energy in an OLED device having an emitting layer of 6 wt%-FIrpic:CBP may be non-radiatively wasted through the back energy transfer to the triplet state of CBP.

In order to confirm this potential mechanism for energy loss in blue phosphorescent devices, thin film PL measurements with FIrpic were determined using the high energy host, *m*-bis(*N*-carbazolyl)benzene (mCP). mCP is shown in the literature to have a T_1 energy of 2.91 eV [20] which is considerably higher than that of CBP (T_1 = 2.53 eV) [19], and should therefore prevent undesired back energy transfer from FIrpic (T_1 = 2.62 eV) [19]. As expected, the η_{PL} of the film of FIrpic:mCP reached 99±1% at 1.4mol% (2wt%), suggesting that the T_1 energy of FIrpic can be completely confined in host materials with a high enough T_1 energy (Fig. 3c).

CONCLUSIONS

We established a simple and accurate system for measuring η_{PL} of luminescent organic materials in submicrometer-thick solid-state thin films using an integrating sphere. The η_{PL} of Ir complexes were measured in solid-state films using an integrating sphere, and the green phosphorescent $Ir(ppy)_3$:CBP film was shown to have an η_{PL} of 97±2% (at 1.5 mol%). This value of almost 100% in the solid state is consistent with previous results for temperature dependency of PL and high EL efficiency in a device (η_{EL} = 19%). Furthermore, the η_{PL} of Btp_2Ir(acac):CBP was 51±1% (at 1.4mol%) and FIrpic:mCP was 99±1% (at 1.2 mol%). These results suggest that the η_{EL} of red phosphorescent OLEDs using Btp_2Ir(acac) as the dopant can be as high as 10%, and blue devices using FIrpic can reach the theoretical limit of 20%.

REFFERENCES

1. C. W. Tang and S. A. VanSlyke, *Appl. Phys. Lett.* **51**, 913 (1987).
2. M. A. Baldo , S. Lamansky, P. E. Burrows, M. E. Thomson and S. R. Forrest, *Appl. Phys. Lett.* **75**, 4 (1999).
3. C. Adachi, M. A. Baldo, M. E. Thompson and S. R. Forrest, *J. Appl. Phys.* **90**, 5048 (2001).
4. M. Ikai, S. Tokito, Y. Sakamoto, T. Suzuki and Y. Taga, *Appl. Phys. Lett.* **79**, 156 (2001).
5. F. Hide, M. A. Diaz-García, B. J. Schwartz, M. R. Andersonn, Q. Pei, and A. J. Heeger, *Science* **273**, 1833 (1996).
6. V. G. Kozlov, V. Bulovic, P. E. Burrows, and S. R. Forrest, *Nature (London)* **389,** 362 (1997).
7. H. Yamamoto, T. Oyamada, H. Sasabe and C. Adachi, *Appl. Phys. Lett.* **84**, 1401 (2004).
8. Y. Kawamura, H. Yamamoto, K. Goushi, H. Sasabe, C. Adachi and H. Yoshizaki, *Appl. Phys. Lett.* **84**, 2724 (2004).
9 T. Sakanoue, E. Fujiwara, R. Yamada and H. Tada, *Appl. Phys. Lett.* **84**, 3037 (2004).
10 S. Lamansky, P. Djurovich, D. Murphy, F. Abdel-Razzaq, C. Adachi, P. E. Burrows, S. R. Forrest and M. E. Thompson, J. Am. Chem. Soc. 123, 4303 (2001).
11 K. Goushi, Y. Kawamura, H. Sasabe and C. Adachi, *Jpn. J. Appl. Phys.* **43**, L937 (2004).
12 K. L. Eckerle, J. W. H. Vanable and V. R. Weidner, *Appl. Opt.* **15**, 703 (1976).
13 N. C. Greenham, I. D. W. Samuel, G. R. Hayes, R. T. Phillips, Y. A. R. R. Kessener, S. C. Moratti, A. B. Holms and R. H. Friend, *Chem. Phys. Lett.* **241**, 89 (1995).
14 H. Mattoussi, H. Murata, C. D. Merritt, Y. Iizumi, J. Kido and Z. H. Kafafi, *J. Appl. Phys.* **86**, 4642 (1999).
15 Y. Kawamura, H. Sasabe and C. Adachi, *Jpn. J. Appl. Phys*, **43,** 7729 (2004)
16. G. Gu, D. Z. Garbuzov, P. E. Burows, S. Venkatesh, S. R. Forrest and M. E. Thompson, *Opt. Lett.* **22**, 396 (1997).
17. S. A. Vanslyke, P. S. Bryan and C. W. Tang, Ext. Abstr. p.195 (The 8th International Workshop on Inorg. and Org. Electroluminescence/EL96 Berlin).
18. Y. Kawamura, K. Goushi, J. Brooks, J. J. Brown, H. Sasabe and C. Adachi, *Appl. Phys. Lett. (in press).*
19 C. Adachi, R. C. Kwong, P. Djurovich, V. Adamovich, M. A. Baldo, M. E. Thompson and S. R. Forrest, *Appl. Phys. Lett.* **79**, 2082 (2001).
20 R. Holmes, S. R. Forrest, Y.-J. Tung, R. C. Kwong, J. J. Brown, S. Garon and M. E. Thompson, *Appl. Phys. Lett.* **82**, 2422 (2003).

Plasmonics

Mater. Res. Soc. Symp. Proc. Vol. 846 © 2005 Materials Research Society DD5.4

Percolation-Enhanced Supercontinuum and Second-Harmonic Generation from Metal Nanoshells

Charles Rohde, Keisuke Hasegawa, Aiqing Chen and Miriam Deutsch
Oregon Center for Optics and Department of Physics,
University of Oregon, Eugene, OR 97403, USA
crohde@uoregon.edu

ABSTRACT

We present results for linear and nonlinear light scattering experiments from percolative silver nanoshells on dielectric silica cores. Using ultrashort pulsed laser illumination we observe strong nonlinear optical (NLO) responses from single metallodielectric core-shell (MDSC) spheres and disordered MDSC sphere aggregates. Finally, combining scaling theory with core-shell Mie scattering formalism we obtain a new model for the observed linear extinction signals.

INTRODUCTION

Metal-dielectric interfaces are known to support the propagation of surface electromagnetic (EM) waves with a broad spectral range.[1] These modes, known as surface-plasmon polaritons (SPPs), are coupled modes of charge-density waves and photons, and are guided at the surface of the metal-dielectric interface. The excitation and propagation of these modes are highly sensitive to the interfacial environment, and may be strongly altered with only slight perturbations to the interface.

The optical response of noble metals changes dramatically when fabricated into nanoparticles.[2] The optical features which evolve with decreasing size of the metal particles may be further controlled through the material's topology. In particular, properly designed noble metal nanoshells allow accurate control of electromagnetic (EM) field distributions and subsequently their surface plasmon resonances (SPRs).[3] Metal nanoshells consist of a nano-scale metal shell (typically 10-30nm) surrounding a dielectric core, thus forming a metallodielectric core-shell (MDCS) structure. These systems exhibit unique extinction spectra, which are geometrically tunable through their core-shell thickness ratios. For core diameters in the sub-micrometer range, the optical response of the composite particles may be tuned over the entire visible and near infrared spectrum. We have developed a method, based on the Tollen's process[4] to fabricate dense, highly uniform nanocrystalline silver shells with thicknesses of 25-100 nm on colloidal silica cores.

The thinnest silver shells grown with this method are highly fractured and form disordered, percolative films. These thinnest films are treated as two dimensional (2d), but may also be grown into thicker 3d shells. Below we model the linear optical response of these thin shells by combining standard core-shell Mie scattering theory [5] with a scaling theory (ST) description[6] of their dielectric response. We show that for thin (2d) percolative shells the ST approach is better suited than the more commonly used Bruggeman effective medium theory (EMT)[7], while thicker (3d) shells are better modeled by the latter.[8]

In addition to their linear response, nanometer-sized metal particles have been the focus of extensive studies owing to their greatly amplified nonlinear optical (NLO) response. These amplifications are attributed mainly to large enhancements of surface-induced electric fields at the nanoparticles' plasma resonance. In a related context, localization of plasmon fields in the plane of percolative metallic films manifests in extremely large local EM fields and giant enhancements of surface NLO phenomena. Examples of such enhancements are surface-enhanced second-harmonic generation (SHG),[9,10] and the enhanced optical Kerr effect.[11] Our particles combine salient features of these two systems : thin, percolative metal films grown in a sub-micron nanoshell geometry.

Under illumination with ultrashort 850nm laser pulses we observe strong frequency-doubled signals at 425nm, and an intense white-light scattered background from these materials. The SHG is attributed to induced surface dipole and quadrupole effects in the metal shells. The continuum generation is related to the enhanced Kerr nonlinearity observed in random metallodielectric films near the percolation threshold.

EXPERIMENTAL MEASUREMENT OF OPTICAL RESPONSES

Silver Shell Characteristics

MDCS particles are typically synthesized using solution-based electroless deposition methods, where a metal precursor is reduced onto colloidal dielectric spheres to produce a nano-scale shell. We have modified conventional electroless deposition techniques to fabricate dense, highly uniform nanocrystalline silver shells.

Using this method we have created MDCS spheres with a controlled variety of shell thicknesses, ranging from 20nm to more than 100nm. Transmission electron microscope (TEM) and scanning electron microscope (SEM) images as shown in Fig. 1 were used to characterize the shell thickness, *t*, and the metal filling fraction of the shells.

Our fabrication process results in a shell which can be characterized by a non-integer fractal dimension, similar to ultra high vacuum (UHV) deposited thin films. Using the SEM images of the thinnest shells as in Fig. 1 (b), we measure the fractal dimension, D, and filling fraction, *p*, of the shells. Using a box counting method on cropped SEM images such as shown in Fig. 1 (c), we find D ~ 1.7-1.8. This agrees well with reported measurements of the fractal dimension of silver films grown under UHV, where D $\cong$ 1.8.[12] The filling fraction of the thin, 2d shells was measured to be in the range $0.5 < p < 0.6$. Using the known value[13] of $p_c = 0.68$ we conclude that for thin shells (~20nm) $p \leq p_c$. As we discuss below, being close to the percolation threshold significantly changes the optical response of the shell.

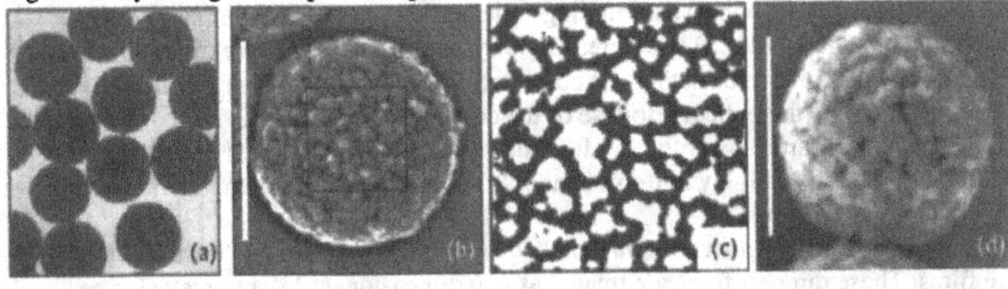

Fig. 1. (a) Transmission electron micrographs of silica spheres 1μm in diameter coated with a thin, ~20nm, silver shell. The particles are highly uniform, with a measured polydispersity of 4%. (b) Scanning electron micrograph (SEM) of a thin-shell silver-silica sphere in (a). (c) Magnified view of the region shown inside the square in (b). It is seen that most metal clusters are not in contact, and therefore this shell is below percolation. (d) SEM of a thickly coated (100nm) silver-silica sphere. The white scale bar in (b) and (d) is 1 μm in length.

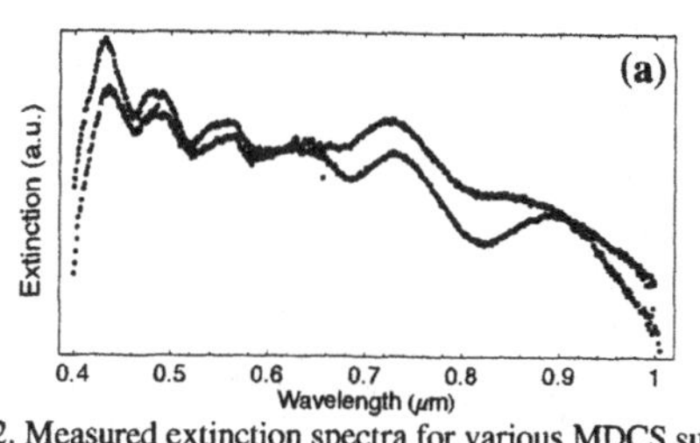

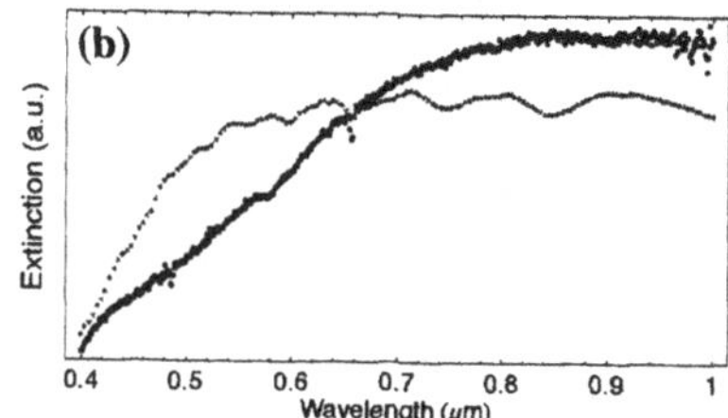

Fig. 2. Measured extinction spectra for various MDCS suspensions. All core diameters are 1 μm. (a) Extinction spectra of two different thin ($t \leq 50$nm) MDCS samples in water. (b) Extinction spectra of two different thick shell samples ($t \geq 100$nm) in water (black) and ethanol (gray).

Near the percolation threshold the dielectric constant can attain either positive or negative values, exhibiting a singularity just below p_c. In this range, standard methods for calculating the dielectric response for a two-component medium, such as Bruggeman EMT, are invalid.

Linear Scattering Measurements

We have collected extinction spectra from colloidal suspensions of silver coated silica spheres suspended in water. Extinction spectra for the two previously shown MDCS sphere species in Fig. 1(b) and 1(d) are shown in Fig. 2. These spectra have two significant robust features: the large number of sharp resonance peaks, and the overall trend of the extinction curves.

The large number of resonance peaks is indicative of coupling to high angular momentum modes, of the order $4 \leq l \leq 7$, of the 1 μm composite core-shell spherical resonator. A change in trend is observed in the extinction spectra when the shell thickness is increased from 20nm to above 100nm. We find that these two shell species differ in their composition; the thin shells are 2d percolative silver films, while the thicker ones are typically 3d metallic shells with dielectric (water) inclusions. Interestingly, we observe that although very thin shells are highly granular and not contiguous, core-shell resonances are clearly observed. This indicates the presence of strong electromagnetic coupling between silver grains in the shell, such that a collective dielectric response is achieved. In what follows we model the dielectric function of the shells.

Nonlinear Light Generation

We have conducted studies to test the NLO response of the MDCS spheres. A concentrated drop of particles was placed on a cleaned silica substrate and allowed to dry. Unless special measures were taken, this resulted in dense disordered aggregates of spheres. In our experiments 120fs, 850nm p-polarized pulse trains from a modelocked Ti:Sapphire laser operating at 76MHz were focused on the samples. The output from the laser was first passed through several filters to eliminate any radiation at wavelengths below 850nm. The pump beam was then focused to a spot ~90μm in diameter, at an incidence angle of 45° on the sample. The reflected signal was collected with a high numerical aperture (NA) microscope objective, filtered to eliminate the pump beam, and dispersed through a spectrometer on a cooled charge-coupled device array.

We observed a frequency-doubled signal at 425nm, and an additional intense supercontinuum (white-light) scattered background, as shown in Figure 3(a).

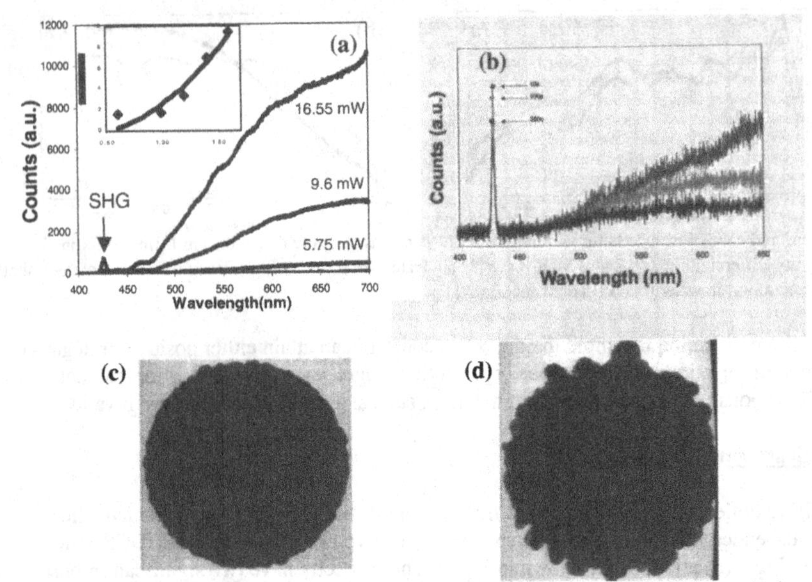

Fig. 3. (a) SHG (marked with arrow) and supercontinuum generated from an aggregate of coated spheres at increasing average pump powers. Inset: SHG counts per second (diamonds) and fitted second order polynomial curve showing quadratic dependence of SHG on average pump power. (b) SHG and supercontinuum from a single sphere. Each scan was taken after exposure to the time noted, using laser pulses of $800MW/cm^2$ peak intensity. The arrows mark the position of the SHG peak. (c) TEM image of 1μm silver-coated sphere before laser illumination. (d) Sphere after exposure to laser for 200 seconds.

The supercontinuum, cutoff by a blocking filter, extends well beyond 700nm into the near infrared range. For control, it was verified that both signals were absent when illuminating uncoated silica sphere aggregates on identical substrates. Due to its centrosymmetric crystalline structure bulk silver does not exhibit SHG in the dipole approximation. This process is therefore attributed to surface induced dipole effects,[14] as well as quadrupole contributions.[15,16]

The continuum generation is related to the enhanced Kerr nonlinearity observed in random metallodielectric films near the percolation threshold.[17] Nevertheless, unlike the latter, our systems are three-dimensional, and therefore require also the consideration of multiple nonlinear scattering from the randomly packed sphere aggregates.

To gain insight into this process, we first address NLO scattering from a single metal-coated sphere. Using a dilute suspension we formed sparsely covered substrates, where single spheres were easily observed. An inverted epi-fluorescence microscope with a high NA objective was used to align and focus the laser beam onto a single sphere. The beam spot diameter was independently measured to be ~1.2μm. The signal was collected in reflection mode and spectrally analyzed as above. Similar SHG and supercontinuum signals were observed. We found that the white-light spectrum generated form a single sphere with a ~25nm silver shell was as broad as that observed from the sphere-aggregate. Figure 3(b) shows the decrease over time in both SHG and white-light signals when continuously pumped for a period of ~3min.

Interestingly, we find that while the continuum signal is significantly lower after several minutes of exposure, the SHG saturates at a fixed value and does not degrade considerably. The reason for these different behaviors is not known. When re-examining the illuminated spheres in TEM, we observe marked photomodification of the silver shell, as shown in Figures 3(c)-(d). We propose that the coalescence and coarsening of the silver grains lead to a shell geometry which is far from percolative. This reduces the possible field localization caused by the fractal structure of the thin shell and weakens the localization induced NLO enhancements.

THEORETICAL MODELING OF SHELL DIELECTRIC CONSTANT

Electromagnetic plane wave scattering from a core-shell sphere system was originally formulated by Aden and Kerker (AK).[5] As an extension to Lorentz-Mie theory, this solution relies on azimuthal symmetry of the particle boundaries. From our SEM images, it is seen that the shells are fragmented, consisting of silver cluster aggregates embedded in a dielectric (in this case it is water.) It is also important to note that using the AK formalism requires knowledge of the dielectric function of both core and shell materials, while in our case the dielectric function of the porous metal shell is not known. The usual approach for modeling the extinction coefficient using tabulated values of the bulk silver dielectric function as the shell material[3] exhibited serious discrepancies and failed to reproduce the experimental data. This is not surprising, since our shells are discontinuous and not uniform in composition, and are likely to be better modeled using an effective medium approach. To implement the AK method we therefore approximate the metal-dielectric composite shell as a homogenous shell with a modified dielectric constant ε_{eff}.

We construct ε_{eff} from first principles, using ST methods.[13] According to this, we first obtain a dielectric function which is associated with smaller sub-sections in the percolative film. The size of these sections is determined by the smallest of length scales characterizing all relevant processes in the film (e.g. dipole coupling, electron scattering,) and is therefore frequency dependent. We then obtain the optical response of the composite shell by taking a weighted average of these dielectric "patches". This dielectric function is then used to calculate the normalized extinction spectrum of thin shells, as shown in Fig. 4(a). The agreement with normalized experimental results is seen to be very good. For comparison, we also plot in Fig. 4(a) the extinction coefficients obtained using both the bulk silver values and the classical Bruggeman EMT model for the shell dielectric function. The latter takes into account geometrical averaging based only on nanoparticle shape and volume fraction in the shell, but neglects the frequency-dependent (AC) coupling existing between silver grains at optical frequencies. It is clear that while EMT qualitatively agrees with the trend of the data, the percolation-based model better reproduces the number of resonances and their positions.

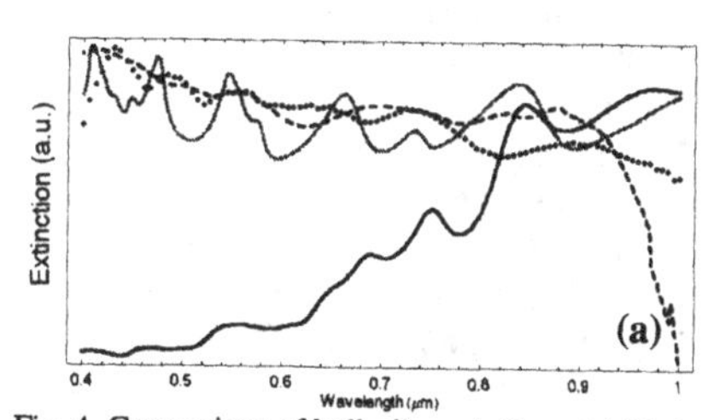

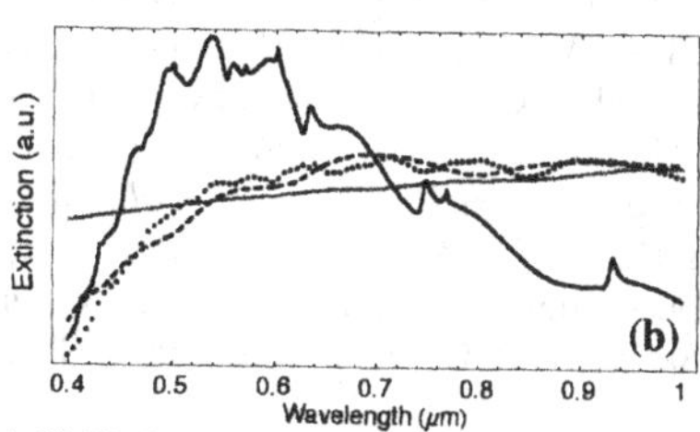

Fig. 4. Comparison of bulk silver shell model (dark trace), EMT (dashed), ST (gray) with measured extinction pectra (points) for (a) thin percolative (t ~20nm) and (b) thick metallic (t ~100nm) silver shells on 1 μm silica cores.

We next examine Fig. 4 (b) where we plot the extinction spectra for thick shells. The shell thicknesses of the MDCS spheres used for this have been measured to be ~100 nm. The SEM image in Fig. 1 (d) shows that this is a porous shell with silver filling fraction well above the 3d percolation threshold of $p_c = 0.3$.[18] In this case we find that the ST model yields a good mean value but cannot reproduce the detailed structure of the spectrum. Instead, EMT provides a closer description of the data.

CONCLUSIONS

We have shown that the presence of a percolative metallic shell in a MDCS system significantly influences the observed optical excitations of the system. The presence of the percolative shell enhances the Kerr nonlinearity of the shell. Altering this shell geometry through laser induced heating, causes a reduction in the observed supercontinuum generation of the shell. Finally, we have presented two models for the dielectric function of the porous shells, based on scaling theory and effective medium theory. When combined with standard core-shell scattering theory these two models respectively reproduce the observed linear extinction results in two separate regimes: near the percolation threshold and well above it.

ACKNOWLEDGMENTS

This work was supported by NSF Grant No. DMR-02-39273 and ARO Grant No. DAAD19-02-1-0286.

REFERENCES

1. H. Raether, *Surface Plasmons*, Springer, Berlin **1988**.
2. U. Kreibig, M. Vollmer, *Optical Properties of Metal Clusters*, Springer, Berlin **1995**.
3. J. B. Jackson, N. J. Halas, J. Phys. Chem. B **105**, 2743 (2001).
4. D. L. Pavia, G. M. Lampman, *et al.*, Saunders College Publishing, USA **1995**.
5. A. L. Aden, M. Kerker, J. Appl. Phys. **22**, 1242 (1951).
6. Y. Yagil, M. Yosefin, *et al.*, Phys. Rev. B, **43**, 11342 (1991).
7. D. Bruggeman, Ann. Phys. (Leipzig) **24**, 6736 (1935).
8. C. Rohde, K. Hasegawa and M. Deutsch, *in preparation* (2004).
9. A. C. R. Pipino, R. P. Van Duyne, G. C. Schatz, *Phys. Rev. B*, **53**, 4162 (1996).
10. C. Anceau, S. Brasselet, *et al.*, *Opt. Lett*, **28**, 713 **(2003)**.
11. D. Ricard, Ph. Roussignol, Chr. Flytzanis, *Phys. Rev. Lett*, **10**, 511 (1985).
12. W. Kim, V.P. Safonov, *et al.*, Phys. Rev. Lett, **82**, 4811 (1999).
13. Y. Yagil, P. Gadenne, *et al.*, Phys. Rev B **46,** 2503 (1992).
14. V. M. Shalaev, A. K. Sarychev, *Phys. Rev. B*, **57**, 13265 (1998).
15. N. Bloembergen, R. K. Chang, *et al.*, *Phys. Rev*, **174**, 813 (1968).
16. J. I. Dadap, J. Shan, *et al.*, *Phys. Rev. Lett.*, **83**, 4045 (1999).
17. S. Ducourtieux, S. Gresillon, *et al.*, *J. Nonlinear Opt. Phys. Mater*, **9**, 105 (2000).
18. A. Sarychev, R. C. McPhedran, and V. Shalaev, Phys. Rev. B, **62**, 8531 (2000).

Mater. Res. Soc. Symp. Proc. Vol. 846 © 2005 Materials Research Society DD5.6

Linear and Nonlinear Transmission of Surface Plasmon Polaritons in an Optical Nanowire

N. C. Panoiu and R. M. Osgood, Jr.
Department of Applied Physics and Applied Mathematics
Columbia University
500 W. 120th Street
New York, NY 10027

ABSTRACT

Polymer-metal composites offer the possibility of strongly enhanced nonlinear optical properties, which can be used for ultrasmall photonic devices. In this paper, we investigate numerically, by means of the finite-difference time-domain (FDTD) method, the propagation characteristics of surface plasmon polariton (SPP) modes excited in an optical nanowire consisting of a chain of either metallic cylinders or metallic spheres embedded in dielectric shells made of polymers (or other material) with optical Kerr nonlinearity. Our FDTD calculations incorporate both the nonlinear optical response of the dielectrics as well as the frequency dispersion of the metals, which is considered to obey a Drude-like model. It is demonstrated that, in the linear limit, the nanowire supports two SPP modes, a transverse and a longitudinal one, separated by $\Delta\lambda = 20$ nm. Furthermore, the dependence of the transmission of these SPP modes, on both the pulse peak power and Kerr coefficient of the dielectric shell, is investigated. Nonlinear optical phenomena, such as power-dependent mode frequency, switching, or optical limiting, are observed.

INTRODUCTION

Until recently, most research work on the optical properties of nanostructured materials has focused on periodic structures consisting of dielectrics, namely photonic crystals. However, recent research studies have shown that it is possible to design metallic photonic crystals with new features: large relative gap width [1,2], unusual transmission properties [3,4], high surface impedance [5], or, arguably the most striking example, structures with negative index of refraction [6-8].

A different class of applications, which is discussed in this paper, is represented by metal-based sub-wavelength-dimension nanodevices used for light guiding. Several approaches to achieve this functionality in passive structures have been proposed and demonstrated. In one case, generally termed "plasmonic" devices, one makes use of the unique optical properties of metallic materials to provide strong optical confinement and eliminate the limitations imposed by strong diffractive effects. For example, cylindrical metallic waveguides with sub-wavelength transverse dimensions [9], plasmon waveguides [10], or optical waveguides consisting of chains of resonantly coupled metallic nanoparticles have been demonstrated [11,12].

The optical properties of these nanoparticle chains can be further improved if their basic building blocks have a more complex geometry, e.g. ellipsoidal particles [12,13], or by combining in the same nanostructure materials with different optical properties. One such example is a metallo-dielectric nanoshell, which consists either of a metallic core surrounded by a dielectric shell [14] or a dielectric core embedded in a metallic shell [15]. An ideal material for such a composite structure would be a polymer, which as a class are known to having very tunable linear optical properties. The frequency of SPP resonances excited in these metallo-dielectric nanostructures depends sensibly on the material parameters of the core and shell, as well as their relative dimensions [16], and therefore, their optical properties are highly tunable.

Despite this work, there have been only limited efforts to explore the questions regarding the *nonlinear* response of these plasmonic light-guiding structures. In this connection, it is known that the basic materials blocks of the nanoparticle chains are a particular efficient medium for nonlinear optical elements. For example, metal nanoparticles have been shown to exhibit strong SPP field enhancement to increase the efficiency of Raman scattering [17], second-harmonic generation [18], and third-order nonlinear processes [19]. In the present paper, we demonstrate that a nanoparticle chain can be used to form a subwavelength nonlinear optical element, which exhibits optical "limiting" and frequency shifting. Further we show that the use of metallo-dielectric nanoparticles leads to lower loss than for pure metal nanospheres.

DESCRIPTION OF THE PLASMONIC NANOSTRUCTURES

In this study we investigate two types of structures: in the first case we consider a chain of infinitely long metallic cylinders coupled through near-field interactions, that is a 2D geometry, whereas in the second case we study chains of metallo-dielectric (i.e. polymers) nanoshells, a full-3D geometry. Since the latter structure better reflects usual experimental settings, we use it to demonstrate that one can achieve dynamic tunability of the optical properties of such nanostructures by incorporating nonlinear (Kerr) optical material in the structure of the metallo-dielectric nanoparticles that form the chain. Specifically, we consider a nanoparticle chain made of metallo-dielectric nanoshells, which contain a metallic core (Ag in our case) embedded in a shell made of a dielectric material with Kerr nonlinearity (see figure 1). The geometrical parameters of the nanostructure are $R = 25$ nm, radius of the metallic spheres, $t = 10$ nm, thickness of the dielectric shell, and $D = 70$ nm, center-to-center distance between adjacent spheres, whereas for the material parameters we chose $n_s = 1.5$, the refractive index of the dielectric shell and $n_b = 1$, the refractive index of the background. The nonlinear properties of the nanostructure are determined by the Kerr coefficient n_2. Finally, we assume that dispersion properties of the metallic core are described by a Drude-like dielectric function

$$\varepsilon(\omega) = \varepsilon_0\left[1 - \frac{\omega_p^2}{\omega(\omega + i\gamma)}\right] \qquad (1)$$

where ε_0 is the permittivity of free space and $\omega_p = 13.7\cdot10^{15}$ rad/s and $\gamma = 2.7\cdot10^{13}$ rad/s are the plasma and damping frequencies, respectively. In the case of metallic cylinders their radius and center-to-center distance are R and D, respectively.

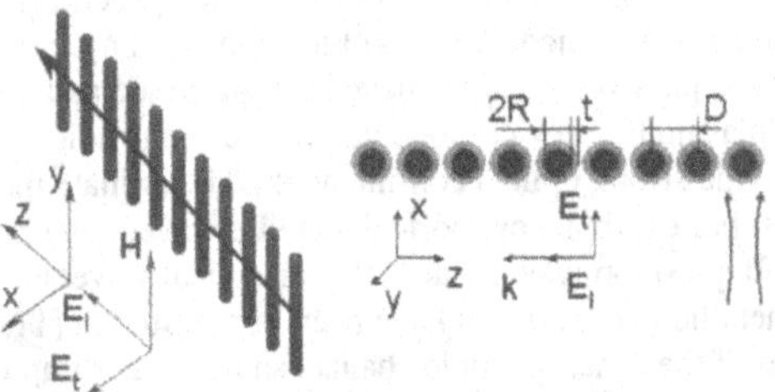

Figure 1. Schematics of the optical waveguides: $R = 25$ nm is the radius of the metallic sphere and cylinders, $t = 10$ nm is the thickness of the dielectric shell, and $D = 70$ nm is the center-to-center distance between adjacent spheres or cylinders.

LINEAR REGIME

Before studying the guiding characteristics of the nanoparticle chain, we briefly discuss the linear optical properties of their building blocks. We consider first an isolated metallo-dielectric nanoshell and use the quasistatic approximation, which is valid if the size of the particle is much smaller than the wavelength, to calculate its extinction and absorption cross-sections. The results show a SPP resonance at $\lambda_r = 285.4$ nm ($\omega_r = 6.6 \cdot 10^{15}$ rad/s). Also, the absorption cross-section is much smaller than the extinction cross section, which means that only a small fraction of the incident light is absorbed by the metallic core. Furthermore, our analytical calculations show that, at the SPP resonance, the electric field is enhanced by more than 300 times, which suggests that the nonlinear optical response of the nanoshell is strongly enhanced at wavelengths close to λ_r. Similar results are obtained if one considers an isolated cylinder, the only difference being that the SPP resonance is excited at a different frequency.

In order to investigate linear and nonlinear optical properties of a nanoparticle chain we employed 2D or a full 3D implementation of the FDTD numerical algorithm. This method consists of discretizing Maxwell equations on a grid and then, starting from a given initial condition, marching the resulting iterative relations in time. Upon choosing a suitably refined grid, the corresponding numerical solution gives an accurate representation of the dynamics of the electromagnetic field. Throughout our simulations, we considered a uniform grid with a 2 nm lattice constant, surrounded by a 25 nm-thick perfectly matched layer. Furthermore, we assumed that the metal dispersion is described by the Drude dielectric function (1) whereas the dielectric material has an instantaneous nonlinear Kerr response. Dispersive and nonlinear optical effects can rigorously be incorporated in the FDTD algorithm. Finally, because of the heavy computational demand imposed by 3D FDTD simulations, we used a parallel implementation of the FDTD algorithm, which has been run on a cluster with 18 Pentium IV processors at 2.8 GHz.

The main results concerning chains made of infinitely long cylinders are summarized in figure 2, which shows the electromagnetic field distribution of a longitudinal SPP mode excited in the chain. Furthermore, our calculations demonstrate that these SPP modes can be used in more complex nanodevices, such as Y-splitters. The field distribution in such a nanodevice, as well as the dependence of its efficiency on the splitting angle between the Y-branches are presented in figure 2(b). Note that although the efficiency of the Y-splitter is ~3%, its size is only a few wavelengths. Since most plasmonic structures incorporated in actual devices are expected to be 3D objects, in what follows we focus on the optical properties of chains of nanoshells.

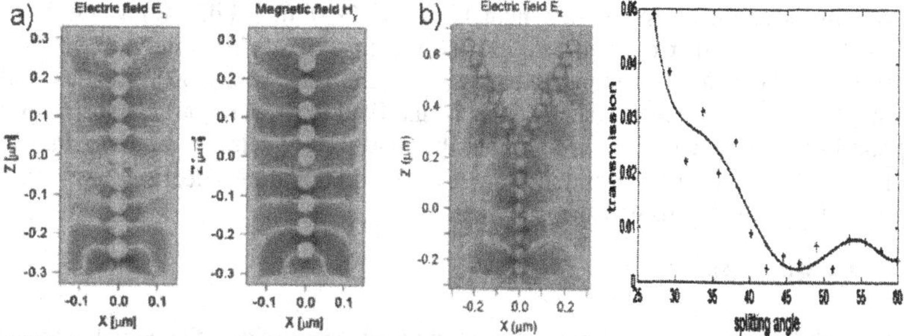

Figure 2. The electromagnetic field distribution of a longitudinal plasmon mode excited in a chain of infinitely long, coupled metallic cylinders, (a); in (b), the field distribution and the dependence on the splitting angle of the efficiency of a plasmonic Y-splitter.

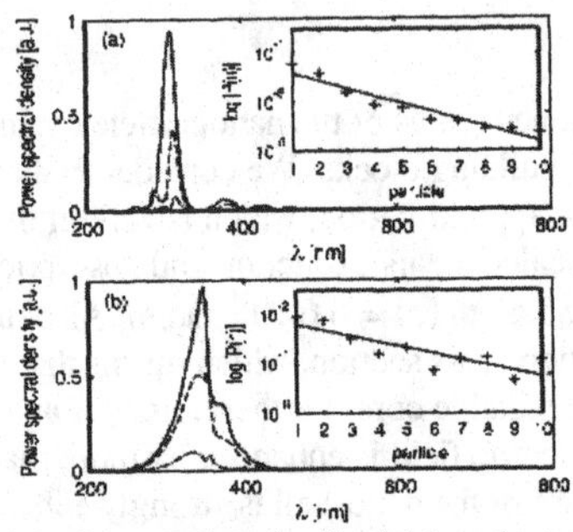

Figure 3. Power spectral density of *linear* longitudinal (a) and transverse (b) SPP modes, calculated after propagation of one (—), two (– – –), and three (– · – · –) interparticle distances. In insets, mode power decay along the particle chain axis.

We started our analysis of nanoshell chains by investigating their linear optical properties. This nanostructure supports two propagating SPP modes, one longitudinal, which corresponds to an induced dipole along the chain axis, and one transverse, with the induced dipole perpendicular to the chain axis. To find these modes we proceeded as follows [20]. An electromagnetic beam with a Gaussian transverse profile with its width comparable to the transverse area of the nanoshell, polarized along the *z*-axis, is excited along the *y*-axis. Due to its close proximity to the first particle in the chain, the beam excites a SPP resonance on the first particle in the chain, with the corresponding dipole along the *z*-axis, which through resonant near-field interactions couples to SPP resonances on subsequent particles in the chain to form a longitudinal SPP propagating mode. To excite a transverse SPP mode, the incoming beam is polarized along the *x*-axis. In both cases, the frequency width of the beam was three times the width of the SPP resonance of an isolated nanoshell. Finally, the values of the electric fields at the middle points between adjacent nanoshels were recorded and then Fourier transformed.

Figure 3 summarizes our calculated results pertaining to the linear optical properties of SPP modes supported by the nanoparticle chain. The graph depicts the two calculated SPP modes with frequencies $\omega_l = 5.98\cdot 10^{15}$ rad/s (longitudinal) and $\omega_t = 5.63\cdot 10^{15}$ rad/s (transverse), both close to the SPP resonance of an isolated nanoshell. Furthermore, from the variation with the propagation distance of the total power in the two modes along the chain we obtained their power absorption coefficients. Our results show that the longitudinal mode experiences larger losses, as compared to the transverse one, namely $\alpha_l = 10.2\cdot 10^{6}$ m^{-1} (3 dB/68 nm) *vs.* $\alpha_t = 7.7\cdot 10^{6}$ m^{-1} (3 dB/90 nm). Note that these absorption coefficients are approximately six times smaller than those in a chain made of metallic spheres with no dielectric shells [12]. This loss reduction is due to the fact that the dielectric shells enhance the field in the interparticle space, increasing the coupling between the particles; in addition, the field in the metallic core is reduced, which leads to lower absorption in the metallic particles. An additional loss decrease, by ~10, can be achieved if particles with elongated shape are used.

NONLINEAR REGIME

When a nonzero Kerr nonlinearity is included in numerical simulations, the optical response of the particle chain becomes dependent on the power coupled into the SPP modes. As previously explained, a beam with peak power P excites a SPP resonance on the first particle in the chain, which then excites a SPP propagating mode; however, when the nonlinear response of the Kerr dielectric is included, the optical properties of the SPP modes depend on both the input power P as well as the Kerr coefficient n_2. To investigate this relationship we proceeded as follows [20]. First, we fix the Kerr coefficient to $n_2 = 1.5 \cdot 10^{-17}$ m^2/W, which is close to Kerr nonlinearities for GaAs, and varied the peak power P of the excitation beam. The results, presented in figure 4 (left panels), show that the frequency of the SPP mode decreases with P, a variation of more than 150 THz being induced when P increases from zero to 46 GW/cm^2. Thus, taking into account the ultra-small transverse area of the SPP mode, one concludes that the frequency switching effect described here is achieved by employing an extremely small amount of optical energy. Furthermore, the fact that the observed frequency shift is due to the nonlinear response of the nanoshells is also illustrated by the inset in figure. 4, which shows the power dependence of the SPP mode frequency shift, calculated at two propagation distances. Thus, due to the losses in the particle chain, the power in the SPP mode decreases with the propagation distance, and therefore the corresponding frequency shift decreases, too.

Figure 4 suggests that the nanoparticle chain can be used as an active nanodevice, namely as an optical limiting device. Thus, the transmission spectra calculated for different powers P, normalized by the peak power, show that at certain wavelengths, the normalized transmission is strongly dependent on the power P. This figure shows that, depending on the operating wavelength, the transmitted power can be either strongly increased or decreased by tuning the input power P.

In a different set of numerical experiments, we fixed the peak power, $P = 100$ MW/cm^2, and varied the Kerr coefficient n_2 from zero to $n_2 = 10^{-12}$ m^2/W. The main results are summarized in figure 4 (right panel), which illustrates that as the Kerr coefficient n_2 varies, the corresponding frequency of the SPP mode changes too, with a steep variation starting at $n_2 \sim 10^{-15}$ m^2/W; for the peak power P shown here, the frequency changes by about 30 THz. Notice also that unlike the linear regime, when optical nonlinearities are included, upon propagation along the particle chain

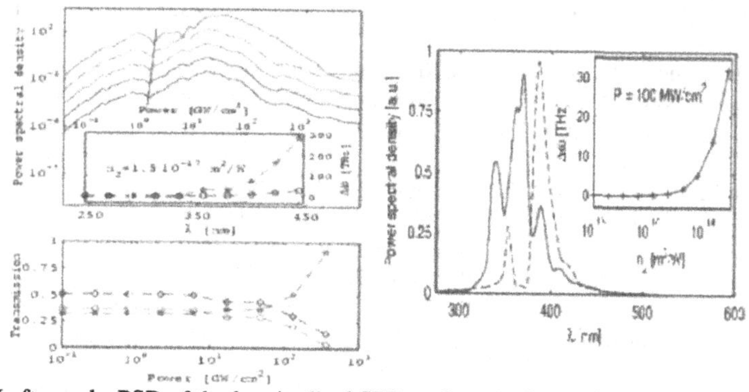

Figure 4. Left panels, PSD of the longitudinal SPP mode, calculated after propagation of three interparticle distances. The curves correspond to input powers of 0.1, 0.77, 6, 46, and 360 GW/cm^2. In inset, the frequency shift of the SPP mode, at three (★) and ten (o) interparticle distances, and, in the lower panel, normalized transmission *vs.* power, calculated at λ=335 (o), 345 (◊), and 376 nm (★). Right panel, PSD of the longitudinal SPP mode, calculated after a propagation of three interparticle distances. The curves correspond to Kerr coefficients of $n_2 = 0$ (---) and $n_2 = 8 \cdot 10^{-14}$ m^2/W (—). In inset, the SPP mode frequency shift *vs.* the Kerr coefficient.

the spectral profile of the SPP mode develops a series of modulations, a phenomenon which is an expression of self-modulation effects.

CONCLUSIONS

In conclusion, we studied both linear and nonlinear properties of a particle chain made of metallic cylinders or metallic spheres embedded into dielectric shells made of Kerr optical material, coupled through near field interactions. We showed that, in the linear limit, these nanostructures support longitudinal and transverse SPP modes, and, for the 3D geometry, calculated the absorption coefficients of both modes. We have also demonstrated that under relatively modest excitation powers strong nonlinear optical response can be observed. The nonlinear optical response of the particle chain includes phenomena such as power dependence of mode frequency, ultra-low power switching, and optical limiting.

ACKNOWLEDGMENTS

This work was supported by the AFOSR STTR, Grant No FA9550-04-C-0022.

REFERENCES

1. A. Moroz, Phys. Rev. Lett. **83**, 5274 (1999).
2. D. F. Sievenpiper, M. E. Sickmiller, and E. Yablonovitch, Phys. Rev. Lett. **76**, 2480 (1996).
3. J.A. Porto, F.J. Garcia-Vidal, and J.B. Pendry, Phys. Rev. Lett. **83**, 2845 (1999.
4. L. Martin-Moreno, F.J. Garcia-Vidal, H. J. Lezec, K. M. Pellerin, T. Thio, J. B. Pendry, and T. W. Ebbesen, Phys. Rev. Lett. **86**, 1114 (2001).
5. D.F. Sievenpiper, L. Zhang, R.F.J. Broas, N.G. Alexopolous, and E. Yablonovitch, IEEE Trans. Microwave Theory Tech. **47**, 2059 (1999).
6. D. R. Smith, W. J. Padilla, D. C. Vier, S. C. Nemat-Nasser, and S. Schultz, Phys. Rev. Lett. **84**, 4184 (2000).
7. N. C. Panoiu and R. M. Osgood, Phys. Rev. E **68**, 016611 (2003).
8. N. C. Panoiu and R. M. Osgood, Opt. Commun. **223**, 331 (2003).
9. J. Takahara, S. Yamagishi, H. Taki, A. Morimoto, and T. Kobayashi, Opt. Lett. **22**, 475 (1997).
10. T. Yatsui, M. Kourogi, and M. Ohtsu, Appl. Phys. Lett. **79**, 4583 (2001).
11. M. Quinten, A. Leitner, J. R. Krenn, and F. R. Aussenegg, Opt. Lett. **23**, 1331 (1998).
12. S. A. Maier, P. G. Kik, and H. A. Atwater, Appl. Phys. Lett. **81**, 1714 (2002).
13. C.J. Chen and R.M. Osgood, Phys. Rev. Lett. **50**, 1705 (1983).
14. L. M. Liz-Marzan, M. Giersig, and P. Mulvaney, Langmuir **12**, 4329 (1996).
15. H. S. Zhou, I. Honma, H. Komiyama, and J. W. Haus, Phys. Rev. B **50**, 12052 (1994); R. D. Averitt, D. Sarkar, and N. J. Halas, Phys. Rev. Lett. **78**, 4217 (1997).
16. S. J. Oldenburg, R. D. Averitt, S. L. Westcott, and N. J. Halas, Chem. Phys. Lett. **288**, 243 (1998).
17. S. Nie and S. R. Emory, Science **275**, 1102 (1997).
18. R. Antoine, P. F. Brevet, H. H. Girault, D. Bethell, and D. J. Schifirin, J. Chem. Soc. Chem. Commun., 1901 (1997).
19. D. Ricard, P. Roussignol, and C. Flytzanis, Opt. Lett. **10**, 511 (1985).
20. N. C. Panoiu and R. M. Osgood, Nano Lett. **4** (12), (2004) (in press).

Mater. Res. Soc. Symp. Proc. Vol. 846 © 2005 Materials Research Society DD5.8

Coherent Oscillations of Breathing Modes in Metal Nanoshells

Arman S. Kirakosyan[1,2], Tigran V. Shahbazyan[1]
[1] Department of Physics, Jackson State University, Jackson, MS 39217, USA
[2] Department of Physics, Yerevan State University, 1 Alex Manoogian St., Yerevan, 375025, Armenia

ABSTRACT

We study coherent oscillations of radial breathing modes in metal nanoparticles with a dielectric core. Vibrational modes are impulsively excited by a rapid heating of the particle lattice that occurs after laser excitation, while the energy transfer to a surrounding dielectric leads to a damping of the oscillations. In nanoshells, the presence of two metal surfaces leads to a substantially different energy spectrum of acoustic vibrations. The lowest and first excited modes correspond to in-phase (n=0) and anti-phase (n=1) contractions of shell-core and shell-matrix interfaces respectively. We calculated the energy spectrum as well as the damping of nanoshell vibrational modes in the presence of surrounding medium, and found that the size-dependences of in-phase and anti-phase modes are different. At the same time, the oscillator strength of the symmetric mode is larger than that in solid nanoparticles leading to stronger oscillations in thin nanoshells.

INTRODUCTION

Acoustic vibrational modes in nanoparticles are impulsively excited by a rapid heating of the lattice that takes place after laser excitation. After initial period of rapid expansion, a nanoparticle undergoes radial contractions and expansions around the new equilibrium. The periodic change in nanoparticle volume translates into a modulation in time of the surface plasmon resonance (SPR) energy that dominates nanoparticle optical absorption spectrum. The spectrum of vibrational modes manifests itself via coherent oscillations of differential transmission at SPR energy measured using ultrafast pump-probe spectroscopy [1, 2]. Since the size of laser spot is usually much larger than nanoparticle diameter, the initial expansion is homogeneous so that predominantly the fundamental ($n = 0$) breathing mode, corresponding to oscillations of nanoparticle volume as a whole, is excited. The lowest excited ($n = 1$) mode has weaker oscillator strength ($\approx 1/4$ of that for $n = 0$), and has also been recently observed [3]. When nanoparticle is embedded in a dielectric medium, the oscillations are damped due to the transfer of latice energy to acoustic waves in surrounding dielectric. In solid particles, the size dependences of eigenmodes energy and decay rate are similar – both are inversely proportional to nanoparticle radius [4].

Here we study the vibrartional modes of metal nanoshells. These recently manufactured metal particles with dielectric core [5] attracted much interest due to unique tunability of their optical properties. By varying the shell thickness during the manufacturing process, the SPR can be tuned in a wide energy interval [6]. Recent pump-probe measurements of vibrational modes dynamics in gold nanoshells submerged in water revealed characteristic oscillation pattern of differential transmission. However, the oscillations period and amplitude as well as their damping were significantly larger than those for solid nanoparticles. We perform detailed analysis of energy spectrum of lowest vibrational modes of a nanoshell in a dielectric medium. We find that the modes eigenenergies exhibit a strong dependence on nanoshell aspect ratio, $\kappa = R_1/R_2$, where R_1 and R_2 are inner and outer radii, respectively. Specifically, for thin nanoshells, the fundamental mode energy is consirerably lower than for solid particles while the damping is significantly larger. At the same time, in the thin shell limit, the fundamental mode carries the *entire* oscillator strength which results in an enhanced oscillations amplitude as compared to solid particles. The analysis also reveals two regimes, where the spectrum is dominated by nanoslell geometry or by surrounding medium, with a sharp crossover governed by the interplay between aspect ratio and impendance.

SPECTRUM OF VIBRATIONAL MODES FOR A NANOSHELL

We consider radial normal modes of a spherical nanoshell with dielectric core extending up to inner radius R_1 in a dielectric medium over outer radius R_2. The core, shell, and medium are characterized by densities $\rho^{(i)}$ longitudinal and tranverse sound velicities $c_{L,T}^{(i)}$ with $i = c, s, m$, respectively. The radial displacement $u(r)$ is determined from Helmholtz equation (at zero angular momentum) [7]

$$u'' + \frac{2u'}{r} + k^2 u = 0, \tag{1}$$

where $k = \omega/c_L$ is the wave-vector with the boundary conditions that the displacement u and the radial component of stress tensor,

$$\sigma = \rho\left[c_L^2 u' + (c_L^2 - 2c_T^2)\frac{2u}{r}\right], \tag{2}$$

are continuous at core/shell and shell/medium interfaces. In the three regions divided by shell boundaries, the solution has the form

$$u^{(c)} \sim \frac{\partial}{\partial r}\frac{\sin k^{(c)}r}{r}, \quad u^{(s)} \sim \frac{\partial}{\partial r}\frac{\sin(k^{(s)}r + \phi)}{r}, \quad u^{(m)} \sim \frac{\partial}{\partial r}\frac{e^{ik^{(m)}r}}{r}, \tag{3}$$

where ϕ is the phase mismatch. The corresponding eigenenergies are, in general, complex due to energy transfer to outgoing wave in the surrounding medium. Matching $u(r)$ and $\sigma(r)$ at $r = R_1, R_2$, we obtain the following equations for eigenvalues $\xi = kR_2$

$$\begin{gathered}\frac{\xi^2}{\xi\cot(\xi+\varphi) - 1} + \frac{\eta_m \xi^2}{1 + i\xi/\alpha_m} + \chi_m = 0,\\ \frac{\xi^2\kappa^2}{\xi\kappa\cot(\xi\kappa+\varphi) - 1} - \frac{\eta_c \xi^2 \kappa^2}{(\xi\kappa/\alpha_c)\cot(\xi\kappa/\alpha_c) - 1} + \chi_c = 0,\end{gathered} \tag{4}$$

where $\kappa = R_1/R_2$ is nanoshell aspect ratio and the parameters

$$\begin{aligned}\alpha_i = c_L^i/c_L^{(s)}, \qquad \eta_i = \rho^{(i)}/\rho^{(s)}, \qquad \chi_i = 4(\beta_s^2 - \eta_i\delta_i^2)\\ \beta_i = c_T^{(i)}/c_L^{(i)}, \qquad \delta_i = c_T^{(i)}/c_L^{(s)}.\end{aligned} \tag{5}$$

characterize the metal/dielectric interfaces. For the ideal case of free boundary conditions, we have $\alpha_c = \eta_m = \eta_c = 0$ and $\chi_c = \chi_m = 4\beta_s^2$. For a thin nanoshell, $1 - \kappa \ll 1$, we then easily recover the known expression for the fundamental mode

$$\xi_0 = 2\beta_s\sqrt{3 - 4\beta_s^2}. \tag{6}$$

The eigenvalue is purely real since no energy leaks through the interface.

In the realistic case of a nanoshell in a medium, the following simplification occurs. The initial laser pulse causes rapid expansion of both dielectric core and metal shell. However, due to a larger value of the metal thermal expansion coefficient, the shell expands to a greater extend than the core, so at the new equilibrium the core and the shell are, in fact, no longer in contact. In this case, the boundary conditions at he core/shell interface should be taken as stress free. For a thin nanoshell, $1-\kappa \ll 1$, Eqs. (4) are then reduced to

$$\lambda_c\left[\lambda_m-\lambda_c+\frac{\alpha_m\eta_m\xi^2}{\alpha_m-i\xi}\right]=(1-\kappa)\left[\left(\lambda_m+\frac{\alpha_m\eta_m\xi^2}{\alpha_m-i\xi}\right)\xi_0^2-\lambda_c\xi^2\right]. \tag{7}$$

Typically, the metal density of the shell is much large that that of the sorrounding dielectric, i.e., the parameter η_m is small. For $\eta_m \ll 1$, using $\chi_m-\chi_c=-4\eta_m\alpha_m^2\beta_m^2$ and $\chi_m/\chi_c = 1-\eta_m\beta_m^2$, we obtain

$$x^2-1+\eta_m\beta_m=\frac{\alpha_m\eta_m}{\xi_0(1-\kappa)}\left[\frac{4\alpha_m\beta_m^2}{\xi_0}-\frac{x^2}{\alpha_m/\xi_0-ix}\right], \tag{8}$$

where $x=\xi/\xi_0$. It can be easily seen that there are two regimes governed by the parameter

$$\lambda=\frac{\alpha_m\eta_m}{\xi_0(1-\kappa)}. \tag{9}$$

For a very thin nanoshell, $\lambda \gg 1$, the lowest eigenvalue is given by

$$\xi \simeq \alpha_m\beta_m - i\frac{\eta_m\alpha_m\beta_m^2}{\xi_0}. \tag{10}$$

In this regime, the energy and damping are comptetely determined by the surrounding medium and do not depend on nanoshell geometry. Note that if the transverse sound speed of the medium is zero (e. g., for water), then both energy and the decay rate vanish. In the opposite case, $\lambda \ll 1$, the solution can be obtained as

$$\omega \approx \frac{c_L\xi_0}{R_2}\left(1+\frac{\alpha_m^2\eta_m}{2\xi_0^2(1-\kappa)}\left[4\beta_m^2-\frac{1}{(\alpha_m/\xi_0)^2+1}\right]\right), \tag{11}$$

$$\gamma \approx \frac{c_L\alpha_m^2\eta_m}{2\xi_0(1-\kappa)(\alpha_m^2/\xi_0^2+1)R_2}. \tag{12}$$

In this case, the spectrum is mostly dominated by nanoshell geometry. Note that in typical experimental situations, the parameter λ is small, and with a good approximation, $\xi \simeq \xi_0$, which is considerably lower that the fundamental mode energy for solid particles, and it depends only weakly on aspect ratio.. At the same time, the damping rate γ is very sensitive to aspect ratio κ and is considerably higher than that for solid particles.

DISCUSSION

Here we present the results of our numerical calculations of vibrational mode spectrum for Au nanoshells in water. In Figs. 1 and 2 we show the energies and damping rates for fundamental ($n = 0$) and first excited ($n = 1$) modes. For fundamental mode, the energy decreases with increasing aspect ratio and, for thin nanoshells, is considerably lower that for nanoparticles, while the damping rate experiences a rapid increase as nanoshell becomes thiner. The sharp crossover for very thin nanoshells corresponds to the transition between geometry and medium dominated regimes, as discussed above. Note that for water ($c_T = 0$) both energy and damping rate vanish in the thin shell limit. In contrast, for $n = 1$ mode, no such transition takes place, and both energy and damping rate increase with aspect ratio.

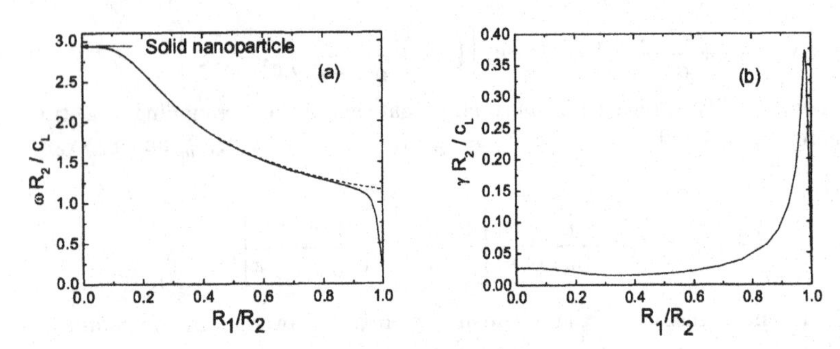

Figure 1: Spectrum for fundamental breathing radial mode in gold nanoshell versus its aspect ratio R_1/R_2. (a) Solid line: eigenfrequency versus R_1/R_2 in the model with free inner boundary and ideal contact between outer shell and matrix. Dashed line: eigenfrequency in the model with free boundaries. (b) Solid line: normalized damping rate versus aspect ratio. Longitudinal sound speed in gold c_L = 3240 m/s, transverse sound speed c_T = 1200 m/s, the density of gold ρ = 19700 kg/m^3. Surrounding matrix is water with c_L= 1490 m/s and $c_T = 0$, ρ = 1000 kg/m^3.

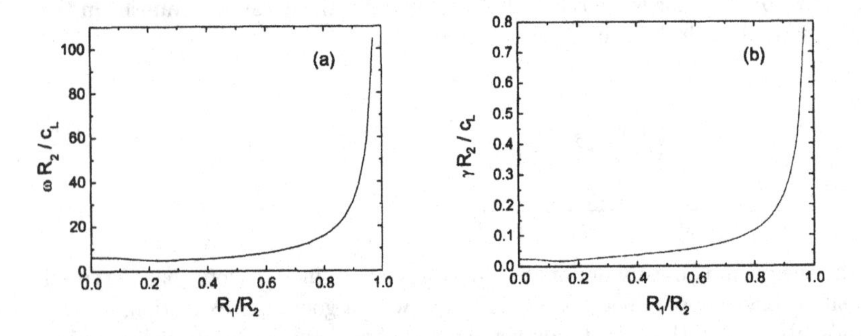

Figure 2: Spectrum for $n = 1$ radial mode in gold nanoshell versus its aspect ratio R_1/R_2. (a) Eigenfrequency calculated with free inner boundary and ideal contact between outer shell and matrix. (b) Normalized damping rate versus aspect ratio.

Let as now turn to the relative contributions of fundamental and excited modes to the pump-probe signal. Since the intial rapid expansion of nanoshell is homogeneous, the oscillator strength of fundamental (symmetric) mode is expected to be larger than that of $n = 1$ (antisymmetric) mode. The expression for oscillator strength of nth mode has the form

$$C_n = \frac{R_2^{-1} \int r\, U_n(r) dV}{V^{1/2} \left[\int U_n^2(r) dV \right]^{1/2}}. \tag{13}$$

In Fig. 3 we show calculated oscillator strengths for $n = 0$ and $n = 1$ modes versus aspect ratio. In contrast to solid particles, where the relative strengths of two modes is constant, $C_1/C_0 = 1/4$, here C_1 vanishes in the $\kappa = 1$ limit, while C_0 reaches it maximal value, $C_0 = 1$. Thus, in thin nanoshells, the fundamental mode carries almost entire oscillator strength. As a result, in nanoshells, excitation of the fundamental mode should result in a greater amplitude of oscillations as compared to solid particles while the $n = 1$ should be practically undetectable.

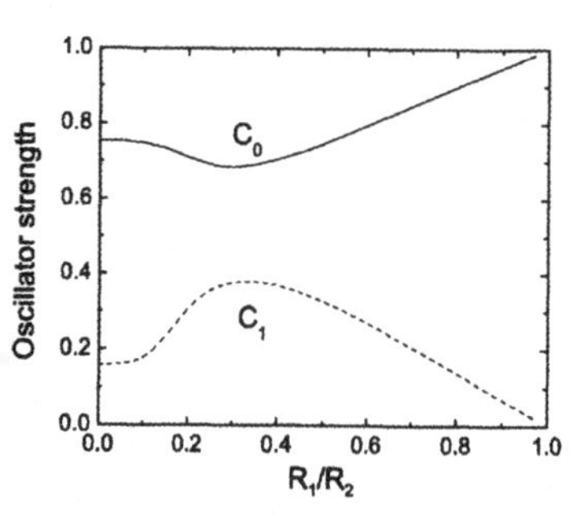

Figure 3: Oscillator strengths for symmetric (solid line) and antisymmetric (dashed line) breathing modes of nanoshell versus nanoshell aspect ratio R_1/R_2. At $R_1 = 0$ oscillator strength coincide with one for nanoparticle.

ACKNOWLEDGMENTS

This work was supported by NSF under grants DMR-0305557 and NUE-0407108, by NIH under grant 5 SO6 GM008047-31, and by ARL under grant DAAD19-01-2-0014.

REFERENCES

1. N. Del Fatti, C. Voisin, F. Chevy, F. Vallée, and C. Flytzanis, J. Chem. Phys. **110**, 11484 (1999).
2. J.S. Hodak, A.Henglein, G.V. Hartland, J. Chem. Phys. **111**, 8613 (1999).
3. C. Voisin, N. Del Fatti, D. Christofilos, and F. Vallée, J. Chem. Phys. **110**, 11484 (1999).
4. V.A. Dubrovskiy and V.S. Morochnik, Izv. Earth Phys. **17**, 494 (1981).
5. R. D. Averitt, D. Sarkar and N. J. Halas, Phys. Rev. Lett. **78**, 4217 (1997).
6. A. L. Aden and M. Kerker, J. Appl. Phys. **22**, 1242 (1951).
7. L.D. Landau and E.M. Lifshitz, *Theory of Elasticity*, (Addison-Wesley, 1987).

Mater. Res. Soc. Symp. Proc. Vol. 846 © 2005 Materials Research Society DD5.10

Surface Plasmon Excitations and Emission Light due to Molecular Luminescence on Metal Thin Film

Futao Kaneko[1,2], Susumu Toyoshima[3], Yasuo Ohdaira[1,2], Kazunari Shinbo[1,2] and Keizo Kato[2,3]
[1]Electrical and Electronic Engineering, Niigata University, Niigata, 950-2181, Japan
[2]Center for Transdisciplinary Research, Niigata University, Niigata, 950-2181, Japan
[3]Graduate School of Science and Technology, Niigata University, Niigata, 950-2181, Japan

ABSTRACT

Emission light due to surface plasmon (SP) excitations has been investigated in Kretschmann configuration of prism/metal thin film/organic dye thin films by means of direct irradiation of a laser beam from air to the films. Emission light through the prism coincided with the resonant conditions of SP excitations in the configuration. It was thought that SPs on the metal surface were excited by luminescence of dye molecules close to the metal surface, and the SPs propagating along the metal surface were converted to the emission light according to the resonant conditions of SPs in the configuration. Intensities and spectra of the SP emission light through the prism strongly depended upon emission angles, nanostructures of dye films, the film thickness and mixture of two kinds of dyes.

INTRODUCTION

Surface plasmons (SPs) that are a coupling mode of free electrons and light can be resonantly excited on metal surfaces in the Kretschmann and the Otto configurations by electromagnetic waves due to the total reflection of a p-polarized laser beam [1, 2]. SPs are two dimensional optical waves and propagate along the surfaces with strong electromagnetic waves, evanescent waves, that decay exponentially away from the surfaces. The attenuated total reflection (ATR or SPR: surface plasmon resonace) method accompanied with SPs is used for evaluations of ultrathin films and sensing [1, 2]. SPs are also utilized in near field optics (NFO) where electromagnetic waves with light frequency are localized in structures smaller than the light wavelength [2]. It is thought that SPs are very useful as a method to connect NFO with three dimensional optical waves, far field optics, and conversions between three and two dimensional optical waves are also important for the future applications [2].

Recently, emission light at a resonant angle region of SP excitations was observed through the prism in the ATR Kretschmann configuration, when metal ultrathin films on the prism or molecular thin films on metal films were directly irradiated from air by a laser beam, that is, reverse irradiation [2-4]. The emission properties corresponded to the resonant conditions of SPs in the Kretschmann configuration, and it is considered that multiple SPs were induced by molecular luminescence of excited dyes in the reverse irradiation and the SPs were converted to the emission light [5-8].

In this study, emission light properties due to multiple SP excitations have been investigated for merocyanine (MC) Langmuir-Blodgett (LB) films and hetero LB films of MC and crystal violet (CV) by means of the reverse irradiation in the Kretschmann configuration. Emission properties have been reported due to propagating directions of SPs and energy interactions between MC and CV molecules.

EXPERIMENTAL DETAILS

MC and CV were used as organic dye molecules in this study [6]. Figure 1 shows chemical structures of MC and CV dye molecules. The MC is one of cyanine dyes showing strong photoluminescence (PL) [8]. The CV is one of triphenylmethyl groups showing n-type conduction. Langmuir-Blodgett (LB) films of the MC and the CV were prepared by LB dipping method. These molecules were mixed with arachidic acid (C20) for excellent LB depositions and the molar ratio in these LB monolayers was [MC or CV]: [C20] =1:2.

Two types of LB films were prepared. The layer structures are shown in figure 2. One is MC LB film with 16 monolayers. Another is MC/CV hetero LB film with 16 monolayers, of which MC bilayers and CV bilayers were deposited alternately. The both LB thin films were deposited on C20 LB films with 2 monolayers on cover glasses coated with vacuum-evaporated Ag thin film. Ag thin film was used as SP active layer and the thickness was approximately 50nm. The C20 LB films were used for the following better depositions.

Figure 3 shows Kretschmann configuration of ATR method (a) and a system for detecting emission light through the prism when the sample was excited by the reverse irradiation in the,

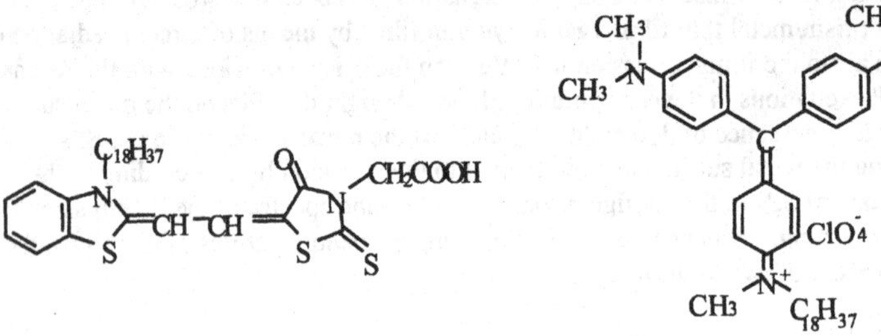

(a) Merocyanine (MC) (b) crystal violet (CV)

Figure 1. Chemical structures of MC (a) and CV dye molecules (b).

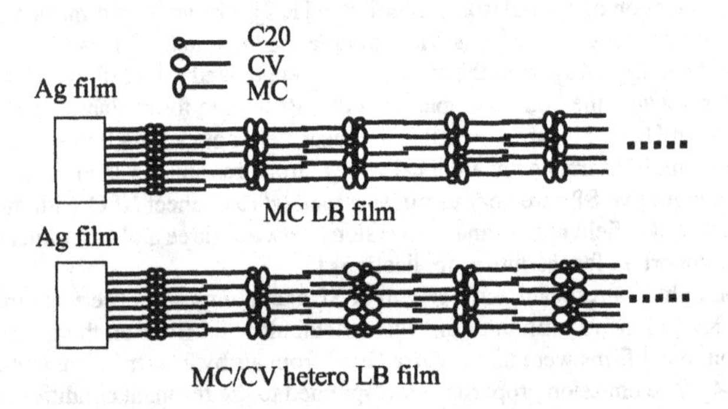

Figure 2. Layer structures of MC LB film (a) and hetero LB film of MC and CV (b).

configuration (b). The sample was put on the back surface of a half-cylindrical prism (BK-7 n=1.512 at λ= 488 nm). Figure 4 shows the measuring system of ATR properties. The sample on the prism was located on a rotating stage and the incident angle of a laser beam was automatically controlled by a computer. Reflectance intensity to incident one Ir/Ii, that is, the ATR signal was measured as a function of the incident angle, θ_i, of the laser beam as shown in figure 3 (a). Lasers at various wavelengths were used in the ATR measurement.

The sample was irradiated at the vertical incident angle by p- and s-polarized Ar^+ laser beams at 488 nm in the reverse irradiation method, as shown in figure 3 (b), in order to induce surface plasmons. The mission light was observed through the prism in the configuration as a function of the emission angle, θe, where the light was observed [5-8]. The spectra of the emission light were also measured at various emission angles.

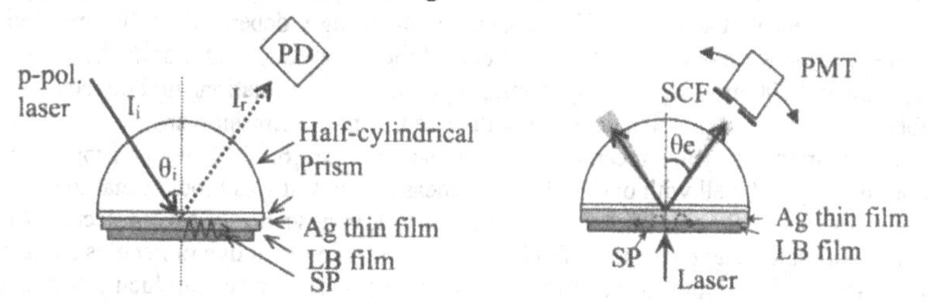

(a) Kretschmann configuration of ATR method. (b) Reverse irradiation and a detecting system.
Figure 3. ATR (SPR) method (a) and reverse irradiation method in the Kretschmann configuration in order to induce surface plasmons (b).

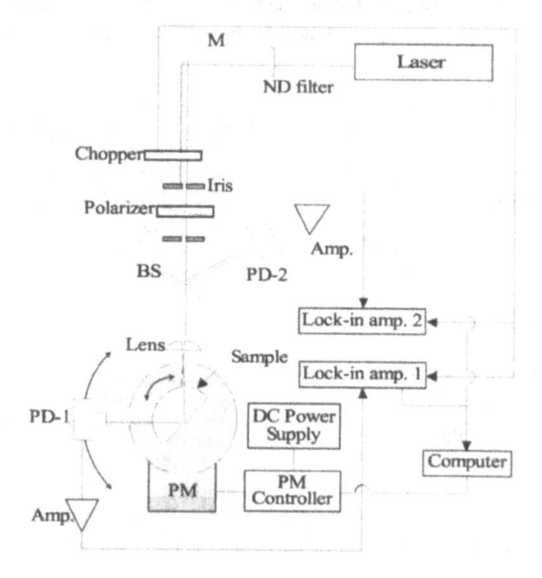

Figure 4. A system to detecting reflection (ATR) signals as a function of incident angles.

RESULTS AND DISCUSSIONS

Figure 5 (a) shows the ATR properties at various laser wavelengths for the Ag/C20 (2 layers)/MC (16 layers) LB thin film. Minima due to resonant excitations of SP were observed in the ATR properties. The resonant angles of the SP (θ_{SP}) were about 61°, 58°, 59° and 55° for the wavelengths of 488.0 nm, 514.5 nm, 594.1 nm and 632.8 nm, respectively. The dip at the resonant angle for 594.1 nm was shallow and observed at a higher incident angle than that for 514.5 nm. It was thought that the property was caused by an anomalous dispersion in the dielectric constants due to the strong and sharp absorption of the MC LB film at approximately 600 nm [6, 8].

Figure 5 (b) shows emission spectra at various emission angles, θe, by the reverse irradiation of p-polarized laser beam at 488.0 nm. The emission light strongly depended on the emission angles. Each spectrum almost corresponded to a part of the PL spectrum of the MC LB film between 600 nm and 700 nm [8]. The dispersion property, that is, a relationship between wavenumber of SP and angular frequency was calculated from the emission angles, θe, and the peak wavelengths in the emission spectra, assuming that θe is θsp [6, 8]. Since the properties of the emission light agreed well with one of the ATR measurement, it was thought that the emission light was due to SP excitations closely related to luminescence of MC molecules [6,8]. Therefore, it is estimated that evanescent fields due to luminescence of dye molecules excited by the laser irradiation induce vibrations of free electrons at the metal surface producing SPs, and the SPs are converted to emission light at the resonant SP conditions in the configuration.

The emission intensities increased with the thickness of the MC LB films [7]. The intensities also strongly depended upon nano-separation between the dye film and became very smaller for the shorter separation [7]. It was estimated to be due to non-radiative energy transfer from the excited MC dyes to Ag [7]. Similar properties have been reported for rhodamine-B LB films [9].

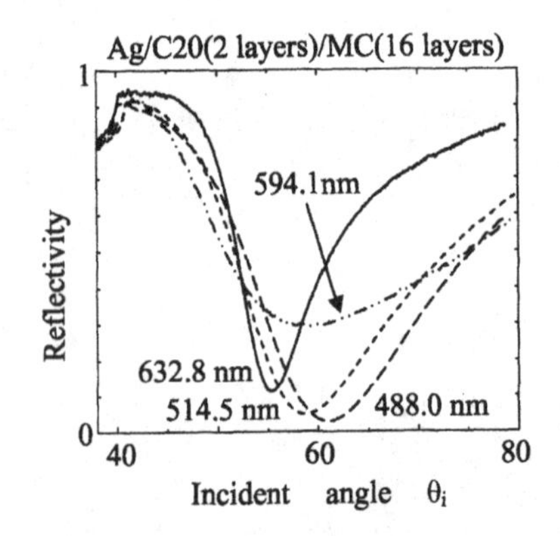

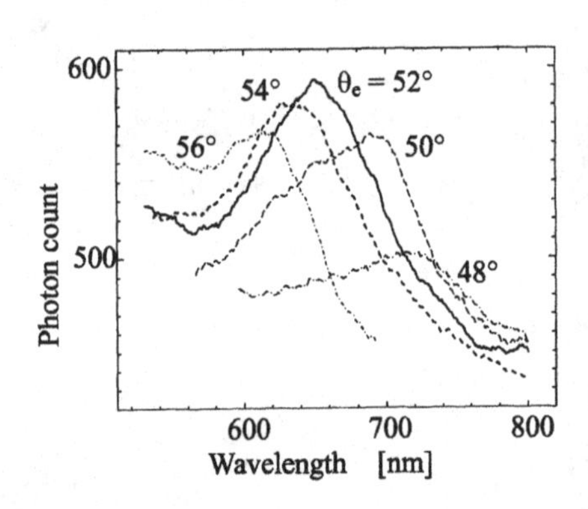

(a) ATR properties. (b) Emission spectra.

Figure 5. ATR properties at various wavelengths (a) and emission spectra at various emission angles by the reverse irradiation of p-polarized laser beam at 488.0 nm (b).

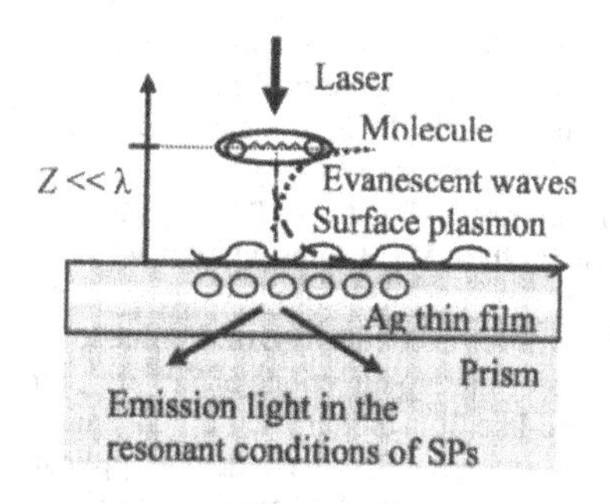

Figure 6. A schematic model of emission light due to molecular luminescence.

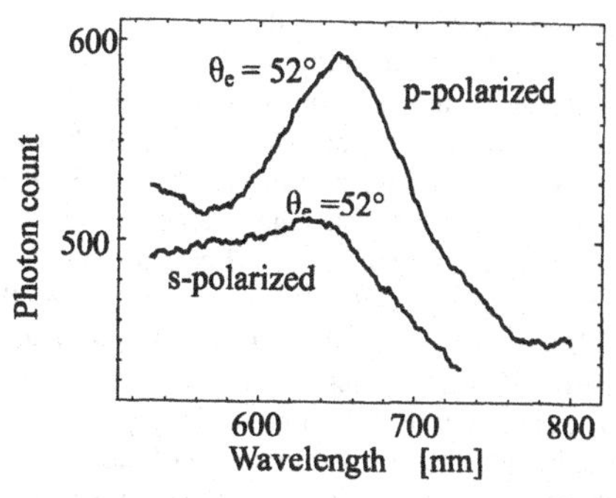

Figure 7. Emission light spectra from prism/ Ag/C20 LB film (2 layers)/MC LB film (16 layers) in the reverse irradiation of p- and s- polarized laser beams at 488 nm.

The schematic model of the emission light due to molecular luminescence is shown in figure 6. The analytical treatment has been reported on light emission properties of electric and magnetic multipoles near dielectric surfaces [10] and the analysis indicates that surface electromagnetic modes localized at the surface are induced when the separation between dipoles and the surface is smaller than the wavelength of light.

Figure 7 shows emission light from Ag/C20 LB film (2 layers)/MC LB film (16 layers) in the reverse irradiation of p- and s- polarized laser beams at 488 nm. The emission peak by the p-polarized laser with the electric fields parallel to the observation plane was larger than that by the s-polarized one with the field perpendicular to the plane. Since orientations of the MC dyes were thought to be almost uniform in the LB film plane, it is tentatively estimated that some MC dyes having the long axis parallel to the polarized plane of the laser were mainly excited by the

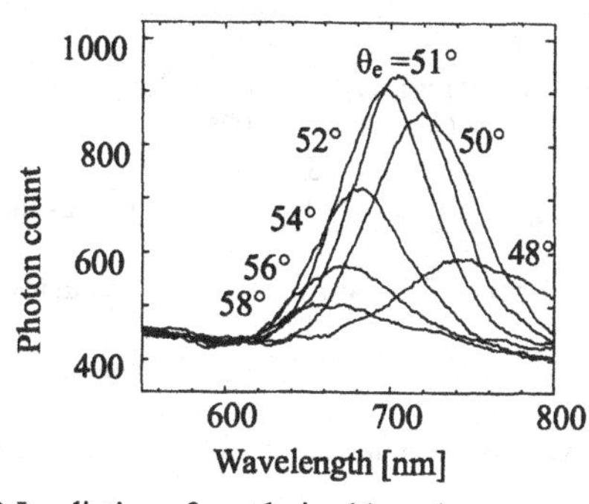

(a) Irradiation of p-polarized laser beam.

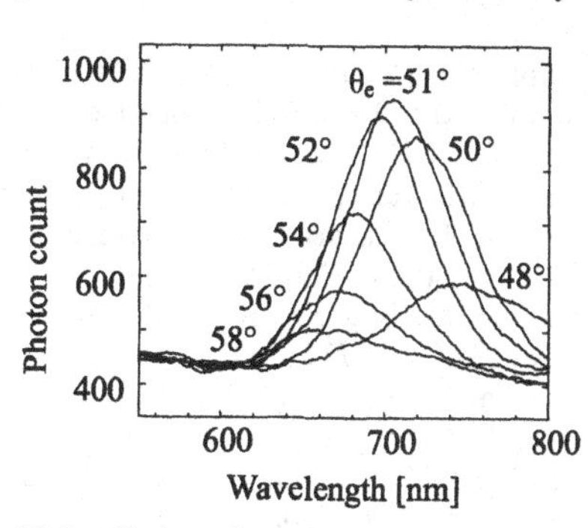

(b) Irradiation of s-polarized laser beam.

Figure 8. Emission light spectra from prism/Ag/C20(2 layers) / MC/CV hetero LB film (16 layers) at various emission angles in the reverse irradiation.

reverse irradiation and the excited dyes induce anisotropic SPs propagating along the Ag surface, which can be observed mainly in the observation plane of the half-cylindrical prism.

Figures 8 (a) and (b) show spectra of the emission light for the MC/CV hetero LB film by the reverse irradiation of p- and s-polarized laser beams at 488.0 nm, respectively.
The emission spectra were observed at around 700 nm corresponding to the PL spectrum of CV molecules and exhibited stronger intensities than those for the CV LB films without MC molecules. There are almost no differences in the spectra and the intensities between p- and s-polarized irradiations. It is tentatively estimated that SPs are excited by PL of CV due to Förster's transition from MC to CV molecules and the propagating directions of SPs are isotropic due to PL of CV molecules that are not rod-like, but disc-like. It is thought that the SP emission properties are very useful for controlling propagating directions of SPs in the device applications.

It is thought that the phenomenon of the SP emission is very useful for applications to nanostructured optical devices, conversion between two- (SP) and three- (light) dimensional optical waves and connections between near field and far field optics utilizing the SP excitations.

CONCLUSIONS

Anisotropic SP emission light was observed due to excited rod-like MC molecules for MC LB film under the reverse irradiation. SP emission light excited by PL of CV was also observed due to Förster's transition from MC to CV molecules and the propagating directions of SPs were isotropic due to the disc-like CV molecules.

ACKNOWLEDGEMENTS

This work was partially supported by Grant for Promotion of Niigata University Research Projects and by a Grand-in-Aid for Scientific Research from the Japan Society of Promotion Science.

REFERENCES

1. W. Knoll, *Ann. Rev. Phys. Chem.* **49**, 569 (1998).
2. S. Kawata (ed.), "Near-Field Optics and Surface Plasmon Polaritons", Springer, Berlin(2001).
3. I. Pockrand, A. Brillante and D. Möbius, *Chemical Physics Lett.* **69**, 499 (1980).
4. S. Hayashi, T. Kume, T.Amano and K. Yamamoto, *Jpn.J.Appl.Phys.* **35**, L331 (1996).
5. T. Nakano, T. Wakamatsu, H. Kobayashi, F. Kaneko, K. Shinbo, K. Kato and T. Kawakami, *Mol. Cryst. Liq. Cryst.* **370**, 265 (2001).
6. T. Nakano, M. Terakado, K. Shinbo, K. Kato, F. Kaneko, T. Kawakami and T. Wakamatsu, *Jpn.J.Appl.Phys.* **41**, 2774 (2002).
7. F. Kaneko, T. Sato, M. Terakado, T. Nakano, K. Shinbo, K. Kato, N. Tsuboi, T. Wakamatsu, and R.C. Advincula, *Jpn.J.Appl.Phys.* **42**, 2511 (2003).
8. S. Toyoshima, F. Kaneko, T. Sato, K. Shinbo, K. Kato, Y. Ohdaira and T. Wakamatsu, *Trans. MRS Japan*, **29**, 755 (2004).
9. F. Kaneko, K. Wakui, W. Saito, H. Hatakeyama, S. Toyoshima, K. Shinbo, K. Kato, T. Kawakami, Y. Ohdaira and T. Wakamatsu, *Jpn.J.Appl.Phys.* **43**, 2335 (2004).
10. T. Inoue and H. Hori, *Opt. Rev.* **5**, 295 (1998).

Electro-Optic and Electronic Materials

Mater. Res. Soc. Symp. Proc. Vol. 846 © 2005 Materials Research Society DD6.1

Functionalized Guanidines for Electro-Optic Materials.

Nicholas Buker, Kimberly A. Firestone, Marnie Haller[1], Lafe Purvis, David Lao, Robert Snoeberger, Alex K.-Y. Jen[1], Larry R. Dalton

University of Washington Dept of Chemistry
Box 351700 Seattle, WA 98195
[1]Univeristy of Washington Dept of Materials Science and Engineering

ABSTRACT

A family of nonlinear optical chromophores has been synthesized containing novel donor systems based on functionalized guanidines. Chromophores utilizing these donor systems display superior transparency and stability properties. The unusual and highly desirable characteristics of these chromophores make them very promising candidates for electro-optic applications. Systematic study of the molecular hyperpolarizabilities and bulk electro-optic properties of polymers containing these chromophores is being used to guide optimization of these systems.

INTRODUCTION

The variable refractive index of electro-optic (EO) materials has led to a great deal of research interest. Application of an electric field to EO materials leads to a linear variance in refractive index known as Pockel's effect. When incorporated into devices, these materials serve as an effective vehicle for rapid modulation of optical signals using electrical inputs.

Nonlinear optical (NLO) chromophores are the most successful family of organic EO materials. These chromophores are typically comprised of a delocalized π-electron system that connects electron rich moieties (donors) to electron deficient functionalities (acceptors). Hyperpolarizability (β), is the component of molecular NLO behavior that is responsible for EO effects. Optimization of NLO chromophores has proven to be a challenging task because enhancement of β typically results in reduced stability, decreased transparency to telecommunication wavelengths and other unwanted consequences.[1,2]

A great deal of effort has been devoted to modification of bridges and acceptors, but donors have been largely neglected. Systems based on oxygen and sulfur have been explored, but the former proved to perform poorly while the latter suffers from stability problems.[1] The standard dialkyl and diaryl amino donor systems have been successful, but evidence suggests that superior systems exist.[3,4] These unconventional guanidine and phosphorous based donor systems demonstrate a significant enhancement of μβ values in small chromophores in conjunction with high degrees of thermal stability. Literature suggests that integration of these improved donors into modern chromophore molecules should yield significant improvements to β resulting in increased r_{33} values and enhanced device performance without sacrificing stability or transparency.[3,4]

RESULTS

The enhanced μβ values, good transparency properties, and high thermal stabilities reported in the literature motivated the decision to further explore these novel donor structures. Reported chromophores consisted only of simple donor-acceptor designs, lacking good bridge systems or advanced acceptors.[3,4] Numerous attempts were made to generate chromophores containing novel donors, but these endeavors failed due to synthetic incompatibility of these donors with advanced chromophore systems. A very mild and efficient method for preparing guanidine-based donors from primary arylamines and imidazolium chlorides was found.[5] Figure 1 illustrates generation of imidazolium chlorides from commercially available materials.

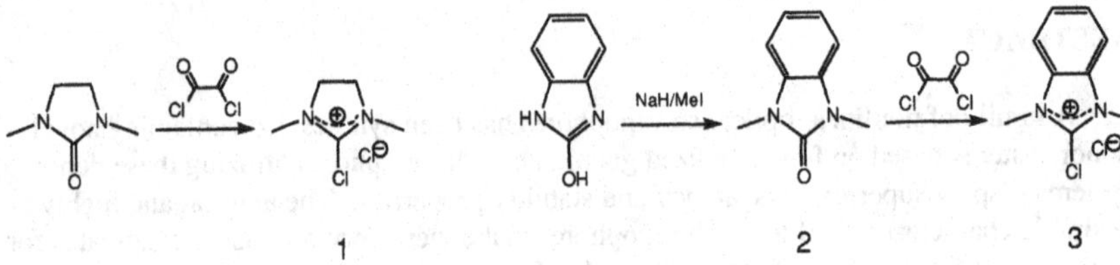

Figure 1. Preparation of Imidazolium Chlorides.

Use of imidazolium chlorides allowed access to various guanidine-containing materials, but it soon become apparent that these substances were incompatible with conventional chromatography techniques. Guanidine-containing structures demonstrated a high affinity for both alumina and silica gel presumably due to protonation. Guanidines were first explored as donors due to their extreme ease of protonation hinting at their capacity as donors in NLO systems.[3] All attempts at running guanidines on silica and alumina columns failed despite the addition of triethylamine to the mobile phase to mitigate protonation problems. It was decided that the guanidine-donor must be generated at the last step to minimize purification difficulties. Preparation of guanidine-containing chromophores using this methodology requires generation of primary arylamines, yet primary amines are incompatible with common chromophore synthetic chemistry. Thus the necessary materials were synthesized from nitro analogs. A selective and mild nitro reduction was found that left vinylic and cyano moieties intact. Figure 2 depicts synthesis of a simple chromophore using this scheme.

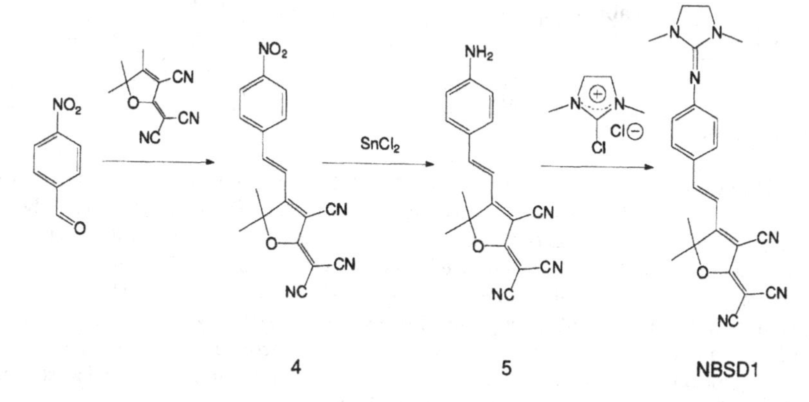

Figure 2. Synthesis of a small chromophore by reduction of a nitro analog and subsequent reaction with imidazolium chloride.

The TCF-1 chromophore, containing a conventional amino donor, was selected as a reference compound for comparison to the NBSD1 chromophore and its guanidine donor. UV-visible spectraphotometry and hyper-Rayleigh scattering (HRS) data for the TCF-1 and NBSD1 chromophores are shown in Table I. Details of the HRS experiment may be found elsewhere.[6] All measurements of β are reported relative to β_{CHCl3} as the hyperpolarizability of chloroform is in debate.[7,8,9] All samples measure in chloroform contained trace triethylamine to prevent protonation of the chromophores by low-level acidity of $CHCl_3$. The NBSD1 chromophore displays the highly desirable and unusual properties of a significantly enhanced β value at 880 nm with a λ max in $CHCl_3$ shifted 24 nm away from telecommunication wavelengths when compared to the conventional donor system. Chromophores with larger β value generally demonstrate a red shifted λ max which can lead to significant absorption problems in devices.[1]

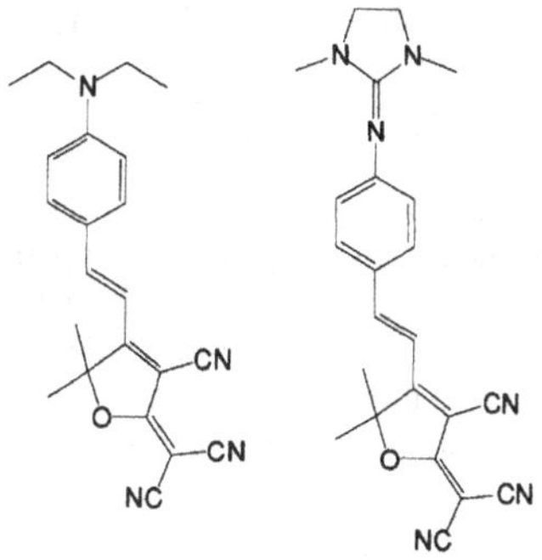

TCF-1 NBSD1

Table I. Spectral Properties of TCF-1 and NBSD1.

Chromophore	λ max in $CHCl_3$ (nm)	β_{HRS} relative to $CHCl_3$*
TCF-1	584	1993 ± 62
NBSD1	560	2501 ± 87

*measured in chloroform at 880 nm

The desirable behavior displayed by the NBSD1 chromophore encouraged efforts to synthesize more potent guanidine-containing chromophores. The stable, synthetically accessible and powerful FTC skeleton was chosen as the base structure for the next family of chromophores to be synthesized. During synthesis of the NBSD1 chromophore, it was noted that compound 4 possessed limited solubility. The decision was made to use a 3,4-dibutyl-thiophene bridge prevent possible synthetic and purification issues related to insolubility of the nitro intermediates. Conventional chromophore chemistry was used to synthesize the NBLP reference compound as shown in Figure 3 and all reactions proceeded without issue.

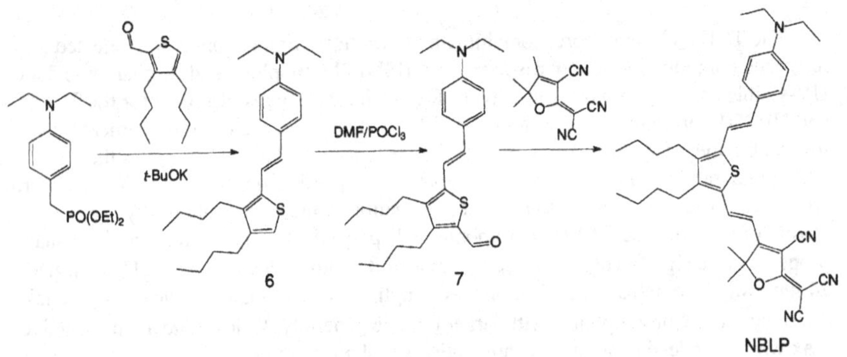

Figure 3. Synthetic Scheme for NBLP reference compound.

The synthetic scheme used to prepare NBSD1 was adapted to a relatively standard FTC synthesis. A nitro-containing analog of the desired chromophore was synthesized by a similar route to NBLP. The synthesis of the nitro materials proceeded without issue and these materials demonstrated good stability and compatibility with conventional chromophore chemistry. This nitro analog was then reduced under similar conditions to those used for the NBSD1 synthesis, and the resulting amine was reacted with the imidazolium chloride salt to yield the desired chromophore. Figure 4 illustrates the scheme used to prepare the guanidine-containing FTC chromophores.

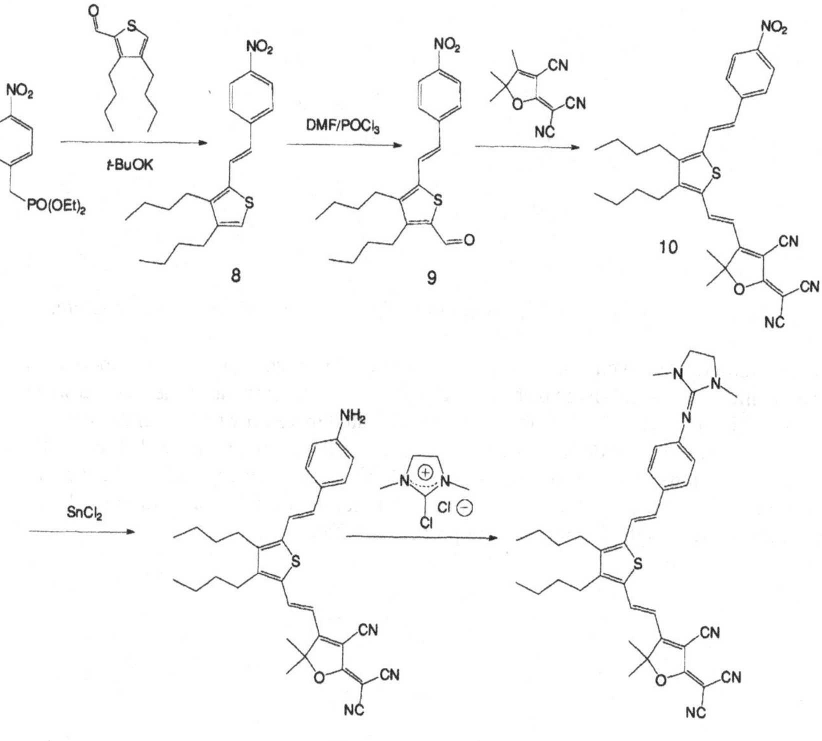

Figure 4. Synthetic scheme used for NBSD1FTC and NBSD2FTC chromophores.

The response of NLO chromophores to an applied electric field may be modeled as a reversible charge transfer from the donor to the acceptor moiety. This simplified model may be used to conceptualize the behavior of high β chromophores. Chromophores with large NLO susceptibilities generally have a very red-shifted λ max. The λ max of these chromophores corresponds to charge transfer, therefore those chromophores with highly red-shifted λ max values have low energy charge transfer states. This knowledge may be used to design systems with increased β value by lowering the energy of the charge transfer state. Guanidine-based donors appeared to provide a good template on which to build systems with stabilized charge transfer states. Figure 5 depicts a benzo-guanidine system in which the donor should possess some degree of aromatic stabilization in the charge transfer state.

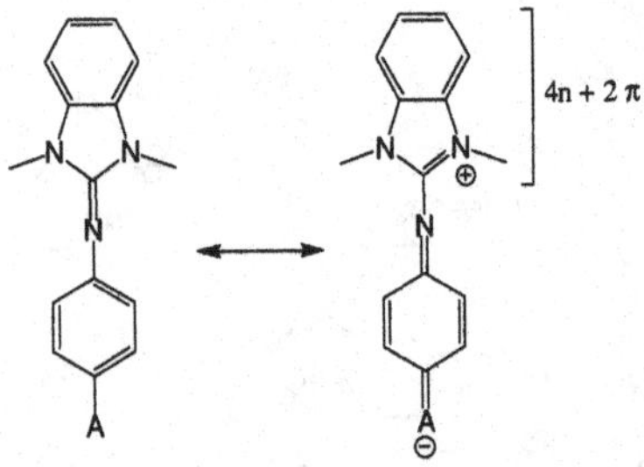

Figure 5. Aromatic stabilization of the charge-separated resonance structure.

The benzimidazolium chloride salt needed to generate benzo-guanidine chromophores was synthesized from 2-hydroxybenzimidazole. This material was dimethylated with sodium hydride and methyl iodide. The dimethylated material was converted to the benzimidazolium chloride salt through adaptation of literature procedure.[5] Formation proceeded in lower yield than the imidazolium chloride presumably due to the increased solubility of the benzimidazolium chloride. Reaction of the salt with compound 11 to produce the NBSD2FTC chromophore proceeded efficiently.

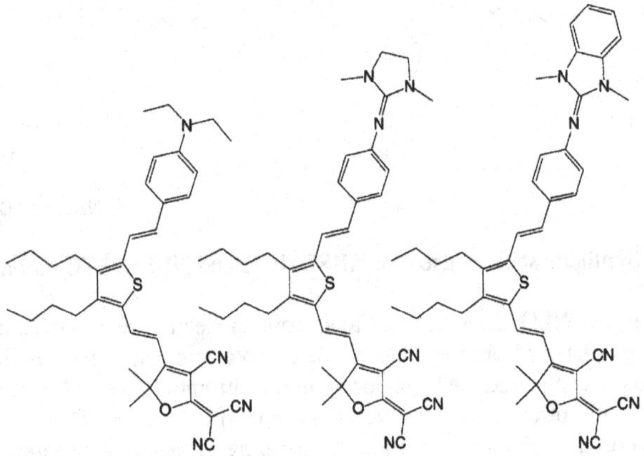

NBLP NBSD1FTC NBSD2FTC

Table II. Data for NBLP, NBSD1FTC and NBSD2FTC chromophores

Chromophore	λ max $CHCl_3$ (nm)	β_{HRS} relative to $CHCl_3$*	r_{33} in pm/V**	T_D (DSC) in ºC
NBLP	700	10769 ± 664	40	220
NBSD1FTC	642	5138 ± 687	15	310
NBSD2FTC	650	6182 ± 477	17	250

*measured in chloroform at 880 nm

**20% in APC measured at 1300 nm by simple reflection technique

CONCLUSIONS

Similar to the NBSD1 and TCF-1 chromophores, the guanidine-containing FTC chromophores were blue shifted relative to the NBLP chromophore. This shift was more pronounced with the NBSD1FTC and NBSD2FTC chromophores being blue shifted 58 and 50 nm respectively. Unlike the smaller chromophores, the guanidine-containing FTC chromophores displayed significantly lower β values than the NBLP chromophore at 880 nm. The NBSD2FTC chromophore displayed a slightly red shifted λ max and somewhat larger β value than the NBSD1FTC chromophore possibly due to aromatic stabilization of the charge transfer state. In line with the β values, NBSD2FTC displayed a slightly larger r_{33} than NBSD1FTC, but both with considerably lower values than the NBLP chromophore.

Clearly the results for the NBSD1FTC and NBSD2FTC chromophores do not correlate with the performance of the smaller NBSD1 chromophore. Both the smaller NBSD1 and the guanidine-containing FTC chromophores demonstrate a blue shifted λ max, but the NBSD1 chromophore has an enhanced β value while the NBSD1FTC and NBSD2FTC molecules have inferior β values when compared to reference compounds. It is possible that the resonance enhancement is responsible for the observed behavior. The λ max of the NBSD1 chromophore is closer to 2ω resonance at 880 nm than that of the TCF-1 chromophore. Conversely, the λ max of the NBLP chromophore is closer to ω resonance at 880 nm that those of the NBSD1FTC and NBSD2FTC chromophores. Despite the fact that β measurements are resonance dependent, the r_{33} data reflects a similar trend despite being measured at a 1300 nm. A plausible explanation for this unexpected behavior is related to the structure of the guanidine donors.

Increasing the length of conjugation between the donor and acceptor in NLO chromophores generally results in enhancement of the β value. These guanidine-based donors possess modestly increased conjugation length compared to simple amino donors. If guanidine-based donors are weaker than amino donors, it is possible that the increased conjugation length of guanidine-containing donors offsets the decreased donor potency, resulting in an overall enhancement of β in short chromophores. In chromophore with lengthy conjugation between the donor and acceptor, the increased conjugation length of the guanidine donor is less significant and presumably would have a less pronounced effect. It is likely that the enhancement of β from slightly increased conjugation length would not be able to overcome the decrease in β due to the reduction in donor strength in longer chromophores.

Despite the decreased β values observed for larger chromophores, guanidine-based acceptors offer promising characteristics. The thermal stabilities observed for guanidine containing chromophores are significantly higher than those offered by similar amino systems. Guanidine-based donors could be valuable for thermally demanding applications. These donors are ideally suited to short chromophores where they can simultaneously offer superior thermal stability, transparency and NLO characteristics compared to amino materials.

REFERENCES

(1) Dalton, L. R. *Advances in Polymer Science*, **2002**, 158, 1-86.

(2) Raimundo, J.-M.; Blanchard, P.; N., G.-P.; Mercier, N.; Ledoux-Rak, I.; Hierle, R.; Roncali, J. *J. Org. Chem.*, **2002**, 67, 205-218.
(3) Boldt, P.; Eisentrager, T.; Glania, C.; Goldenitz, J.; Kramer, P.; Matschiner, R.; Rase, J.; Schwesinger, R.; Wichern, J.; Wortman, R. *Adv. Mater.* **1996**, *8*, 672-675.
(4) Katti, K. V.; Raghuraman, K.; Pillarsetty, N.; Karra, S. R.; Gulotty, R. J.; Chartier, M. A.; Langhoff, C. A. *Chem. Mater.* **2002**, 14, 2436-2438.
(5) Isobe, T.; Fukuda, K.; Ishikawa, T. *Tetrahedron: Asym.* **1998**, 9, 1729-1735.
(6) Firestone, K. A.; Reid, P. J.; Lawson, R.; Jang, S.-H.; Dalton, L. R. *Inorg. Chim. Acta.* **2004**, 357, 3957-3966.
(7) F., K.; I., L.; Zyss, J. *Phys. Rev. A.* **1987**, 36, 2210-2219.
(8) Clays, K.; Persoons, A. *Phys. Rev. Lett.* **1991**, 66, 2980-2983.
(9) Kaatz, P.; Shelton, D. P. *Opt. Commun.* **1998**, 157, 177-181.

Mater. Res. Soc. Symp. Proc. Vol. 846 © 2005 Materials Research Society DD6.2

Highly Ordered Pseudo-Discotic Chromophore Systems for Electro-Optic Materials and Devices

Nishant Bhatambrekar[1], Scott Hammond[1], Jessica Sinness[1], Dr. Olivier Clot[1], Harry Rommel[1], Dr. Antao Chen[2], Dr. Bruce Robinson[1], Dr. Alex K-Y. Jen[3], Dr. Larry Dalton[1]
[1] Department of Chemistry, University of Washington, Seattle, WA-98195-1700
[2] Applied Physics Laboratory, University of Washington, Seattle, WA 98105-5640
[3] Department of Material Science and Engineering, University of Washington, Seattle, WA-98195-2120

ABSTRACT

In order to achieve the near-ferroelectric order desired in organic electro-optic (EO) chromophore systems, a pseudo-discotic chromophore is under investigation. Calculations suggest head-to-tail inter-chromophore dipole-dipole interactions should drive chromophores with an appropriate aspect ratio into ferroelectric columns similar to those seen in discotic liquid crystals (DLCs). Therefore, the liquid crystalline properties of these chromophores are being examined by differential scanning calorimetery (DSC), polarized optical microscopy (POM), and X-ray diffraction (XRD). Furthermore, the effect of this discotic behavior on the order and EO properties of the system are being examined both dynamically by second harmonic generation (SHG) and statically by attenuated total reflection (ATR). Additionally, these chromophores are being incorporated into waveguide-based photonic devices.

INTRODUCTION

In the past couple of decades, there has been a great attention towards improving materials for electro-optic devices. Organic based EO materials have been proven to be a much alternative to traditional inorganic crystals such as $LiNbO_3$. One of the main advantages of the organic EO materials over inorganic crystals is the exceptionally high band widths.[1] Other advantages of organic materials include low dielectric constant, easy processability into electro-optic devices, and low operating voltages.[2] Despite of all the aforementioned advantages, their temporal instability is the main impediment to commercialization of organic EO materials and devices.

The fundamental unit of the organic EO materials is a π-conjugated network in which an electron rich donor is connected to an electron withdrawing acceptor through a π-conjugated bridge. The electron density is transferred from the donor to the acceptor via the π-conjugated bridge in the presence of an external electric field. This changes the refractive index of the medium and hence the speed and the phase of light in the material. For example, in a typical Mach-Zender type waveguide device a phase shift can be introduced between the two arms by applying an external electric field to one of the arms. In devices made of high electro-optic coefficient (r_{33}) organic materials, the operational voltages can be reduced to less than one volt.[2]

THEORY

One challenging problem in improving electro-optic activity of these devices is to achieve unidirectional ordering of the EO materials such that all the dipoles point in the same direction. As in most organic systems, the dipoles of these compounds tend to arrange themselves in a centerosymmetric fashion.[3] This leads to decrease in the order parameter $<cos^3\theta>$ of the system, and hence a decrease in the EO activity. The typical procedure to orient the molecules in one direction is called 'poling', in which the EO material is heated close to its glass transition temperature in the presence of the external bias voltage. However, poling efficiencies of the EO materials is generally quite low (on the order of 30%) due to their dielectric nature. While applying higher voltages could achieve greater efficiencies, dielectric breakdown is a fundamental limit to the bias electric potential. Although many synthetic attempts are being made to improve the hyperpolarizabilty of the individual chromophore, one must also consider the macroscopic ordering of the material in order to achieve higher r_{33} values for the system.

In the past, Bruce Robinson *et al.* have done pioneering work in rationalizing the effect of inter-chromophoric interactions on the decreased EO activity.[3] As mentioned previously, the organic EO molecules tend to orient opposing the external applied field and this alignment efficiency is related to the acentric order, which is mathematically expressed as the statistical average $<cos^3\theta>$ of these molecules, inC which θ is the angle subtended by the dipole with an applied electric field **E** (Figure 1).

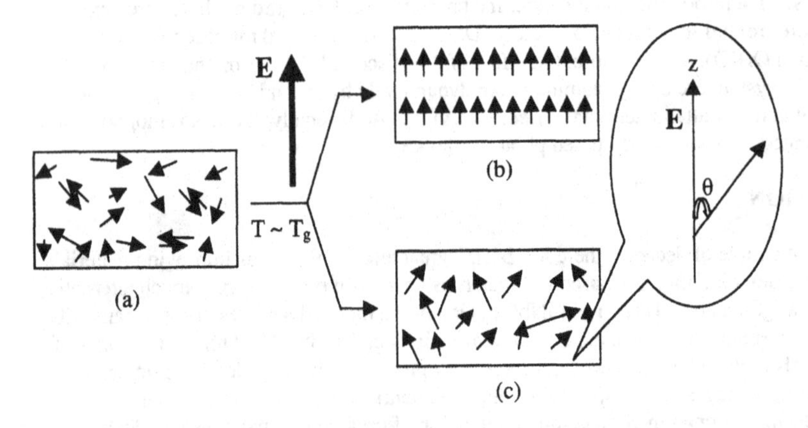

Figure 1. (a) Random orientation and the interchromophore aggregates (b) The perfect alignment of EO chromophores and (c) actual arrangement of EO chromophores under the poling

The expression relating the electro-optic coefficient, r_{33}, to the molecular hyperpolarizability, β, is given by

$$r_{zzz} = r_{33} = -2\frac{F^{local}\beta_{zzz}(-\omega,\omega,0)}{n_{zz}^4}N < \cos^3\theta > \quad (1)$$

where n_{zz} is The bulk refractive index along the z-axis and N is the number density of the chromophores measured in number of dipoles in unit volume. F^{local} is the field factor and

$\beta_{zzz}(\omega,-\omega,0)$ is the molecular hyperpolarizability.[3] While $\beta_{zzz}(\omega,-\omega,0)$ and F^{local} are essential for the EO activity, they will not be discussed here. In the current paper, the design strategy for the molecular architecture for the optimum acentric ordering is discussed.

It was argued by Robinson *et al.* that the entire material can be visualized as a collection of the crystalline domains in the form of columns with uniformly random macroscopic orientations.[4] In the case of low chromophore concentration mataerial, a linear dependence of the loading density on the r_{33} was observed.[3,4] However, as the chromophore concentration in the system increases, the inter-chromophore electrostatic interactions start to dominate, which leads to a decrease in the order and thus a decrease in the EO activity. In other words, the acentric order reaches its maximum and eventually starts to wane with the increase in the concentration. However, Monto Carlo calculations suggest that as the aspect ratio of dipolar molecules approaches and exceeds 1.5:1, i.e. the width exceeds the dipole-axis length, noncentrosymmetric head-to-tail dipole interactions dominate over the centrosymmetric side-to-side interactions. Chromophores possessing this aspect ratio and having a disk like structure shown in Figure 2(a) may tend to stack effectively in noncentrosymmetric columns (b) or to pack together in an inter-digitated fashion as shown in (c). In either case, the weak centrosymmetric side-to-side interactions should allow efficient poling of the individual columns or packed domains to produce large enhancements in EO activity. Additionally, the weak centrosymmetric interactions should enhance the temporal stability of the ordering in these materials.

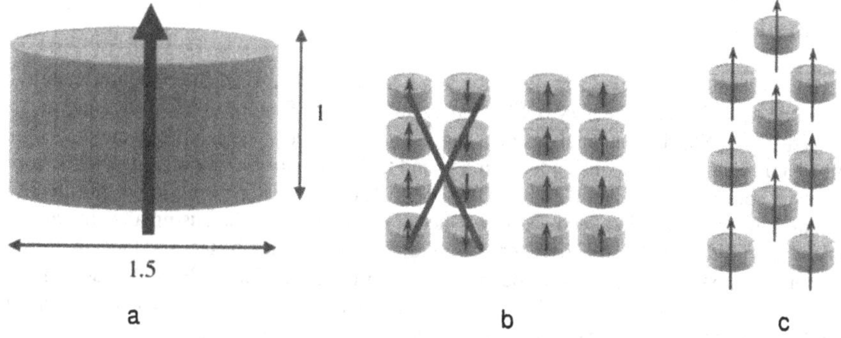

Figure 2. (a) Aspect Ratio of the molecule (b) Columnar stacking of the molecules (c) Interdigitation

The chromophore used for this study is based on the OLD chromophores, which contain a dialkylaniline donor is connected to the TCF acceptor through two 3,4-dialkoxythiophenes fused together at their α-positions. The chromophore will allow 3-dimentional functionalization around the center of the bridge. To increase the bulk around the chromophore and provide a disc like structure, gallic ester based dendrons are attached to the thiophenes through the ester bond as shown in the Figure 3. For a very simple structure, ethylhexyl chains are attached via an ether linkage to the gallic ester. Synthetic details for this dendrimeric chromophore are given elsewhere.

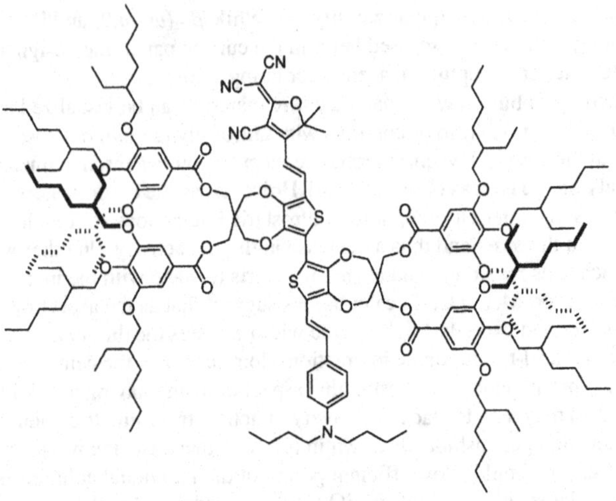

Figure 3. The complete structure of the dendrimeric OLD-type chromophore

EXPERIMENT

UV-visible spectroscopy was performed on a Shimadzu UV 1601 Spectrophotometer. Differential scanning calorimetery (DSC) was performed on a Shimadzu DSC-60 using a Shimadzu TA-50WS thermal analysis data acquisition interface. The samples were placed in crimped aluminum sample pans under nitrogen and the temperature was ramped at 10 °C/min. Films were spun cast on a Laurell WS-400A-6NPP/LITE spin coater from 5-6 wt. % filtered solutions onto commercial ITO coated glass slides that were half-etched using a chemically resistant polyimide tape in 36 N hydrochloric acid, washed with the water, and then sonicated in chloroform, methanol and acetone respectively for 20 min. each. The slides were additionally cleaned immediately before depositing the chromophore solution with acetone and isopropanol at 5000 revolutions per minute (RPMs). Films were initially "soft-baked" on a hot plate, followed by overnight storage in a vacuum oven at room temperature. Variable temperature and polarized optical microscopy (POM) were performed on an Instec STC100 hot stage and an Olympus BH-2 optical microscope equipped with polarizing filters.

RESULTS

The UV-visible absorption spectroscopy data shows the chromophore to have maximum absorption (λ_{max}) at 734 nm, with an extinction coefficient (ε) of 68700 L/mol*cm. These values are comparable to those for the previously-reported parent chromophore based on the OLD bridge without the bulky dendrons.[5] This suggests that the bulky dendron groups do not adversely affect the planarity or the hyperpolarizability of the chromophore backbone.

The DSC data indicates this system has complex thermal behavior. The initial sample showed no obvious peaks up to about 340 °C at which point the chromophore decomposed. A second sample again showed no obvious peaks up to 300 °C, and the sample was then allowed to slowly cool back to room temperature. This sample was run a second time, and it exhibited a distinct glass transition temperature (T_g) at 48 °C. This indicates that the system is extremely sensitive to previous thermal treatment, and the macroscopic order of the system is expected to depend heavily on the detailed sample treatment. This behavior was examined in more detail using variable-temperature microscopy and polarized optical microscopy.

A small sample of as-prepared dendrimer was placed between microscope slide cover slips and heated initially to 100 °C, at which point it was a freely flowing liquid, to clear the sample of previous thermal treatment effects. This was then slowly cooled to room temperature, and then re-heated to 45 °C. At this temperature the sample was rigid and resisted shearing. On heating to 50 °C the sample still would not flow, but was much less resistant to shearing, suggesting it was above the glass transition temperature. The sample was seen to fully melt around 80 °C, at which time the sample was free flowing and readily sheared. This transition was not apparent from the DSC data. On rapid cooling from 150 °C to room temperature and subsequent heating the sample displayed no differences in behavior. The sample did not exhibit birefringence at any stage in the heating/cooling cycles, indicating no discotic liquid crystalline behavior.

Surprisingly, the dendrimer displays only moderate solubility in a variety of organic solvents. It is sufficiently soluble in solvents such as acetone, methylene chloride, and chloroform to permit routine manipulation, purification, and characterization, but it doesn't exhibit the degree of solubility expected from the large aliphatic dendrons. This reduced solubility has significantly hampered the formation of high-quality amorphous films for electro-optic characterization.

Simple drop-cast tests were performed with dilute solutions of the dendrimer in 1,1,2-trichloroethane (TCE), 1,2-dichloroethane (DCE), toluene, cyclopentanone (CP), and dioxane, all common spin-casting solvents, to examine the film forming properties. Tests indicated that DCE, CP, and dioxane all gave cloudy, poor-quality films, while TCE and toluene gave shiny, potentially good-quality films. The dendrimer has limited solubility in TCE and toluene, however, with 5-6 wt. % solutions exhibiting particulates. Attempts were made to spin cast films from these solutions after filtering through a 0.2 μm filter. Due to the low concentration and viscosity of these solutions, relatively low speeds (300-700 RPMs) were required to obtain films of sufficient thickness for electro-optic characterization. Early attempts resulted in wide-spread aggregation throughout the samples. Subsequent attempts demonstrated that a brief high-temperature (100 °C, as opposed to 45°C) soft-baking step immediately after spin casting results in improved amorphous films, probably due to rapid removal of the residual solvent preventing the dendrimer from aggregating. However, these films still exhibit small local aggregates, likely due to the saturated solutions from which they were cast. The aggregates cause inhomogenaities in film thickness, which, combined with the relatively thin films (~0.6 μm), results in electrical shorts after sputtering of the top gold electrode. This prevents contact poling and EO characterization of the samples.

The strong tendency for the dendrimer to aggregate is contrary to the initial design concepts, which sought to prevent dipole-dipole induced side-on-side chromophore aggreagation. The nature of the aggregation, whether due to dipole-dipole interactions, inter-digitation of the alkyl chains, or π-π stacking of the dendron phenyl rings is of utmost importance for the design of future generations of dendrons for this system. Two of the primarily-amorphous "soft-baked" films spin cast from toluene were annealed at 70 °C overnight in a vacuum oven. Surprisingly, examination of these films in an optical microscope revealed the formation of a large number of crystalline growth sites out of the amorphous film. The formation of crystals at 70 °C is feasible, as it is below the melting temperature but above the T_g, allowing for facile rearrangement of the dendrimer. These crystals were seen to melt back into amorphous material at temperatures above 80 °C, and were not seen to re-crystallize readily on cooling. The ability for the dendrimer to crystallize, while surprising, allows for the possibility of growing single crystals and obtaining X-ray diffraction data. This would provide detailed information about the intermolecular interactions leading to aggregating and could be used to evaluate the design concepts of the system This information could then be used to design future generations of dendrons that may potentially increase solubility, decrease aggregation, and yield higher glass transition temperatures.

ACKNOWLEDGMENTS

We would like to thank NRO and DARPA for the funding for this work.

REFERENCES

1. L.R. Dalton *Synthetic Metals* **124**, 3 (2001).
2. Y. Shi, C. Zhang, H. Zhang, J.H. Bechtel, L.R. Dalton, B.H. Robinson, W.H. Steier, *Science* **288**, 119 (2000).
3. B.H. Robinson, L.R. Dalton *J. Phys. Chem. A* **104**, 4785 (2000).
4. R. D. Nielsen, H. L. Rommel, and B. H. Robinson, *J. Phys Chem. B.* **108**, 8659 (2004).
5. O. Clot, *in press*.

Mater. Res. Soc. Symp. Proc. Vol. 846 © 2005 Materials Research Society DD6.3

Synthesis of Dendridic NLO Chromophores for the Improvement of Order in Electro-optics

[1]Jessica Sinness, [1]Olivier Clot, [1]Scott R. Hammond,[1]Nishant Bhatambrekar, [1]Harrison L. Rommel, [1]Bruce Robinson, [2]Alex K-Y. Jen, [1]Larry Dalton.
[1]University of Washington, Department of Chemistry, PO Box 173500, Seattle, WA 98195.
[2]University of Washington, Department of Material Science and Engineering, PO Box 352120, WA-98195.

ABSTRACT

Previous research in organic electro-optics has shown dramatic increases in the hyperpolarizablity of NLO chromophores. However, this large microscopic activity has not been translated to the macroscopic domain. The polymeric electro-optic (E-O) materials continue to lack the high noncentrosymmetric order of the poled chromophores within the matrix necessary for high E-O response (r_{33}). This deficiency of order represents one major obstacle that must be overcome before E-O device commercialization can be achieved. This lack of order is partially due to the large dipole moments of high μβ chromophores, which cause the chromophores to align in a centrosymmetric fashion through intermolecular electrostatic interactions. However, quantum calculations show that when the aspect ratio between the width and length of the chromophore system is adjusted to be greater than 1.4:1 by adding bulky side groups around the center of the chromophore, it would prevent side on pairing of the chromophores. This would cause a decrease in the large areas of centrosymmetric aggregation and thus allow for easier poling of the system. Here we report the synthesis of a nanoscale NLO architecture in which dendritic moieties have been incorporated around the center of the chromophore to give a three dimensional structure in order to achieve the 1.4:1 aspect ratio and maximize the macroscopic order of the system.

INTRODUCTION

Electro-optic modulators currently have the potential to have profound effect on the telecommunications industry and the speed with which we can send and receive information. These modulators allow for an electrical signal to be translated to a photonic signal that will carry this information significantly faster than its electronic counterpart. The key component of these modulators is that their optical properties vary under an electrical field. Lithium niobate ($LiNbO_3$) is the current commercial standard but its speed limitations and high cost are forcing researchers to look for improved materials to convert electrical to optical signals more efficiently. One area that shows much promise is conjugated organic materials.[1]

These conjugated organic chromophores generally consist of an electron donating group and an electron accepting group connected by a π-conjugated bridge. In the presence of an electrical field the π-electron density shifts from the donor to the acceptor, thus changing the index of refraction of the material. The ease of which the electrons can be shifted from the donor to the acceptor is measured by the hyperpolarizablity, or β.

In order to fabricate an E-O device or characterize the E-O chromophore, guest-host system approach is followed, in which the chromophore is mixed into a polymer matrix solution and thin films are made. However, the activity of these films is based not only on β but also on the macroscopic order of the system, meaning all the dipole moments of the molecules must be oriented in the same direction. To achieve this order the films are heated near the glass transition point of the molecule and an electric field is applied. Unfortunately, this method does not achieve a well ordered system because the poling field cannot always overcome the strong dipole-dipole interaction and after poling the system is also prone to relaxation back to its centrosymmetric form.

Recent calculations done by Robinson *et al.* suggest that by designing systems in which the distance between the chromophores is greater than 1.5 times the length of the chromophores, a significant increase in the order of the chromophores and thus the electo-optic activity can be achieved.[2] Because dipole-dipole interactions fall off proportional to $1/r^3$, the order of this system would be driven by head to tail interactions rather than dipole pairing. These systems should form long columns of order that should then pole much easier than traditional systems in which the dipole-dipole interactions dominate. To achieve this, a system was designed in which the width of the chromophores is 1.5 times the length (Figure 1a). These chromophores may line up in distinct ferrioelectric or antiferroelectric columns (1b) or may interdigitate to form small pockets of order within the film(1c). In both systems the macroscopic order will remain centrosymmetric and the system will need to be poled.

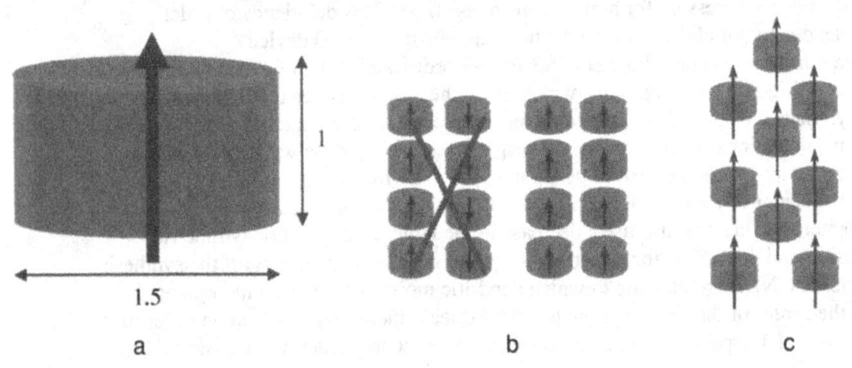

Figure 1. Interchromophore interactions (a) Aspect Ratio of the molecule (b) Columnar stacking of the molecules (c) Interdigitation

To achieve this type of system, this work focuses on synthesizing a chromophore with bulky dendrons surrounding the bridge of the chromophore (**1**). This should decrease the centrosymmetric dipole-dipole interactions, while allowing for noncentrosymmetric head-to-tail interactions between chromophores. The OLD bridge was used due to its ease in functionalization and the 3-D structure of the bridge that should help give a more cylindrical shape, preventing side by side interactions. The dendrons will be added to the system at the end of the synthesis to allow for easy manipulation of the properties of the material through changing these dendrons.

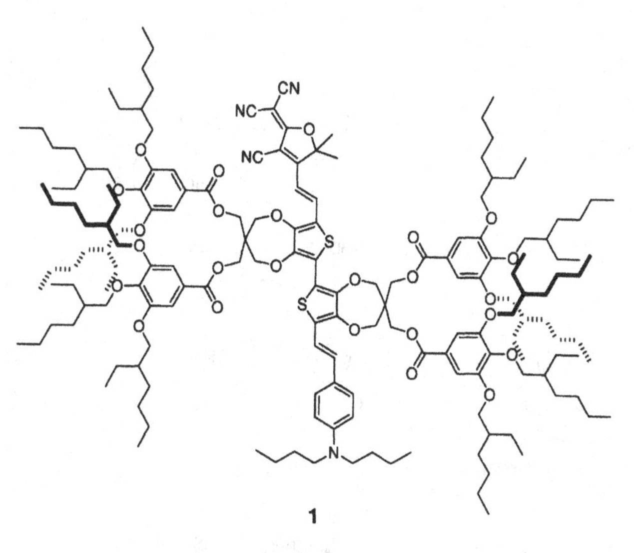

Figure 2. Proposed chromophore system

EXPERIMENTAL DETAILS

The ProDOT-OTBDMS bridge was synthesized as shown in Scheme 1. Diethyl malonate was first benyl protected to give **3**, which was reduced with $LiAlH_4$ to give the diol, **4**. **4** was then reacted in a transetherification reaction to give the benzyl protected ProDOT monomer **5**. Due to complications in debenzylation later in the synthesis, the benzyl group was removed and the TBDMS group was attached. At this point the ProDOT was dimerized using $Fe(acac)_3$ and formylated using ^{n}butyl lithium and N,N-dimethyl formamide. Once the bridge was synthesized, the donor and acceptor could be attached to give the desired chromophore (Scheme 4). The donor (**9**) was attached to the donor via a horner-emmons reaction to give **10**. The donor bridge (**10**) was then formylated (**11**) and the acceptor (**12**) was attached at high concentrations in the presence of diethyl isopropyl amine to give the desired chromophore (**13**).

The dendron (**16**) was synthesized according to the procedure shown in Scheme 3. First, ethyl hexyl chains were attached to the hydroxyl groups of methyl-3,4,5-tri-hydroxy benzoate (**14**) to give **15**. **15** was then hydrolyzed in alkaline condition to give the final dendron, **16**.

In order to obtain the 1.4:1 aspect ratio of this chromophore, dendrons were required around the center of the bridge. Although the TBDMS group is generally removed with 1 M HCl at room temperature, the ProDOT ring system showed some decomposition due to ring opening when reacted under these conditions. In order to alleviate this problem the TBDMS groups were removed using TBAF in THF (Scheme 4). The dendrons (**16**) were then attached to **17** using DMAP/DCC in a mixture of dichloromethane and DMF to give the desired product (**1**) in 68% (Scheme 5).

Scheme 1

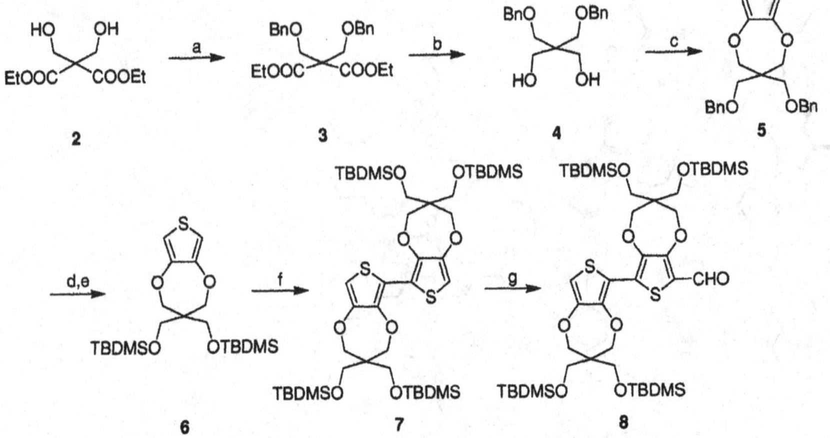

a) benzyl 2,2,2-trichloroacetimidate, triflic acid, ether, 99% b) $LiAlH_4$, ether, 95% c) 3,4 dimethoxythiophene, *p*-toluenesulfonic acid, toluene, 63% d) Pd/C, EtOH e) TBDMS-Cl, diisopropylethyl amine, DMF, 80% f) ^nBuLi, $Fe(acac)_3$, THF g) ^nBuLi, DMF, THF, 53%

Scheme 2

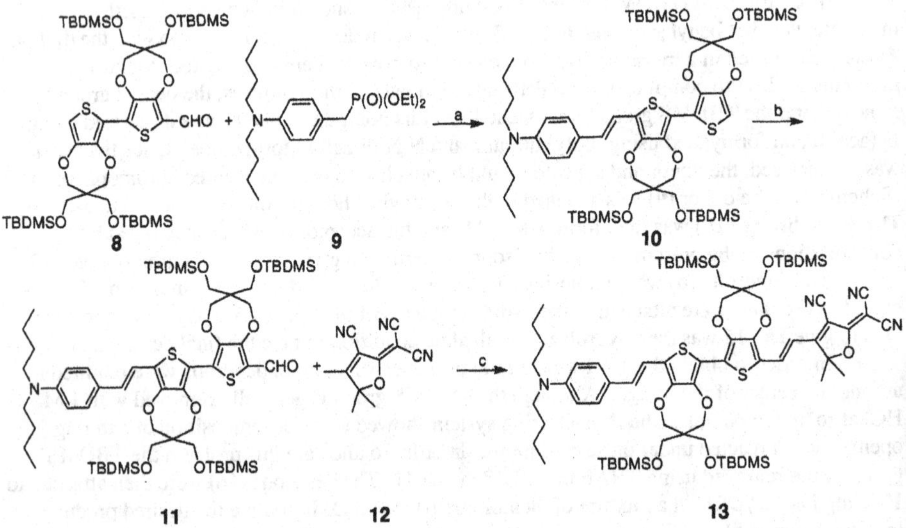

a) ^tBuOK, THF, 84% b) ^nBuLi, DMF, THF, 51% c) triethyl amine, $CHCl_3$, 93%

Scheme 3

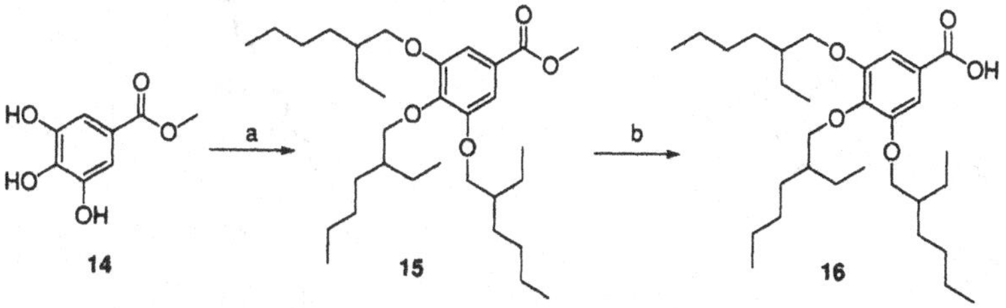

a) 1-Bromo 2-ethylhexyl, K_2CO_3, Acetone/DMSO, 83% b) MeOH/KOH

Scheme 4

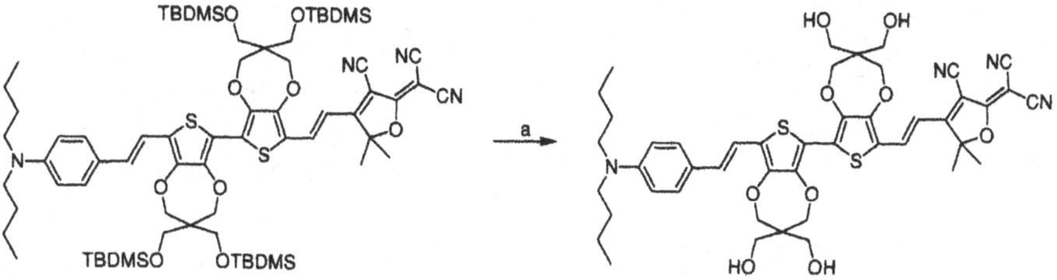

a) TBAF, THF

DISCUSSION

The synthesis of the ProDOT bridge structure can be accomplished in fairly good yields. The benzyl protecting group was selected due to its stability in the presence of hot *p*-toluene sulfonic acid which needed to close the ring. They were exchanged for the TBDMS groups due to the need for a group that could be easily removed in high yields without decomposing the completed chromophore. They synthesis of the chromophore was straightforward and followed the procedures generally employed to make the ProDOT chromophores.[3] The attachment of the dendrons went in high yields suggesting the alcohols are not significantly experiencing steric hindrance and preventing the chromophore from fully reacting. The attachment of dendrons to all four positions around the bridge was confirmed by MALDI-MS, which gave a peak at 2823.89 and was also characterized by ^{13}C and ^{1}H-NMR. The partially reacted product could also be isolated and reacted again to increase the total amount of material obtained. UV-Vis spectra show a λ_{max} of 734 nm and a ε of 68700 L/mol*cm suggesting similar microscopic activity to the original ProDOT chomophores.[3] This indicates that the electronic properties of the chromophore backbone have not been compromised with the addition of the dendrons. This suggests that the structure remains planer, at least in solution, and that the communication between the donor and the acceptor is still relatively strong.

Scheme 5

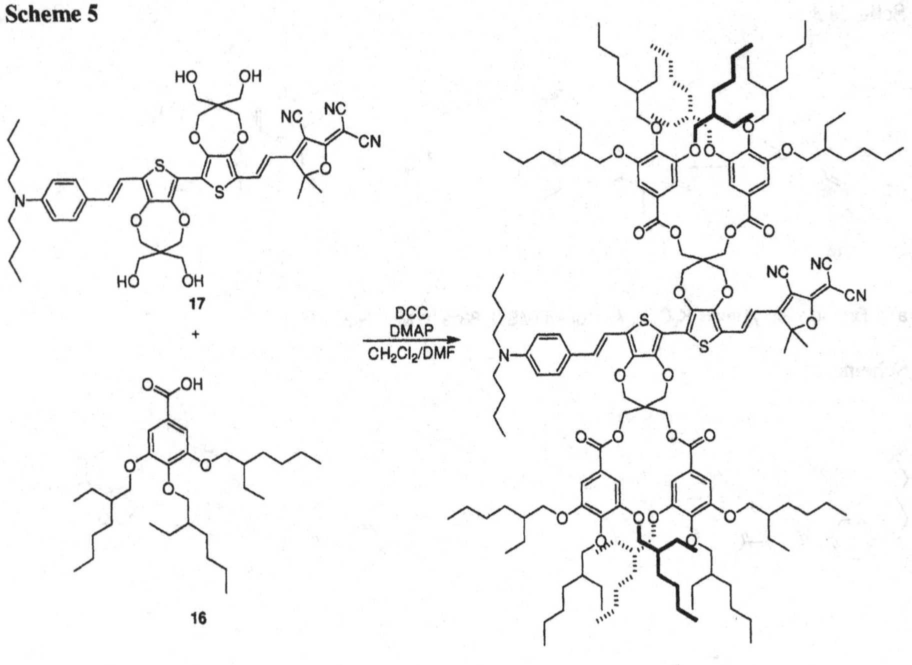

HO OH NC NC CN N S O O S O O HO OH 17 + O OH O O O 16

DCC DMAP CH_2Cl_2/DMF

1

The T_g of the material was measured at 48 °C, which is low for these general types of systems. However, there is evidence that films can still be made and poled at these low temperatures. The processes of film making, poling, and the electro-optic activity will be reported in a later paper.

In the future, new systems will be made that incorporate different dendrons to alter the physical properties of the material. The optimal material will have a higher T_g and cross linkable functional groups to add to the long term stability of the system.

ACKNOWLEDGEMENTS

We would like to thank NRO and DARPA for the funding for this work.

REFERENCES

1. L. Dalton, *Adv. in Poly. Sci* **158**, 1 (2002).

2. R. D. Nielsen, H. L. Rommel, and B. H. Robinson, *J. Phys Chem. B.* **108**, 8659 (2004).

3. O. Clot, *in press.*

Mater. Res. Soc. Symp. Proc. Vol. 846 © 2005 Materials Research Society DD6.4

Optical Switching Devices Using Redox Polymer Nanosheet

Jun Matsui, Kenichi Abe, Masaya Mitsuishi, Atsushi Aoki[1] and Tokuji Miyashita
Institute of Multidisciplinary Research for Advanced Materials, Tohoku University,
2-1-1, Katahira, Aoba-ku, Sendai 980-8577, Japan,
[1]Nagoya Institute of Technology
Gokis-cho, Showa-ku, Nagoya, 466-8555, Japan.

ABSTRACT

Optical switching devices, which control the current direction by light stimuli, were constructed using redox polymer nanosheet assemblies. Two kind of nanosheet-photodiode, which can operate by different light irradiation wavelength were assembled in a series so that each current direction becomes opposite. In the nanosheet photodiodes, one photodiode contains ruthenium complex as a sensitizer and the other contains anthracene. When 460 nm light was irradiated to the nanosheet assemblies, only the photodiode containing the ruthenium complex becomes active and an anodic photocurrent was observed. On the other hand, when 390 nm light was irradiated to the photodiode assemblies, only the photodiode containing the anthracene chromophore becomes active and a cathodic photocurrent was observed. This means that we can control the photocurrent direction by 460 nm and 390 nm irradiation as input signals.

INTRODUCTION

Recently, development of molecular devices for information processing applications becomes one of the major research fields in material science. Such devices include molecular wires, switches and diodes, which can be designed for use in photonic, electronic or optoelectronic systems[1-4]. However, most of those molecular devices work only in a single molecular level. To fabricate three-dimensional nano-devices, those molecules should be organized and integrated.

For this purpose, we have studied the fabrication of organized polymer sheet assembly using Langmuir-Blodgett (LB) technique. Recently we found that an alkylacrylamide polymer forms a condensed monolayer on water surface, and the monolayer can be transferred onto a solid support by the LB technique, yielding two-dimensional polymer nano-sheet and the three dimensional assembly by repeat deposition. Moreover, the polymer nano-sheet is functionalized by a copolymerization method with photo and/or electro functional monomers. We have developed several organic photofunctional devices using the functional polymer nanosheet assemblies. For example, it was confirmed that the photodiode where a polymer nano-sheet containing a photoredox chromophore is deposited onto a polymer nano-sheet containing an electron donor, works an efficient photoenergy conversion nanodevice[5]. Moreover, we have succeeded to create a new type of molecular logic device from the photodiode[6,7]. With assembling two kind of nanosheet photodiode, which can operate, individually by different wavelength irradiation, we have constructed optical AND, exclusive OR (EXOR) and switch gates. In this report, we describe the application of the redox polymer nanosheet assembly to a molecular switching device operating at visible region wavelength. The switching device was constructed from an assembly of functional polymer nanosheets, which include photoredox active molecules, and electron carrier molecules. With the heterodeposition of different polymer nanosheets, the electron flow direction can be controlled by light stimuli.

EXPERIMENTAL SECTION

All copolymers used in the current study, p(DDA/Ru), p(DDA/Fc), p(*t*PA/DNB), and p(nPMA/AMMA) was prepared as described previously (Figure 1) [5-7]. The measurement of pressure (π)- area(A) isotherms and the deposition of monolayers were carried out with an automatic Langmuir trough (USI, LB lift Controller FSD-51 using a Wilhelmy-type film balance). Stable monolayers were prepared on the pure water subphase (>18 MΩ cm, CPW-101, Advantec) by spreading chloroform solution (1 mM) of the polymers at 15 °C and compressed at a rate of 15 cm^2 min^{-1}. ITO and quartz substrates were cleaned as described previously. Photocurrent was measured by using a three electrode cell. We used 0.1 M $NaClO_4$ as supporting electrode and the solution was initially purged with N_2 for 30 min to remove oxygen. A 500-W xenon lamp was used

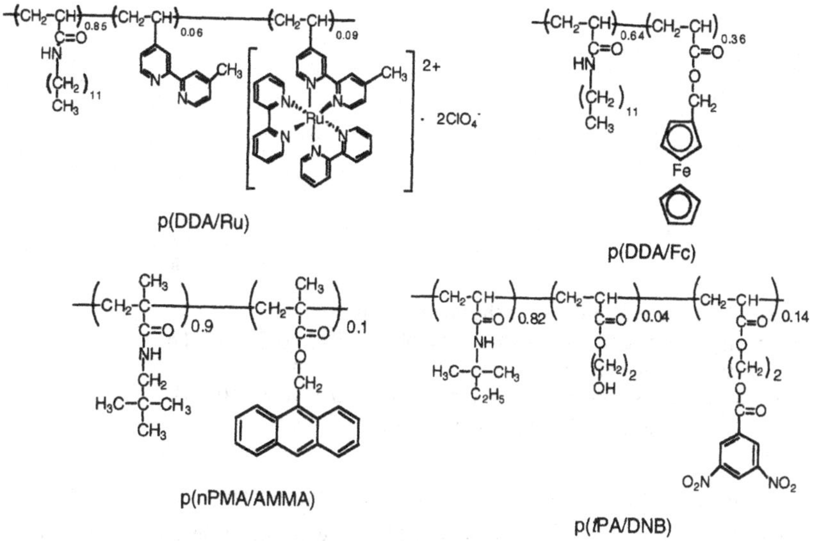

p(DDA/Ru)

p(DDA/Fc)

p(nPMA/AMMA)

p(*t*PA/DNB)

Figure 1 Chemical structures of the polymers

as a light source, and the interference filters were used to obtain monochromatic light.

RESULTS AND DISCUSSION

First, we have investigated the monolayer behavior formation of the polymers on water surface by measurement of the pressure(π)-area (A) isotherms. The isotherms of all the polymers show a steep rise with high collapse pressure, indicating the formation of stable condensed monolayer at the air/water interface. Moreover, those monolayers can be transferred regularly onto a solid substrate in a tailor-made manner by using the LB technique. Therefore, the polymer nanosheets were heterodeposited onto solid substrates using the LB technique to construct the polymer nanosheet photodiodes. We used p(DDA/Ru) and p(DDA/Fc) pair to construct first polymer nanosheet diode and p(nPMA/AMMA) and p(*t*PA/DNB) pair to construct second polymer nanosheet diode. Photoinduced electron transfer between the each pairs, that is the p(DDA/Ru) and the p(DDA/Fc) nano-sheet pair or p(nPMA/AMMA) and p(*t*PA/DNB) nano-sheet

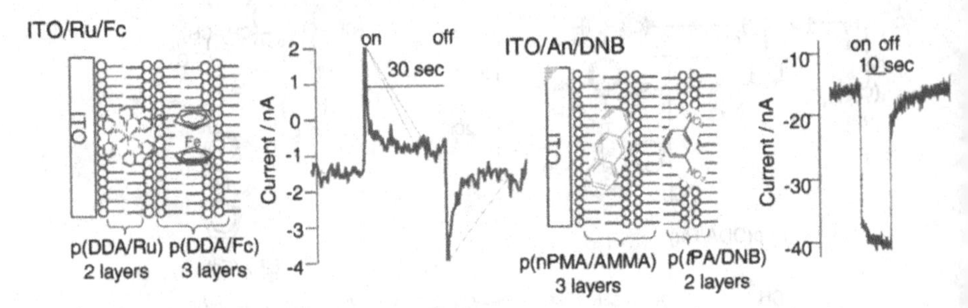

Figure 2 Photocurrent response of the nanosheet photodiode.
(a) nanosheet photodiode constructed from p(DDA/Ru)/p(DDA/Fc) pair
excitation wavelength 460 nm, excitation power 2.5 mW
(b) nanosheet photodiode constructed from p(nPMA/AMMA)/p(tPA/DNB) pair
excitation wavelength 390 nm, excitation power 2.5 mW

pair was confirmed by effective fluorescence quenching measurements. The photocurrent response of the nanosheet assemblies was measured by three-electrode electrochemical cell using $NaClO_4$ as a supporting electrolyte. Figure 2 shows film structure of nanosheet photodiodes and the photocurrent of the films. In the p(DDA/Ru)/p(DDA/Fc) pair, a steady state anodic photocurrent was observed during light irradiation. Because in this nanosheet assemblies, p(DDA/Ru) is an inner layer and p(DDA/Fc) is an outer layer of the heterodeposited film, the excited ruthenium accepts an electron from outer electron donating layer, that is p(DDA/Fc) and then give an electron to electrode (vide infra), resulting in an anodic photocurrent. On the other hand, in the p(nPMA/AMMA)/p(*t*PA/DNB) pair, a steady state cathodic photocurrent was observed. In this case, the chromophore layer i.e. p(nPMA/AMMA), was also located at an inner layer of the heterodeposited film and the excited anthracene transfer an electron to the outer electron accepting layer, p(*t*PA/DNB) and then accept electron form electrode (vide infra), result in a cathodic photocurrent. As can be shown in figure 2 it is demonstrated that a photocurrent direction is controllable by the order of alignment of the polymer nano-sheets in heterodeposited assemblies.

Next, for the construction of molecular switching device, we connected the two polymer nanosheet photodiode in a series. Figure 3 shows the film structure for the switching devices. The nanosheets were assembled in a tailor-made manner onto an ITO electrode using the LB technique as shown in Figure 3. Figure 4 shows a typical photocurrent response on light irradiation (stimuli). When 460 nm wavelength light was irradiated onto the assemblies a steady state anodic

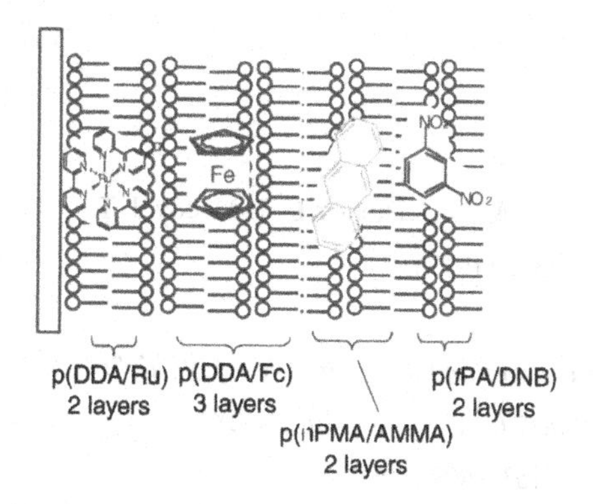

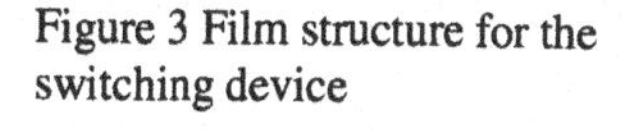
Figure 3 Film structure for the switching device

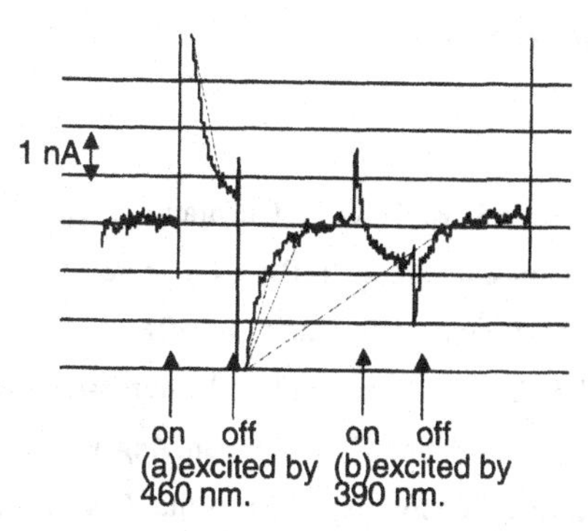

Figure 4 Photocurrent response of the switching device.
(a) ruthenium excitation (460 nm, 2.5 mW)
(b) anthracene excitataion (390 nm, 2.5 mW)

photocurrent about 0.7 nA was observed because only the nano-sheet diodes constructed from p(DDA/Ru)/p(DDA/Fc) pair is active by the excitation of the wavelength. On the other hand when it was excited by 390 nm, the p(nPMA/AMMA)/p(*t*PA/DNB) pair nanosheet photodiode became active and a steady state cathodic photocurrent about 0.9 nA was observed. Conclusively the nanosheet assemblies shown in Figure 3 act as a molecular switching device where a direction of photocurrent can be switched by changing a stimuli by different wavelength irradiation.

In conclusion, photocurrent switching device was constructed by a tailor-made alignment of redox active polymer nano-sheet diode. With the selective excitation of the chromophore by different operation wavelength, photoinduced electron transfer direction can be controlled in regular alignment of the nano-sheets assembly. This means that multi-wavelength operative devices can be constructed by the photofunctional polymer nano-sheet diodes assemblies.

ACKNOWLEDGMENTS

This research was supported by Grant-in-AIds (No. 16750107) from Ministry of Education, Culture, Sports, Science and Technology (MEXT), Japan. J.M. also thanks to Rikougaku shinnkoukai for their finical help and Dr. Go Hirai for his advice for synthesis.

REFERENCES

1. A. P. de Silva, H. Q. N. Gunaratne, T. Gunnlaugsson, A. J. M. Huxley, C. P. Mccoy, J. T. Rademacher and E. T. Rice, *Chem Rev*. **97**, 1515-1566 (1997).
2. F. M. Raymo, *Adv. Mater*, **14**, 401-414 (2002)
3. V. Balzain, A. Credi and M. Venturi, *Molecular Devices and Machines* (Wiley-VCH, 2003).
4. L. R. Caroll and C. B. Gorman, *Angew. Chem., Int. Ed.* **41**, 4378-4400 (2002).
5. A. Aoki, Y. Abe and T. Miyashita, *Langmuir*, **15**, 1463-1469 (1999).
6. J. Matsui, M. Mitsuishi, A. Aoki and T. Miyashita, *Angew. Chem., Int. Ed.* **42**, 2272-2275 (2003).
7. J. Matsui, M. Mitsuishi, A. Aoki and T. Miyashita, *J. Am. Chem. Soc*. **126**, 3708-3709 (2004)

Mater. Res. Soc. Symp. Proc. Vol. 846 © 2005 Materials Research Society DD6.6

Size-dependence of the Linear and Nonlinear Optical Properties of GaN Nanoclusters

Andrew C. Pineda[1,2] and Shashi P. Karna[3]
[1]US Air Force Research Laboratory, Space Vehicles Directorate, 3550 Aberdeen Ave, SE, Bldg. 914, Kirtland Air Force Base, New Mexico 87117-5776
[2]The Center for High Performance Computing and Department of Chemistry, The University of New Mexico, MSC01 1190, 1 University of New Mexico, Albuquerque, NM 87131-0001
[3]US Army Research Laboratory, Weapons and Materials Directorate, ATTN: AMSRD-ARL-WMBD, Bldg. 4600, Aberdeen Proving Ground, MD 21005-5069

ABSTRACT

In this paper, we present the results of our first-principles quantum mechanical studies of the electronic structure, geometry, and linear and nonlinear optical (NLO) properties of tetrahedral Ga_mN_m (m=1, 4, 7, 17) atomic clusters. Our calculated results suggest that the linear and NLO properties both exhibit a strong dependence upon cluster size and shape (geometry). However, the size- and the geometry-dependences are more pronounced for the NLO properties than for the linear optical properties. For clusters containing equal numbers of Ga and N atoms, an open-structure with no network-forming ring has a much larger second-order NLO coefficient than a cluster with a closed ring structure.

INTRODUCTION

The linear and nonlinear optical properties of III-V binary semiconductors have been the subject of active research since the late 1960s [1,2]. Recent advancements in (a) experimental techniques to fabricate/produce stable isolated nanometer-size binary atomic clusters composed of group III and group V elements and (b) techniques and tools to probe response properties of nanoscale objects, have attracted a great deal of attention to the linear and nonlinear optical properties of III-V nanoclusters due to their potential applications in future communications technologies [3]. An important issue in a bottom-up approach to fabricating nanoclusters for future technological applications is an understanding of the evolution of their response properties with cluster size and shape. Theoretical work on atomic clusters of III-V semiconductors have examined the electronic structure and stability of small clusters using density functional theory [4,5] and their linear and non-linear optical properties using the *ab initio* time dependent Hartree-Fock formalism [6,7,8] as well as time-dependent density functional theory [9]. Work by Kandalam et al. suggests that as cluster sizes increase a transition from low energy planar to bulk-like 3-D structures occurs in some nitrides [4]. As non-linear optical properties of molecules are particularly sensitive to their 3-D structure, this suggests a strong dependence of the NLO properties on cluster size. Indeed, work by Natarajan et al. on the atomic and electronic structures of IV-VI semiconductor structures (Ge_mTe_n, $m+n \leq 10$) suggest that the polarizabilities of the clusters vary non-monotonically with cluster size [10].

In order to better understand the dependence of optical properties on geometrical structure, we have undertaken a systematic study of the electronic and geometrical structures and the optical properties of stable GaN nanoclusters. We present an *ab initio* time-dependent Hartree-Fock (TDHF) study of the (hyper)polarizabilities of three-dimensional Ga_mN_m (m=1,4,7,17) clusters with bulk-like bonding. The technical details of the calculations are discussed in the first section. We then present the results for both linear and non-linear properties. Issues relating to cluster size and shape, such as approach to the bulk limit, are discussed.

CALCULATIONAL DETAILS

All optical property calculations were performed on Ga_mN_mclusters using the GAMESS electronic structure package [11]. Surface hydrogens were added to the clusters in order to eliminate surface electronic effects by terminating dangling bonds. The geometrical structures of the clusters were optimized at the *ab initio* Hartree-Fock level using a large polarized basis set, DZP(1*p*,1*d*), as well as Stevens-Basch-Krauss-Jasien-Cundari (SBKJC) effective core potentials [12] for the Ga and N atoms. The exponents of the basis sets employed are the defaults in GAMESS. The components of the polarizability and the 1st hyper-polarizability tensors were computed using the time-dependent Hartree-Fock (TDHF) method [13] using the same DZP basis set as used for the geometry calculation. The orientationally averaged linear polarizability $\langle\alpha(\omega)\rangle$ is presented for the static limit ω=0 as well as the optical wavelength of λ= 1064nm. The hyperpolarizability tensors corresponding to the electroptic Pockels effect (EOPE), $\beta(-\omega;0,\omega)$, second harmonic generation (SHG),$\beta(-2\omega;\omega,\omega)$, and optical rectification (OR), $\beta(0;\omega,-\omega)$, were computed at λ= 1064nm.

The sensitivity of the optical properties to the computed geometry was studied by comparing the properties of clusters optimized with SBKJC and DZP(1*p*,1*d*) basis sets. The convergence of the optical properties with respect to basis set size (including number of polarization functions employed) on the calculated optical properties was also studied by recomputing the properties of the DZP(1*p*,1*d*) structures with additional *p*, *d*, and *f* polarization functions.

RESULTS

Figures 1 and 2 show the clusters employed in this study arranged in order of increasingly bulk-like character. The bonding in these clusters is generally tetrahedral. The clusters in Figure 1 are what we term "open" in the sense that they do not contain ring structures. By contrast, the clusters in Figure 2 are composed of "closed" network-forming rings. The clusters in Figure 1 were taken from Korambath et al.[7] who employed an even-tempered Gaussian (ETG) basis set for their calculations on clusters up through m=4. The Ga_4N_4 clusters, Figure 1(B) and Figure 2(A), are related in that they represent the simplest open and closes GaN structures that can be carved out of a periodic arrangement of GaN atoms like that in Figure 2(C). They were chosen to provide insight into the effect that small changes in atomic arrangements can have on optical properties.

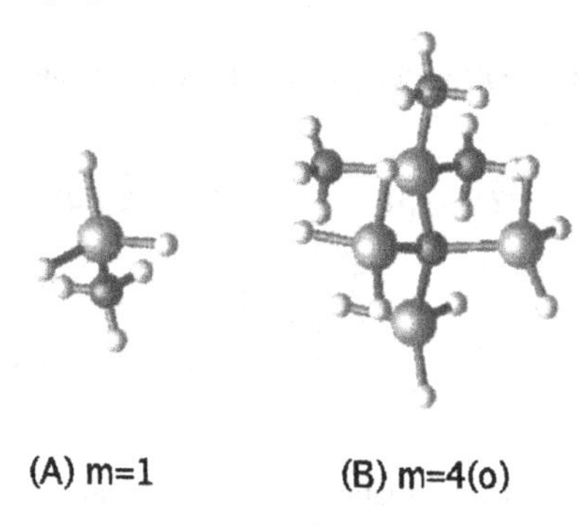

Figure 1. Open GaN structures: (A) $GaNH_6$ and (B) $Ga_4N_4H_{18}$. Nitrogen atoms are shown in blue, gallium in silver and hydrogen in white.

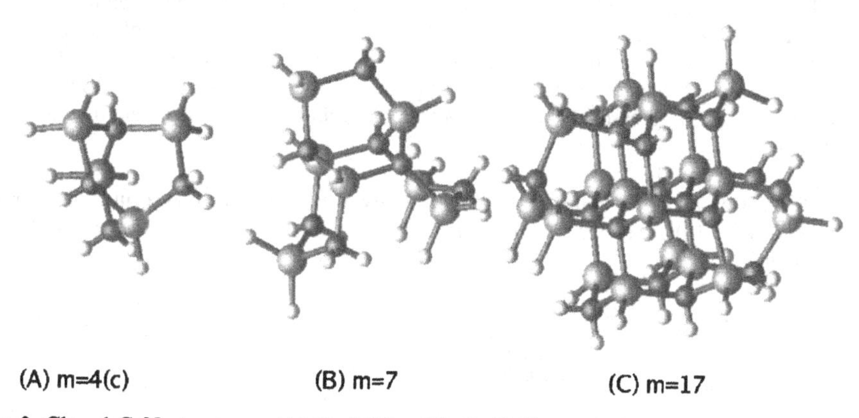

Figure 2. Closed GaN structures: (A) $Ga_4N_4H_{14}$, (B) $Ga_7N_7H_{22}$, and (C) $Ga_{17}N_{17}H_{38}$. Nitrogen atoms are shown in blue, gallium in silver and hydrogen in white.

Linear Optical Properties

Table I shows the calculated dipole moments of the clusters as a function of basis set size. The calculated dipole moment appears to be reasonably well converged even for the smallest basis set. The magnitudes of the computed dipole moments are not well correlated with cluster size, but are more generally related to the overall symmetry of the cluster.

Table I. Dipole moments in Debye for optimized clusters computed at the DZP(1*p*,1*d*) geometry as a function of basis set size.

***m*, Ga$_m$N$_m$**	**DZP(1*p*,1*d*)**	**DZP(3*p*,3*d*)**	**DZP(3*p*,3*d*,1*f*)**
1	5.21	5.15	5.15
4(o)	18.60	18.34	18.34
4(c)	5.92	5.80	5.80
7	2.66	2.62	2.62
17	13.19	-	-

Tables II and III show the variation of the isotropically-averaged static and frequency dependent polarizabilities as functions of cluster size and the basis set employed in the properties calculation. Polarizabilities for basis sets larger than DZP(1*p*,1*d*) for the *m*=17 cluster were not computed owing to the large computational requirements. However, we can still draw conclusions from the trends in the data since (1) the polarizabilities were generally found to be well-converged at the DZP(3*p*, 3*d*) level and (2) the relative differences between results of the relatively inexpensive DZP(1*p*,1*d*) polarizability calculations and the results of the more expensive calculations rapidly decrease from 14% to 7% as *m* increases from 1 to 7. This suggests that the computed value of the linear polarizability computed at the DZP(1*p*,1*d*) level for *m*=17 should be a reasonable estimate of the polarizability one would obtain from a larger calculation.

Table II. Isotropically-averaged static polarizability, α(0) (10^{-24} cm^3), as a function of basis set at the DZP(1*p*,1*d*) geometry. The final column gives our best value for the polarizability per GaN unit.

***m*, Ga$_m$N$_m$**	**DZP(1*p*,1*d*)**	**DZP(3*p*,3*d*)**	**DZP(3*p*,3*d*,1*f*)**	**α(0)/*m***
1	5.38	6.26	6.26	6.26
4(o)	20.00	22.34	22.35	5.59
4(c)	17.68	19.25	19.25	4.81
7	29.79	31.93	31.94	4.56
17	63.97	-	-	3.76

Table III. Isotropically-averaged frequency dependent polarizability, $\alpha(\omega)$ (10^{-24} cm^3 ≡ 10^{-24} esu^2 cm^2 erg^{-1}) at λ = 1064nm, as a function of basis set at the DZP(1*p*,1*d*) geometry. The final column gives our best value for the polarizability per GaN unit.

***m*, Ga$_m$N$_m$**	**DZP(1*p*,1*d*)**	**DZP(3*p*,3*d*)**	**DZP(3*p*,3*d*,1*f*)**	$\alpha(\omega)/m$
1	5.42	6.32	6.32	6.32
4(o)	20.18	22.58	22.59	5.65
4(c)	17.83	19.43	19.44	4.86
7	30.03	32.22	32.23	4.60
17	64.48	-	-	3.79

The values we compute for the dipole moments and linear polarizabilities for *m*=1 and 4(o) are in rough agreement with the results obtained by Korambath et al. [7] who noted that the polarizability of clusters depended in an approximately linear fashion on the number of GaN units in the cluster. Our value for the polarizability per GaN unit is slightly larger however. In the final columns of Tables II and III, we compute the polarizability per GaN unit using our best available value for each cluster. Our results suggest that the polarizability per GaN unit slowly decreases with cluster size and that it may already be near the bulk limit (~3.2-3.9 x 10^{-24} cm^3) by *m*=17.[15]

The apparent convergence toward the bulk limit observed in the polarizability is not seen in other properties. Table IV shows the trends in energy difference between the highest occupied molecular orbital (HOMO) and the lowest unoccupied molecular orbital (LUMO) as a function of basis set and the sizes of our clusters. It is expected that this quantity should decrease as a function of cluster size, and in the bulk limit approach the band gap of the crystalline material. Table IV suggests that this approach to the bulk limit is slow and depends on the shape of the cluster in a non-trivial fashion.

Table IV. HOMO-LUMO energy differences in eV as a function of basis set for optimized clusters at the DZP(1*p*,1*d*) geometry.

***m*, Ga$_m$N$_m$**	**DZP(1*p*,1*d*)**	**DZP(3*p*,3*d*)**	**DZP(3*p*,3*d*,1*f*)**
1	13.58	11.99	11.99
4(o)	10.85	9.13	9.13
4(c)	13.03	11.21	11.21
7	13.41	11.86	11.86
17	11.64	-	-

Non-linear optical properties

The first hyperpolarizability, $\beta=\beta(-\omega_\sigma;\omega_1,\omega_2)$ is a rank 3 tensor quantity, and depends upon the symmetry of the atomic system studied. For centrosymmetric systems, β vanishes. When β is non-zero, a vector component β_μ, directed along the applied electric field, is usually observed. A related and frequently reported theoretical quantity is

$$\beta_{vec} = \sqrt{(\beta_x^2 + \beta_y^2 + \beta_z^2)} \quad (2)$$

where the β_i ($i = x, y, z$) are determined from the tensor components, β_{ijk}, of β as

$$\beta_i = \frac{1}{3} \sum_{j=x,y,z} (\beta_{ijj} + \beta_{jij} + \beta_{jji}). \quad (3)$$

The former quantity (Eq. 2) evaluated for frequencies corresponding to the static hyperpolarizability, the electroptic Pockels effect, and second harmonic generation is shown in Tables V-VIII. The calculated values have also been divided by a factor of 2 so that the values we report are consistent with the conventions found in the experimental literature instead of those in the quantum chemical literature [14].

Tables V and VI show the effect of the basis set used for the geometry optimization as well as the basis set used on the NLO properties on the static hyperpolarizabilities, $\beta(0;0,0)$, computed for our clusters. As was seen in the calculation of the linear polarizabilities, the smallest basis set results are not completely converged. From tables V and VI we see that the geometry used in the calculation (SBKJC or DZP) has a weak effect on the computed hyperpolarizability.

Table V. Static hyperpolarizability, $\beta(0;0,0)$, of the clusters in 10^{-32} esu^3 cm^3 erg^{-2} as a function of basis set at SBKJC optimized geometry.

m, Ga_mN_m	DZP(1*p*,1*d*)	DZP(3*p*,3*d*)	DZP(3*p*,3*d*,1*f*)
1	30.87	31.89	32.05
4(c)	5.80	14.15	13.99
7	44.75	38.15	37.99
17	119.27	-	-

Table VI. Static hyperpolarizability, $\beta(0;0,0)$, of the clusters in 10^{-32} esu^3 cm^3 erg^{-2} as a function of basis set at the DZP(1*p*,1*d*) optimized geometry.

m, Ga_mN_m	DZP(1*p*,1*d*)	DZP(3*p*,3*d*)	DZP(3*p*,3*d*,1*f*)
1	27.16	29.43	29.61
4(o)	153.83	89.97	90.94
4(c)	0.93	11.93	11.74
7	32.29	33.59	33.32
17	116.14	-	-

Table VII. EOPE hyperpolarizability, β(-ω;0,ω), of clusters in 10^{-32} esu^3 cm^3 erg^{-2} at λ = 1064nm as a function of basis set using DZP(1*p*,1*d*) optimized geometry. The results for OR (optical rectification), β(0;-ω,ω) are identical.

***m*, Ga$_m$N$_m$**	**DZP(1*p*,1*d*)**	**DZP(3*p*,3*d*)**	**DZP(3*p*,3*d*,1*f*)**
1	27.71	30.17	30.35
4 (o)	158.67	95.01	96.02
4 (c)	1.05	12.18	11.99
7	33.08	34.21	33.93
17	119.93	-	-

Table VIII. Second harmonic generation hyperpolarizability, β(-2ω; ω, ω), of clusters in 10^{-32} esu^3 cm^3 erg^{-2} at λ = 1064nm as a function of basis set using DZP(1*p*,1*d*) optimized geometry.

***m*, Ga$_m$N$_m$**	**DZP(1*p*,1*d*)**	**DZP(3*p*,3*d*)**	**DZP(3*p*,3*d*,1*f*)**
1	28.87	31.72	31.91
4(o)	169.15	106.89	107.98
4(c)	1.31	12.68	12.48
7	34.76	35.47	35.16
17	128.10	-	-

Tables VI – VIII show that the variation of hyperpolarizabilities with cluster size and shape are pronounced, but perhaps not as large as suggested by the previous results of Korambath, et al. who computed hyperpolarizabilities for *m*=4(o) that are more than twice what we found. The hyperpolarizabilities in our study vary over roughly an order of magnitude and appear to be strongly correlated with both the size of the clusters and degree of compactness of the cluster shapes. These size and shape effects can either reinforce one another or cancel. This correlation suggests that experimentally observed hyperpolarizabilities will exhibit oscillatory behavior as a function of cluster size. The degree of oscillation will depend upon the relative stabilities of the various possible shapes of a cluster of a given size and their point symmetry. The magnitudes of the various second-order effects show the expected trend of β(-2ω; ω, ω) > β(-ω;0,ω) > β(0;0,0) as was reported previously [7].

SUMMARY

We have calculated the (hyper)polarizabilites of a series of Ga$_m$N$_m$ clusters of varying size and shape. The convergence of these properties with respect to basis set size was studied and found to be well converged at the DZP(3*p*, 3*d*) level. Our results also suggest that the number of polarization functions included in the calculation becomes less important for larger clusters. The linear polarizability of a GaN unit in a cluster is found to slowly vary with cluster size and appears to quickly approach the bulk value of 3.9 Å^3. The linear polarizability exhibits small frequency dispersion. Other properties appear to converge more slowly with cluster size.

The hyperpolarizabilities are found to oscillate strongly with cluster size owing to competing size and shape effects. For equal numbers of Ga and N atoms (e.g., Ga_4N_4), the structure with reduced symmetry gives a substantially enhanced value of β as compared with the more symmetric structure. As the range of energetically accessible structures depends upon temperature, this suggests that the variation of hyperpolarizabilities with cluster size will depend strongly on temperature.

ACKNOWLEDGEMENTS

We thank Dr. Prakashan Korambath and Prof. Mark Gordon for useful discussions. We also thank the Center for High Performance Computing at the University of New Mexico for access to the Los Lobos Linux Supercluster.

REFERENCES

1. B. F. Levine, Phys. Rev. Lett. **22**, 787 (1969).
2. J. Ch. Phillips and J. A. Van Vechten, Phys. Rev. **183**, 709 (1969).
3. R. Schaefer, S. Schlecht, J. Woenckhaus, and J. A. Becker, Chem. Phys. Lett. **76**, 471 (1996).
4. A.K. Kandalam, M.A. Blanco, and R. Pandey, J. Phys. Chem. B **105**, 6080-6084 (2001); **106**, 1945-1953 (2002).
5. A. Costales, A.K. Kandalam, R. Franco, and R. Pandey, J. Phys. Chem. B **106**, 1940-1944 (2002).
6. P. P. Korambath, B. K. Singaraju, S. P. Karna, Int. J. Quant. Chem. **77**, 563 (2000).
7. P. P. Korambath and S. P. Karna, J. Phys. Chem. A **104**, 4801-4804 (2000).
8. Y.-Z. Lan, W.-D. Cheng, D.-S. Wu, X.-D. Li, H. Zhang, Y.-J. Gong, Chem. Phys. Lett. **372**, 645 (2003).
9. I. Vasiliev, S. Öğüt, and J.R. Chelikowsky, Phys. Rev B. **60**, R8477 (1999); Phys. Rev. Lett. 78 (25), 4805-4808 (1997).
10. R. Natarajan and S. Öğüt, Phys. Rev. B **67**, 235326 (2003).
11. M. W. Schmidt, K. K. Baldridge, J. A. Boatz, S. T. Elbert, M. S. Gordon, J. H. Jensen, S. Koseki, N. Matsunaga, K. A. Nguyen, S. J. Su, T. L. Windus, M. Dupuis, and J. A. Montgomery, J. Comput. Chem. **14**, 1347-1363 (1993).
12. W. J. Stevens, H. Basch, M. Krauss, J. Chem. Phys. **81**, 6026-6033 (1984). W. J. Stephens, H. Basch, M. Krauss, P. Jasien, Can. J. Chem. **70**, 612-630 (1992). T. R. Cundari, W. J. Stevens, J. Chem. Phys. **98** (7), 5555-5565 (1993).
13. S. P. Karna and M. Dupuis, J. Comput. Chem. **12**, 487-504 (1991). S. P. Karna, Chem. Phys. Lett. **214** (2), 186-192 (1993).
14. S. P. Karna, P. N. Prasad, and M. Dupuis, J. Chem. Phys. **92**, 7418 (1990).
15. V. Bougrov, M.E. Levinshtein, S.L. Rumyantsev., and A. Zubrilov, in *Properties of Advanced Semiconductor Materials GaN, AlN, InN, BN, SiC, SiGe* . edited by M.E. Levinshtein, S.L. Rumyantsev, and M.S. Shur., (John Wiley & Sons, Inc., New York, 2001), p. 1-30.

Mater. Res. Soc. Symp. Proc. Vol. 846 © 2005 Materials Research Society DD6.8

Current-voltage characteristics in organic semiconductor crystals: space charge vs. contact-limited carrier transport

J. Reynaert*, V. I. Arkhipov, J. Genoe, G. Borghs, P. Heremans
IMEC, Kapeldreef 75, B-3001 Leuven, Belgium
*also with ESAT, Katholieke Universiteit Leuven, Leuven, Belgium

ABSTRACT

Numerous experimental studies, mostly based on the time-of flight (TOF) technique, showed that the conductivity in organic crystals can be analysed in terms of (trap-controlled) band transport. However, recent comparative studies of TOF signals and space charge limited currents (SCLCs) in tetracene crystals revealed a striking difference in carrier mobilities estimated from TOF current transients and from SCLC curves. The analysis of the SCLC curves yielded the mobilities wildly varying within 6 orders of magnitude. Therefore, it is not always clear whether the measured current-voltage (IV) device characteristics are controlled by charge injection or by transport in the bulk. In this work, we formulate a model of dopant-assisted carrier injection across a metal/organics interface and use this model for the analysis of IV curves measured on a tetracene and perylene crystal. The model suggests the occurrence of an energetically disordered layer at the surface of an organic crystal. This might be either an amorphous phase of the same material or a crystalline layer with a high density of defects and/or impurities. Since, at variance with bulk properties, the surface of an organic crystal is poorly controlled and can be strongly modified upon the contact deposition, the model of injection-controlled IV characteristics can explain the striking difference between the TOF mobility and the apparent 'SCLC mobility' measured in tetracene crystals. In order to give more credence to the role of surface defects states in the dark charge transport, we compare IV characteristics measured on sandwich and coplanar structures. In the latter structure, surface states show a major contribution to the conductivity.

INTRODUCTION

Single crystals of small molecule organic semiconductors form an interesting case of study as it is expected that their electrical properties are not subjected to grain boundaries or other defects due to morphological disorder, which are present in thin films of these same materials. Due to the structural order in well-grown crystals, the carrier transport is expected to happen through (trap-controlled) hopping of weakly localized carriers which is virtually equivalent to band transport [1, 2]. Therefore, it is interesting to derive the intrinsic properties of these organic materials such as carrier mobility or impurity (trap) concentration through their electrical characterization.

A modest attempt is made in this introduction to give an overview of the opto-electrical methods that could eventually lead to a characterization of the electrical transport in an organic crystal.

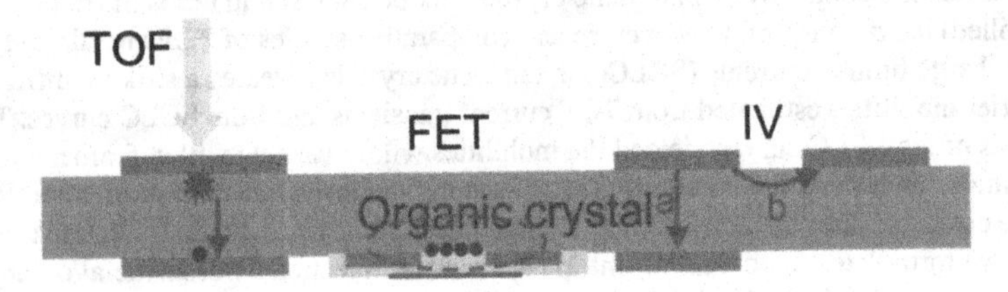

Figure 1: Opto-electronic methods for transport property characertization

The time-of-flight (TOF) method has been used extensively in the past for the determination of the carrier mobility in organic crystals [1, 2]. As illustrated on the left of Fig.1, TOF is based on the photogeneration of a carrier close to a (semi-)transparent contact, which then selectively propagates by the applied bias through the crystal to the opposite contact. Another technique to determine the carrier mobility is to provide the crystal surface with source and drain contacts, and a gate contact separated from the crystal surface by a thin dielectric material. In this way, a field effect transistor (FET) is created as shown in the middle part of Fig.1. FET-mobilities can be derived by applying the appropriate voltages to the contacts. Recent work on tetracene, pentacene and rubrene [3, 4] has revealed nice mobility results, nevertheless, one can not be sure that the obtained mobility is a real intrinsic bulk mobility rather than a surface mobility subjected to surface defects. A last method to determine electrical properties is by two-probe current-voltage (IV) measurements. This can be implemented either by sandwich-contacting the organic crystal, either by coplanar contacting, shown by *a* and *b* respectively in the right part of Fig. 1. There again the latter method might be stronger influenced by the crystal surface [5]. The importance of possible defects at the crystal surface will be highlighted further on in this work. IV-measurements are typically interpreted in terms of space-charge limited current (SCLC), for which we refer to Ref[6].

Table 1 gives an overview of the formulas used to determine the organic crystal mobility for each method presented here.

Table 1: Overview of mobility characterization methods and formulas used thereto.

Method	Formula
TOF	$\mu_{TOF} = \frac{V}{L^2 \tau_{tr}}$
FET	$\mu_{FET,sat} = \frac{2.L}{W.C}\left(\frac{\partial\sqrt{I_{ds}}}{\partial V_{gs}}\right)^2$
Trap-free SCLC[6] for sandwich contacted crystals	$\mu = \frac{8\varepsilon_0 \varepsilon L^3}{9}\left(\frac{\partial\sqrt{J}}{\partial V}\right)^2$

where L is the crystal thickness, τ_{tr} the carrier transit time (for TOF).
W, L, C, I_{ds}, V_{gs} are the transistor channel width, length, capacitance per unit area, drain-source current and gate-source voltage (for FET).
L, ε_0, ε, J are the crystal thickness, permittivity of free space, dielectric constant and current density (for trap-free SCLC).

EXPERIMENTAL RESULTS AND DISCUSSION

A description of the experimental setup and sample preparation can be found elsewhere [7]. Below, we will restrict ourselves to information strictly necessary for the further understanding of this work.
Organic crystals of small molecules can be grown in a variety of ways [8], one of which is from the convective flow of an inert carrier gas [9]. The growth system, represented in Fig. 2, consists of an inlet of preheated carrier gas close to the material to be purified and a heating system maintaining a temperature gradient between the gas inlet and the cooled end of the growth tube.

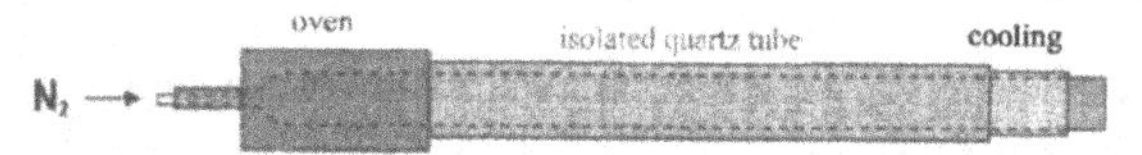

Figure 2: Growth tube for organic crystals

Perylene and tetracene crystals obtained by such growth are shown in Fig.3a and b, respectively.

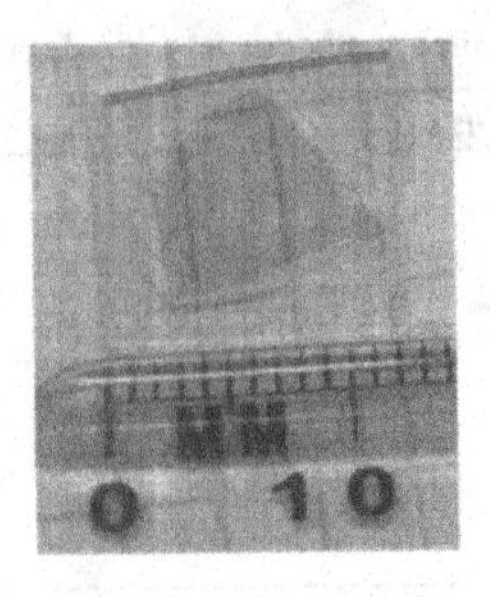

a.

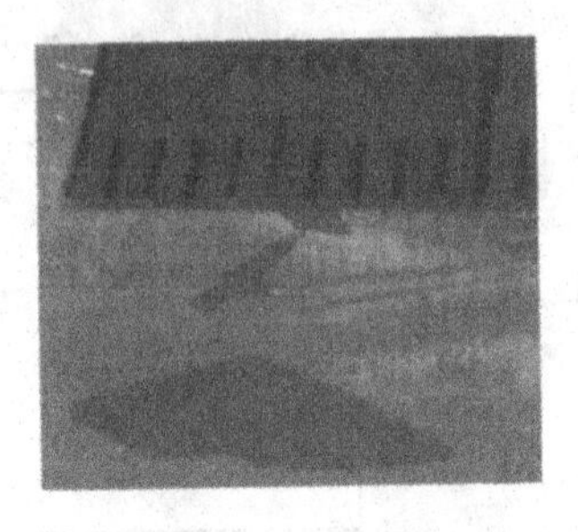

b.

Figure 3a, b: Perylene and tetracene crystals grown in a crystal growth system as in Fig. 2.

In this work, we refer to crystals, grown with poor impurity evacuation and taken from the growth tube after the first growth, as to 'lower' quality crystals. Crystals, made by at least two subsequent growths after intensive impurity evacuation for at least half an hour at 3 mTorr, will be referred as to 'higher' quality crystals. This quality difference will only be made for crystals out of perylene.

IV-curves measured on sandwich- or coplanar contacted crystals, can have various origins, namely:

1. Ohmic conductivity, obeying $I \sim \frac{V}{L}$ law
2. Bulk-limited current, obeying $I \sim \frac{V^{r+1}}{L^{2r+1}}$ behaviour

   for a sandwich-contacted crystal, r=1 in the trap-free regime [6].
3. Contact-limited current

Each of these possible mechanisms will be further addressed in this work.

The IV-curves shown in Fig. 4a were measured on a lower quality perylene crystal, for several contact interdistances L. At first sight, these curves reveal an Ohmic voltage dependence. This is nevertheless contradicted by the fact that the current at 30V is not inversely reciprocal to the contact interdistance, as seen from Fig. 4b. The temperature-dependent IV-curves of Fig. 5, measured on a perylene crystal from the same growth run, show a very weak temperature dependence. This could indicate the presence of only shallow traps if we assume the current is bulk-limited. This would nevertheless invoke a strong super-quadratic IV-dependence, which is not the case as seen from Fig. 4a. Moreover, a weak temperature-dependence is more likely in the case of contact-limited current.

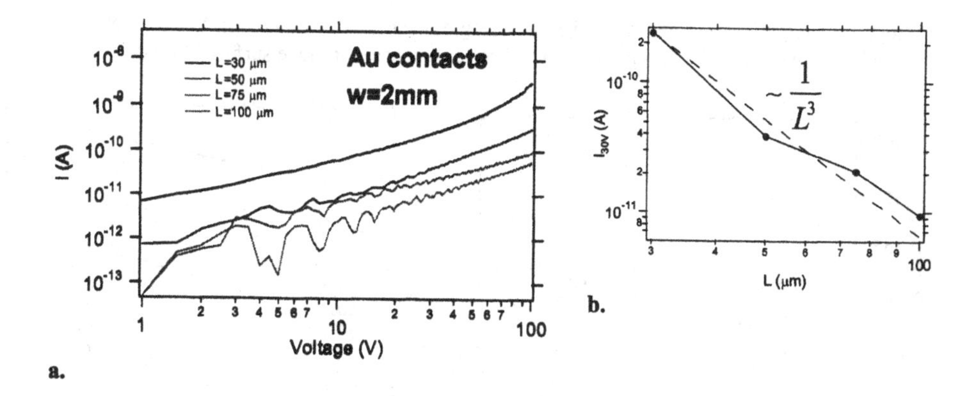

Figure 4: a. Results of IV-measurements on a lower quality perylene crystal with coplanar Au contacts, with contact width 2 mm and various contact interdistances L, b. Current at 30V of the same measurements vs. contact distance L. The red line indicates $1/L^3$-dependency with d the crystal thickness.

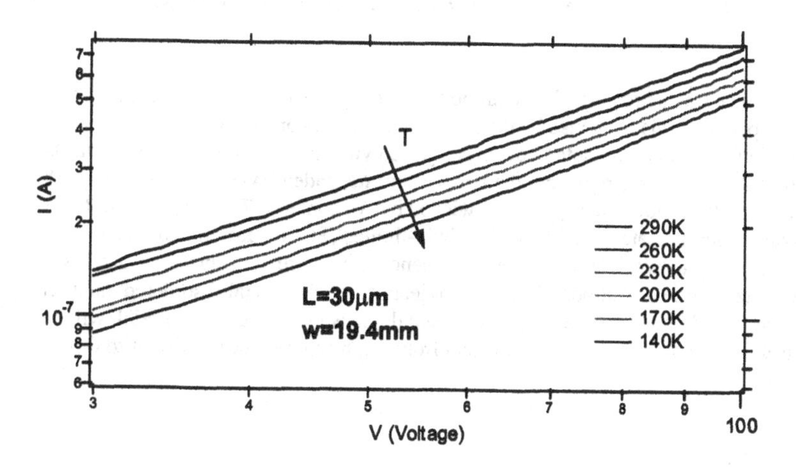

Figure 5: Temperature dependence of a lower quality perylene with coplanar contacts.

The importance of surface states for conductivity between coplanar contacts is illustrated by the measurement from Fig. 6a. There, we start from a higher quality perylene crystal, which shows a very low conductivity. The conductivity is nevertheless increased by orders of magnitude after 20 seconds of oxygen plasma treatment, and increases upon another 20 seconds of oxygen plasma treatment. In Fig. 6b, the temperature dependence of the IV-curves for the oxygen plasma treated sample is shown. For these samples, the temperature dependence is stronger although the respective activation energy is not larger than 0.1 eV. During oxygen plasma treatment, free oxygen radicals can react with the organic crystal surface and hence introduce surface defects.

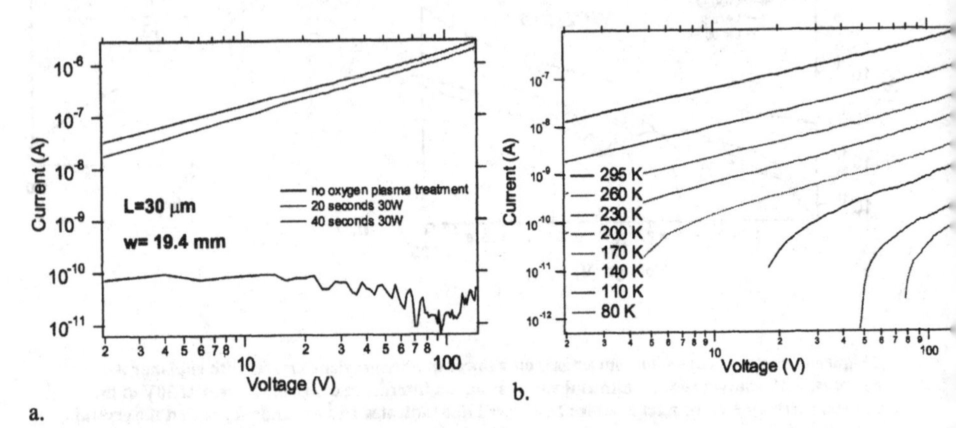

Figure 6: a. IV on a higher quality perylene crystal, before, after 20" and after 40" of oxygen plasma treatment, b. Temperature-dependent IV on the same crystal after 40" oxygen plasma treatment.

If surface states prove to play an important role in coplanar contacted crystals one could expect that this must also be the case for sandwich-contacted crystals, as the current in a sandwich-contacted device is injected via the metal/organic crystal interface. The datapoints in Fig. 8 represent the temperature-dependent IV-measurements on a 8 micrometer thick tetracene crystal sandwiched by Au contacts. The supra-quadratic IV-behavior could indicate a trap-controlled bulk-limited (SCLC) current, but this is contradicted by its weak temperature-dependence. Therefore, this current is also assumed to be contact-limited. A model for current injection across a metal/disordered material interface, based on the presence of inadvertent dopants in the organic material, will be outlined in the following section. The data from Fig.8 can be successfully fit to this model [7].

MODELING

Figure 7a represents the energy diagram at the metal/organic interface. In this figure, Δ is the zero-field contact barrier, F_0 the externally applied electrical field, $U(x)$ the potential energy of an injected carrier, taking into account barrier lowering by image charge. We assume the organic crystal surface to have a Gaussian density-of-states (DOS) distribution, given by

$$g(E)=\frac{N_t}{\sqrt{2\pi}\sigma}\exp\left(-\frac{E^2}{2\sigma^2}\right) \tag{1}$$

The injected current is determined by the amount of carriers that overcome the barrier by thermal activation, without falling back through surface recombination, as illustrated in Figure 7b. The exact mathematical formulation is worked through in Ref.[10], the resulting injection current density j is given by

$$j = e\nu \frac{\int_a^\infty dx_0 \exp(-2\gamma x_0) \int_{-\infty}^{\infty} dE \ \mathrm{Bol}(E) g[E - U(x_0)] \int_a^{x_0} dx \exp\left[-\frac{e}{kT}\left(F_0 x + \frac{e}{16\pi\varepsilon_0 \varepsilon x} \right) \right]}{\int_a^\infty dx \exp\left[-\frac{e}{kT}\left(F_0 x + \frac{e}{16\pi\varepsilon_0 \varepsilon x} \right) \right]} \tag{2}$$

with ν being the attempt-to-jump frequency, γ the inverse localization radius, *Bol(E)* the Boltzmann distribution function, *a* the distance from the contact to the closest hopping site and x_0 the jump distance from the contact.

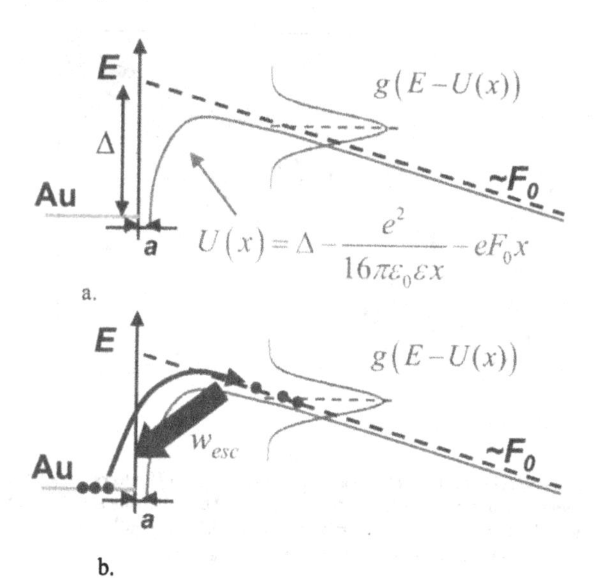

Figure 7: a. Energy diagram at a metal/organic interface with initial barrier height Δ, Gaussian surface DOS g(E) and externally applied field F_0, b. current injection by thermally activation (thin arrow) and carrier backflow by surface recombination (thick arrow). The probability to avoid carrier backflow is defined as w_{esc}.

The relation between the current and the applied field is given by

$$j = e\mu p(x) F(x) \tag{3}$$

The field F and the carrier density in the bulk are related by Poisson's equation

$$\frac{dF(x)}{dx} = \frac{e}{\varepsilon_0 \varepsilon}\left[p(x) - p_0 \right] \tag{4}$$

The relation between the external applied voltage and the field is

$$V = -\int_0^L dx\, F(x) \tag{5}$$

where μ is the carrier mobility, p the carrier density, p_0 the dopant concentration and L the crystal thickness.

The model according to Eqs. (1-5) is applied to fit the data shown in Fig. 8, and the resulting fit lines are added to Fig. 8. The fit parameters can be found back in the figure caption.

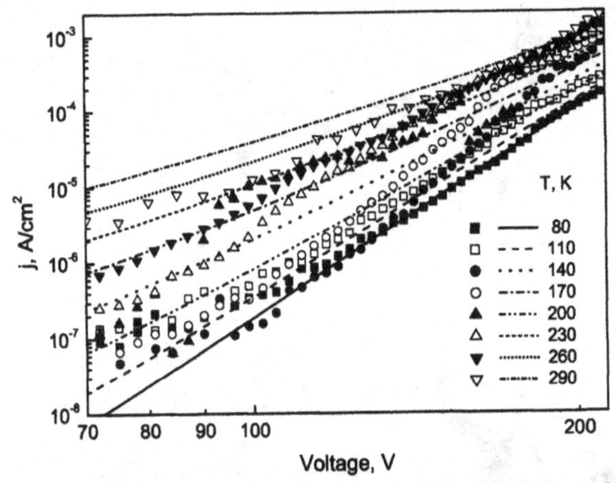

Figure 8: Experimental (symbols) and theoretical (lines) IV characteristics of a gold/tetracene/gold structure at different temperatures. The following set of parameters has been used in the calculation: Δ=0.55 eV, σ= 49 meV, N_t = 3.5×10^{20} cm^{-3}, ν_0 = 10^{11} s^{-1}, γ= 9 nm^{-1}, μ = 0.9 cm^{2}V^{-1}s^{-1}, ε = 3.5, L = 8 μm, p_0 =5×10^{16} cm^{-3}.

One should note that the theoretical curves are only very weakly dependent on the mobility fit parameter. By consequence, our results show that contact-limited IV-measurements on organic crystals do not reveal any information about the carrier mobility. The applied model does allow to extract quantitative information about the surface DOS width σ and the density of localized states N_t.

CONCLUSION

In this work, we have shown the importance of surface states for the conductivity of coplanar contacted organic crystals. By applying oxygen plasma treatment to a perylene crystal, the conductivity was increased by orders of magnitude. We assume this is due to

an improved conductivity via surface states created upon the oxygen plasma treatment. If so, these states should also control injection across the metal/organic interface in sandwich-type samples. IV-measurements on a sandwich-contacted tetracene crystal indeed show characteristic features of a contact-limited current. The obtained data are fit to the proposed model of dopant-assisted current injection.
The results of the present work suggest that deliberate doping would favor the current injection process in organic crystals, as recently seen by the authors of Ref. [11]. Therefore, increasing dopant concentration could eventually lead to the current in sandwich-contacted crystals being bulk- rather than contact-limited. In such case, one would be able to determine the bulk carrier mobility by mere IV-measurements on sandwich-contacted crystals.

ACKNOWLEDGEMENTS

The authors want to acknowledge E. Emelianova for help with numerical calculations and M. Van der Auweraer for the interesting discussions. One of the authors (JR) wants to thank the Flemish IWT-fund for his scholarship.

REFERENCES

1. Pope, M. and C.E. Swenberg, *Electronic Processes in Organic Crystals and Polymers*. 2nd ed. 1999, New York: Oxford University Press.
2. Karl, N., Synthetic Metals, 2003. **133**: p. 649-657.
3. Podzorov, V., V.M. Pudalov, and M.E. Gershenson, Applied Physics Letters, 2003. **82**(11): p. 1739-1741.
4. Goldmann, C., S. Haas, C. Krellner, K.P. Pernstich, D.J. Gundlach, and B. Batlogg, Journal of Applied Physics, 2004. **96**(4): p. 2080-2086.
5. de Boer, R.W.I., M. Jochemsen, T.M. Klapwijk, and A.F. Morpurgo, J. Appl. Phys., 2004. **95**(3): p. 1196.
6. Lampert, M.A. and P. Mark, *Current injection in Solids*. 1970, New York: Academic.
7. Reynaert, J., V.I. Arkhipov, G. Borghs, and P. Heremans, Applied Physics Letters, 2004. **85**(4): p. 603-605.
8. Karl, N., Molecular Crystals and Liquid Crystals, 1989. **171**: p. 157-&.
9. Laudise, R.A., C. Kloc, P.G. Simpkins, and T. Siegrist, Journal of Crystal Growth, 1998. **187**(3-4): p. 449.
10. Arkhipov, V.I., E.V. Emelianova, Y.H. Tak, and H. Bassler, Journal of Applied Physics, 1998. **84**(2): p. 848.
11. Hiramoto, M., A. Tomioka, K. Suemori, and M. Yokoyama, Applied Physics Letters, 2004. **85**(10): p. 1852-1854.

Mater. Res. Soc. Symp. Proc. Vol. 846 © 2005 Materials Research Society DD6.9

Photoinduced Memory Effects based on Poly(3-hexylthiophene) and Al Interface

Keiichi Kaneto, Mitsuru Ujimoto and Wataru Takashima
Graduate School of Life Science and Systems Engineering, Kyushu Institute of Technology, 2-4 Wakamatsu-ku, Kitakyushu, Fukuoka, 808-0196 JAPAN

ABSTRACT

Organic memory cells have been constructed utilizing conducting polymer, poly(3-hexylthiophene), PHT. The PHT film with Au electrodes at the bottom was coated with ultra thin Al having the thickness of 1 nm on the top. The cell showed the large photoconductivity, which was measured between Au bottom electrodes. The photoconductivity lasted for several hours after turn off the illumination, indicating photo induced memory effect. The phenomenon is discussed in terms of the persistent photoconduction, which is resulted from the electron trapping at Al donor sites in the depletion layer formed near the top of film.

INTRODUCTION

Organic electronics devices like electro luminescence (EL) displays, field effect transistors (FET) and solar cells have been intensively interested and studied, because of the lightweight, thin film and low cost manufacturing process. For the fabrication of all organic electronic circuits, organic memory devices are required. Though a large number of studies on the improvement of EL and FET have being carried out, organic memory devices have not been studied. A cell consisting of polyethlenedioxythiophene (PEDOT) and polystyrene sulfonic acid has revealed a write-once, read-many-times (WORM) memory[1]. Bistable memory devices using organic materials of 2-amino-4,5-imidazoledicabonitrile (AIDCN)[2] have shown that the conductivity of the films switched on and off by the application of voltages at certain threshold. The reports have stimulated further interests to realize the rewritable organic memories. In this paper, a preliminary study on the memory effect utilizing a persistent photoconduction in conducting polymers, poly(3-hexylthiophene), (PHT) film with thin Al layer will be mentioned.

EXPERIMENTALS

Typical structures of memory devices are shown in Fig.1(A). Au electrodes with the thickness of 30 nm were firstly deposited in vacuum on glass substrate. The distance between Au electrodes and the width of electrodes were 25 μm and 2 mm, respectively. Then PHT was spin-coated with 2000 rpm for several minutes. The thickness of the PHT films was tuned by using various concentrations of PHT in chloroform ranging 0.125–1.0 wt%. The thickness of PHT film was ranging 50–200 nm. Al with thickness of approximately 1 nm was deposited on it for memory cells. Another type cell with thin Al layer deposited on Au electrode underneath PHT was also prepared as shown in Fig.1 (B). The electrical characteristics of the cell were measured in vacuum using the Keithley 6517 electrometer. Light source was a monochromated light of 550nm with the intensity of 50 $\mu W/cm^2$ from the Xe lamp. The intensity of light was tuned by using neutral optical density filters.

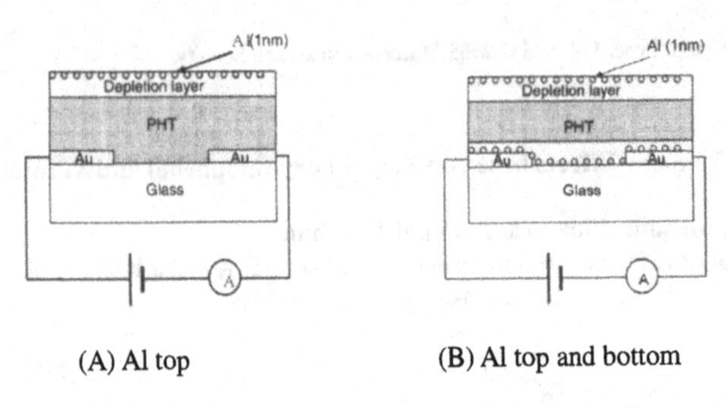

(A) Al top (B) Al top and bottom

Fig. 1 Schematic drawing of the memory cells.

RESULTS AND DISCUSSION

Curves in Fig. 2 show the resistances of PHT for the cell (A) shown in Fig.1, as a function of Al thickness [3,4], which was obtained during deposition of Al. The resistance increased at the initial stage of Al deposition, followed by decrease above the thickness of 10 nm. The result indicates that the increase of resistance at the first stage is due to the formation of depletion layer with much higher resistance. This is resulted from compensation of p-type PHT by donated electrons from Al. The resistance of depletion layer is much higher than that of pristine PHT by several orders of magnitude. The decrease of resistance at the second stage originates from the build up of conductive path by Al at the top. The red line in the inset shows the current path above the Al thickness of 20 nm. From the magnitude of the increased resistance at the first stage, the thickness of depletion layer was estimated to be several tens nm (20 - 100 nm depending on time after deposition). It should be noted that the depletion layer plays the role of Schottky type junction, resulting in a rectification with the combination of ohmic contact by Au electrode.

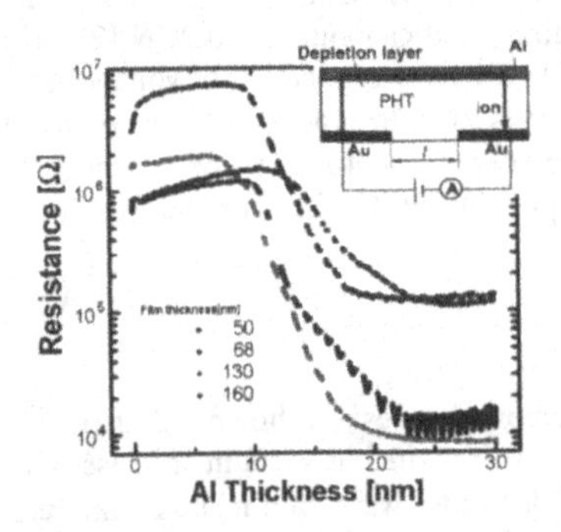

Fig.2 Resistance change of the cell (A) as the function of Al thickness.

The light of 550 nm with the intensity of 50 μW/cm^2 was illuminated onto the top of cell. The action spectrum of the photocurrent showed similar to the absorption spectra peaking at around 550 nm. Curves in Fig.3 show the time response of the current for the cell (B). The current increased slowly by the light illumination with the time constant of approximately 100 s and saturated in 1,000-2,000 s. The photo-induced current lasted for long time after turn off the light illumination as shown in Fig.3. The life time of memory effect typically consisted with two components of approximately 20 s and 13 hrs. The similar photo-induced memory effect was also observed in the cell (A). The characteristic of memory effect, such as the magnitude of photocurrent and the time constant depended on samples.

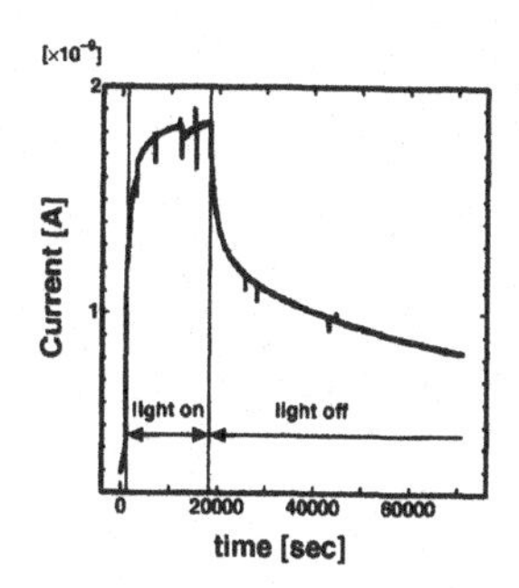

Fig.3 Typical photoresponse of the current upon light on and off in the cell (B).

Figure 4 (a) shows time response of the current as the parameter of applied bias. The *I-V* characteristic of the cells in the dark were linear and had ohmic behavior. Photo-current and memory current were defined by the current at 60 s after turn on the light and 120 s after turn off the light as shown in Fig.4 (a) at this stage of experiment. The bias dependencies of photo-current and memory-current are shown in Fig.4 (b), indicating approximately the linear dependence.

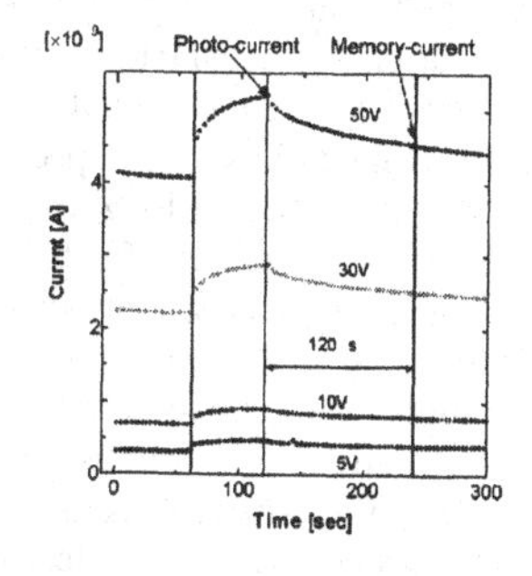

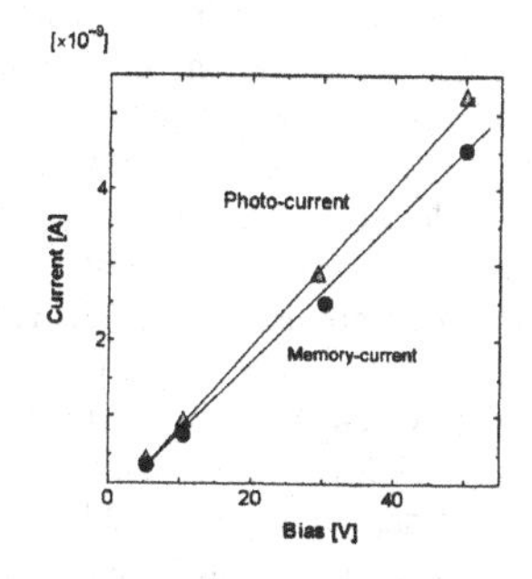

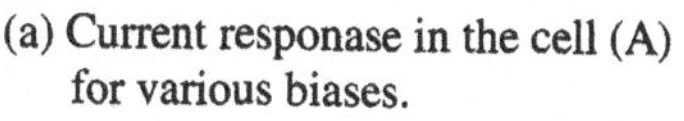

(a) Current responase in the cell (A) for various biases.

(b) Bias dependence of photo-current and memory-current in the cell (A).

Fig.4 Bias dependencies of photo-current and memory current, which are defined in (a).

It was found that the decay of memory-current was not affected by disconnection of bias and the polarity of bias. The facts indicate that the memory state did not disturved by the change of bias below of 20 kV/cm. The light intensity dependencies of photo-current and memory-current were measured and the results are shown in Fig.5 (a) and (b). It is interesting to note that the built up response of photocurrent varied by the light intensity. And also the magnitude of memory current tended to saturate at higher light intensity, even weak light intensity of 10 $\mu W/cm^2$.

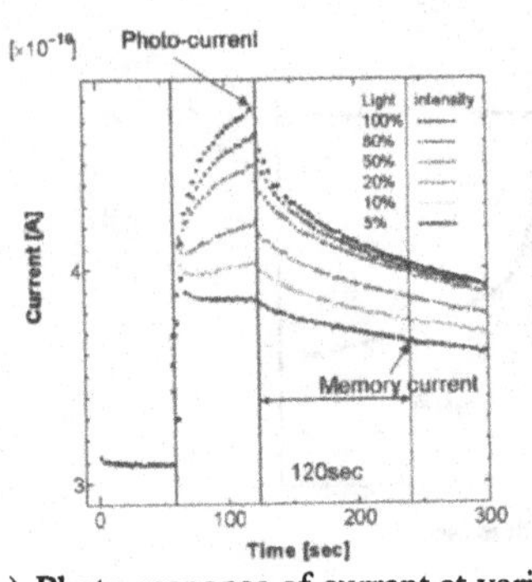

(a) Photo-response of current at various light intensity.

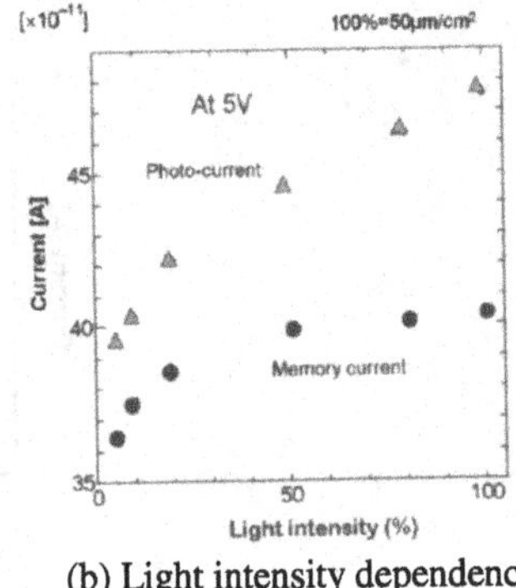

(b) Light intensity dependence of Photo-current and memory current.

Fig.5 Current responses of the cell (A) as the function of light intensity.

The mechanism of photo induced memory effect shown here could be accounted by taking the depletion layer into consideration. Due to the smaller work function of Al compared to PHT, electrons transfer from Al to acceptors in PHT, resulting in loose of conductivity and formation of depletion layer. The origins of acceptors are associated with the trace of catalyst or oxygen. Hence the internal field are built between Al particle and PHT in the depletion layer as shown in Fig.6(a). Upon light illumination, the photocarrier generation was satisfactorily explained by the dissociation of excitons with the assistance of the internal field in the photovoltaic cell [5].

The photo carriers of electrons and holes migrate to Al and bulk sides, respectively, by the internal field. If electrons, which are trapped at the Al particles, stay for long time, the excess holes also remain and give rise to persistent photoconduction or memory effect. This idea shown by Fig.6 (b) is one of possible mechanisms to explain the photo induced memory effect. The resistance of depletion layer must be large enough for the long term memory. This may be supported by the evidence that the on/off current ratio for forward and reversed bias in Al/PHT/Au diode exceeded more than 10^5 [5]. The on/off ratio roughly corresponds to the resistance ratio of depletion layer and bulk regions of PHT film. The large resistance of depletion layer isolates charges trapped at Al particles from the bulk region and electrodes. There could be the other possible mechanisms besides the present model.

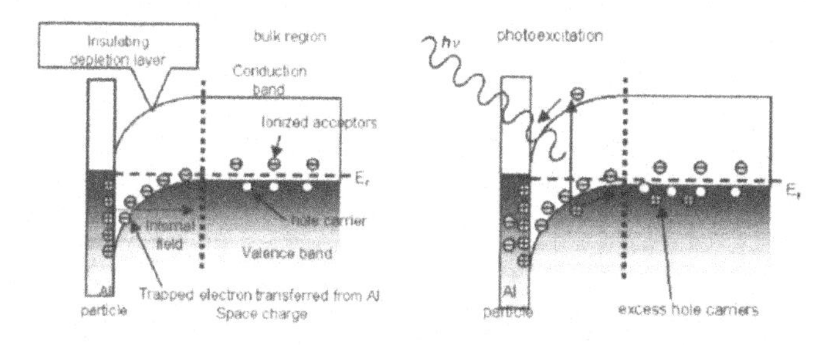

(a) Interface of Al and PHT. (b) Excess carrier generation by light illumination.

Fig.6 Mechnisms of photo induced memory effect.

CONCLUSION

The organic memory devices using conducting polymer, PHT and reducing particles of Al are mentioned. The trapped charges at Al particles play the important role for memory effect. Further investigations to clarify the detailed mechanisms and to improve the durability of memory effect are under study. It is also under consideration to write and erase the memory by electronic means.

REFERENCES

1. S. Muller, C. Parlor, W. Jackson, C. Taussig, S.R. Forrest, *Nature* **426** 166 (2003).
2. L. Ma, S. Pyo, J. Quyang, Q. Xu and Y. Yang, *Appl. Phys. Letts.*, **82** 1419 (2003).
3. K. Kaneto, K. Takayama, W. Takashima, T. Endo and M. Rikukawa, *Jpn. J. Appl. Phys.*, **41** 675 (2002).
4. Koichi Rikitake, D.Tanimura, Wataru Takashima, Keiichi Kaneto, *Jpn.J. Appl.Phys.* **42** 5561 (2003).
5. K. Kaneto and W. Takashima, *Curr. Appl. Phys.*, 1 355 (2001).

Organic and Hybrid Light Emitting Devices

Mater. Res. Soc. Symp. Proc. Vol. 846 © 2005 Materials Research Society DD7.3

Large Area Microcontact Printing Presses for Plastic Electronics

Hee Hyun Lee[1,2], Etienne Menard[1], Nancy G. Tassi[2], John A. Rogers[1] and Graciela B. Blanchet [2]
[1]University of Illinois at Urbana-Champaign, Department of Materials Science and Engineering, Urbana, IL 61801, U.S.A.
[2]DuPont, Central Research, Wilmington, DE 19880, U.S.A.

ABSTRACT

Low cost fabrication is key to the successful introduction of organic electronics and roll to roll manufacturing processes. We propose here that extending flexography into the micron size resolution regime may provide an economical commercialization path for plastic devices. Flexography is a high-speed technique commonly used for printing onto very large area flexible substrates.[1] Although low resolution and poor registration are characteristics of today's flexographic process, it has many similarities with soft lithographic techniques. This work shows that large, (12"x12") high-resolution printing plates appropriate for use on small tag and label flexographic presses can be prepared using simple and inexpensive flexographic compatible processes. We illustrate the use of these plates for three representative soft lithographic processes: microcontact printing, replica molding, and phase shift lithography.

INTRODUCTION

Microcontact printing (μCP) [2,3], a representative soft lithographic technique, has a demonstrated ability to print sub-micron features of Au and other metals, typically over small areas. It is well suited for the fabrication of conducting layers in high performance electronic devices such as backplanes for electrophoretic displays.[4] Numerous groups around the world have demonstrated that several techniques are viable for printing electronic devices[5-7]. Source/drain levels of thin film transistors (TFT) with small channel lengths have been printed using μCP [4], thermal [5], ink jet printing [6] and photolithography [7]. However, commercializing high performance flexible displays manufactured in a roll to roll process, would perhaps entail fabricating large area μCP plates and adopting a *flexography-like* process.[1] A starting point in such ambitious path, and the heart of this work, is learning how to produce flexible plastic printing plates that offer high resolution over large areas at a reasonable cost. In this work, we demonstrate that 12" x 12" polydimethylsiloxane (PDMS) stamps with micron size features can be fabricated via a simple procedure. The size was chosen to fit the plate size requirements of small commercial tags and labels flexographic presses. Stamps were used to print micron-size Au lines onto 12" x 12" Mylar substrates using standard thiol chemistry. We show also that these same stamps can be used as molds for replica molding of photopolymers and as phase masks for exposing photoresist layers.[8]

EXPERIMENTAL DETAILS

The large PDMS stamps were constructed via a straightforward and inexpensive process. A standard photoresist (PR) coating on 12" x 14" kapton sheet (thickness = 125μm) was used to create relief structure in a master. The PR was exposed through a glass photomask using a standard exposure unit which is nearly collimated over a 30" x 40" area and commonly available in printing shops. Standard PR developing chemistry followed exposure. A PR master with 1~10 micron wide lines and spaces (L/S) was then used to produce the 12" x 12" PDMS stamps with Mylar backing on vacuum plate. Pulling vacuum at the edges of the slightly oversized Mylar film causes the elastomer to spread evenly over the whole area of the master. (Figure 1) The stamps were inked using standard thiol chemistry and micron size Au lines were then printed onto a Mylar substrate.

Atomic force microscopy (AFM) images of the developed PR on Kapton master are shown in Figures 2 (a). These stamps can be use also as molds for creating surface relief structures in photopolymers. In this case, a photopolymer (Norland optical adhesive (NOA 73), Norland Products Inc.) is meyer bar coated onto a 12"x14" glass sheet (Corning 1737, thickness = 0.7mm) with thickness of 25μm and a mold was placed on top. After curing the laminated NOA 73 for 1hr with the same exposure unit used for the photolithography, the mold was delaminated yielding replica relief features in the NOA surface. AFM analysis is shown in Figure 2 (b). Similar measurements on the printed Au lines are shown in Figure 2 (c).

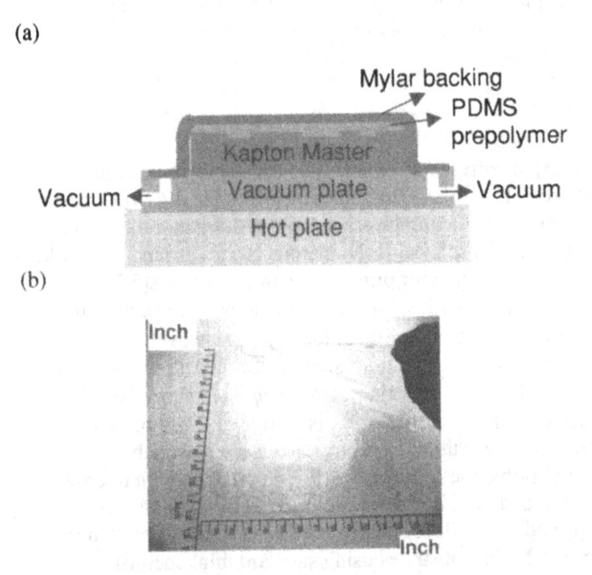

Figure 1. (a) Schematic illustration of the vacuum plate setup used to fabricate 12"x12" PDMS stamps with Mylar backing layers; (b) 12"x12" stamp after separation from the master.

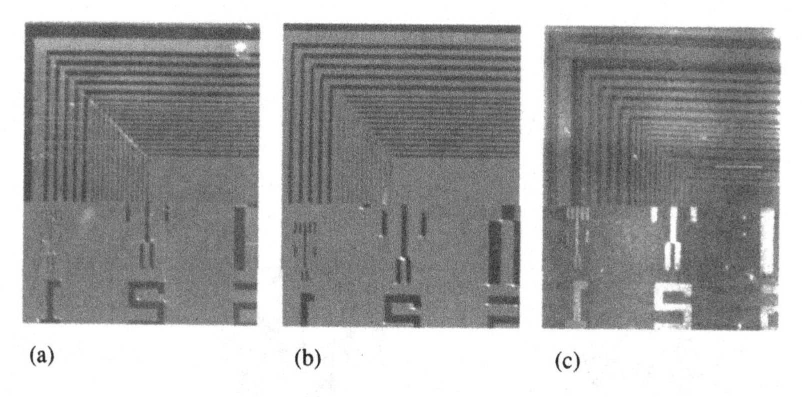

Figure 2. (a) AFM images of 1, 2, and 5 micron line and spacing of photoresist on kapton that served as a master; (b) AFM images of 1, 2, and 5 micron line and spacing of a photopolymer molded using a PDMS mold generated from this master; (c) 1, 2, and 5 micron line and spacing of etched Au after microcontact printing with a PDMS stamp generated from this master.

In addition, the stamps are used for phase masks that can be used to produce sub-micron features of photoresist via near field exposure.[8] In this case, a 12"x12" plastic photomask on 10 mil thick Mylar sheet having half mil lines and spacings throughout was purchased from Advance Reproductions Corp. A 12"x12" PR layer was coated on 12"x14" glass sheet (Corning 1737, thickness = 0.7mm) using a Meyer rod (measured thickness of 220nm), followed by a prebake at 115 °C for 2min. The developed PR was used to fabricate the PDMS stamp with 12.5 microns lines and spaces. The adhesion of photoresist to the 12" x 14" glass plate was improved by priming the glass with hexamethyl-disilazane (HMDS). The patterned stamps were placed in contact with the unexposed PR coated glass (thickness of 500nm) and exposed for 5 sec in the ultraviolet exposure system. After post-baking at 115°C for 2min, exposed regions were developed with MF-319 developer for 30sec (Figure 3).

Deformations in the stamps can cause difficulties in alignment and registration. To quantify these deformations, the positions of features on the stamp were compared to those on the PR master.[9] Although absolute distortions are important, relative distortions between multiple imprints of the same pattern are enough to prove the feasibility of using μCP technique for printing multilayer structures. The measurement begins by placing displacing the PDMS stamp and the PR master by a slight amount. After relaminating the stamp against the master, the offsets between the positions of 36 features of the master and of stamp were recorded/measured. Figure 4 shows distortions that correspond to measurements of the positions of features on the stamp relative to those on its master. The displacement offset vectors were measured, looking through the stamp, with an optical microscope equipped with a high resolution CCD camera. The patterns on the master consisted of an array of 6x6 patterns equally spaced across the 12"x12" master surface area.

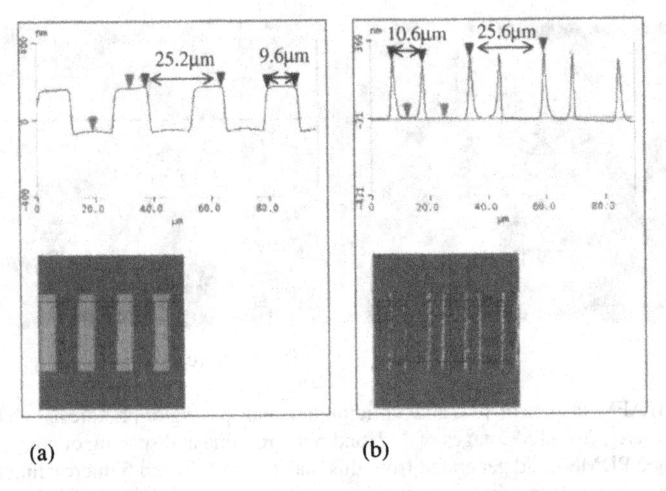

Figure 3. (a) AFM image of photoresist on a 12"x14" glass master made using a 12"x12" plastic photomask; (b) AFM image of photoresist on a 12"x14" glass plate patterned using a PDMS phase mask generated from the structure shown in (a).

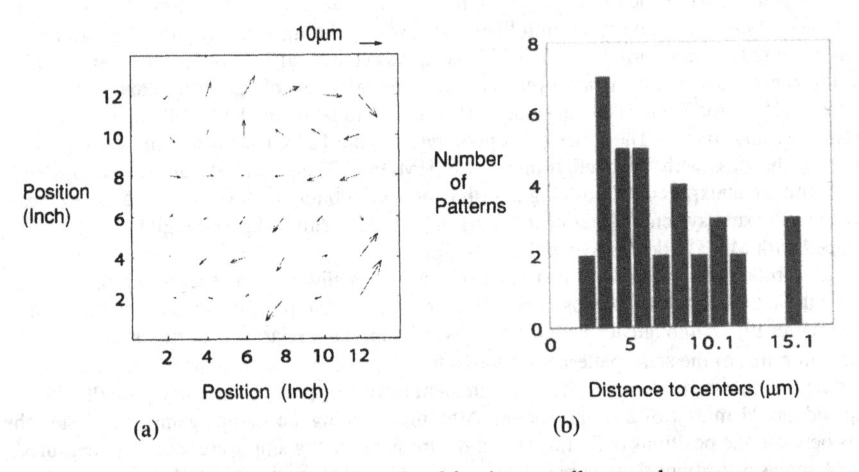

Figure 4. Distortion measurements collected at 36 points equally spaced across a 12x12 inches stamp supported by a 250 μm thick Mylar backing. (a) A vector diagram of misalignments between the stamp and its master. (Overall translational and rotational misalignments are subtracted.) (b) A histogram plot of the lengths of the vectors illustrated in the top frame.

The stamp used for this experiment consisted of a ~ 50μm thick PDMS layer supported by a 10mil thick Mylar sheet. These residual distortions results include corrections for overall translational and rotational misalignment and isotropic shrinkage (rms = 22 ppm (6.5μm over 30cm) for a stamp cured using vacuum plate at 67 °C hotplate). The algorithm used to perform these corrections was similar to the one used by E. Menard *et al.* in the case of thin PDMS stamps backed with 2 Kapton layers.[9] The residual distortions have a median value of 6.5 microns. The estimated accuracy of the measurement method is estimated to be ± 1.5 microns.

DISCUSSION

One of the factors that may facilitate the successful introduction of plastic electronics would be to provide a clear differentiation between any new approach being brought into the marketplace and the well-established Si technology. Positioning plastic electronics as technologically capable of producing electronic devices over large areas, with high throughput, onto flexible substrates and perhaps with a roll to roll printing process would clearly emphasize the distinction.

Our attempt with this work was to support this avenue by exploring whether μCP could be the starting point for the development of μ-flexography. The μCP plates described here are micro-molded. While some flexo plates are made out of molded rubber materials, the majority are photopolymer base systems.[1] Our hope is to move forward evaluating whether a photopolymer plate in which exposure is followed by wet develop is more desirable than developing a master which is then used to mold a plate. The main goal is to demonstrate the manufacturability μ-flexo plates with a 30X increase in resolution relative to the flexo-plates currently available. The focus here was to demonstrate that this resolution could in fact be achieved, over significantly large areas and to demonstrate that the designed plates are compatible with standard tag and label flexographic presses.

CONCLUSIONS

μCP has been widely used in organic electronics to print high- resolution features over fairly small areas used as source/drain levels for thin film transistors. This work represents an attempt to demonstrate that μCP maybe a viable path to manufacture large area plates with micron size resolution. This path may perhaps lead to the development of μ-flexography on press. The process used to produce 12" x 12" PDMS plates is simple, inexpensive and scalable to 30" x 40", the most common plates size in the flexographic market. The master was fabricated with a commercial positive photoresist that was imaged to produce 1-micron features over large areas using a nearly collimated UV exposure system standard in the proofing industry. Similarly, the manufacturing of the large 12" x 12" stamp was achieved in our laboratory version using vacuum plate.

ACKNOWLEDGMENTS

The authors thanks Adrian Lungus and Lyla El-Sayed in Dupont Cyrel for useful and stimulating discussions. This work was supported by the Grainger foundation.

REFERENCES

1. http://www.linkisd.org.uk/PDF/20.flexolithographic_print.pdf
2. A. Kumar and G. Whitesides, *Appl. Phys. Lett.* **63**, 2002 (1993)
3. Y. Xia and G. Whitesides, *Annu. Rev. Mater. Sci.* **28**, 153 (1998)
4. J. A. Rogers, Z. Bao, K. Balwin, A. Dodabalapur, B. Crone, V. R. Raju, V. Kuck, H. Katz, K. Amundson and P. Drzaic, PNAS, 98, 4835 (2001)
5. G. B. Blanchet, Y-L Loo, J. A. Rogers, C. R. Fincher and F. Gao, *Appl. Phys. Lett.* **82**, 463 (2003).
6. H. Sirringhaus, T. Kaswase,R. H. Friend,T. Shimoda, M. Inbasekaran, W. Wu, E. P. Woo, *Science*, **290**, 2123 (2000)
7. C. J. Drury, C. M. Mutsaers, C. M. Hart, M. Halters and D. M. de Leeuw, *Appl. Phys. Lett.* **73**, 108 (1998)
8. J. A. Rogers, K. Paul, R. Jackman and G. Whitesides *Appl. Phys. Lett.* **70**, 2658 (1997)
9. E. Menard, L. Bilbaut, J. Zaumseil, and J. A. Rogers *Langmuir* **20**, 6871 (2004).

Mater. Res. Soc. Symp. Proc. Vol. 846 © 2005 Materials Research Society DD7.7

New Light Emitting Polymers and High Energy Hosts for Triplet Emission

Chris S. K. Mak[1], Scott E. Watkins[1,2], Charlotte K. Williams[3], Nicholas R. Evans[1], Khai Leok Chan[1], Sung Yong Cho[1], Andrew B. Holmes[1,2,3], Clare E. Boothby[4], Richard H. Friend[4], Anna Hayer[4] and Anna Köhler[4]
[1]Melville Laboratory, Department of Chemistry, University of Cambridge, Lensfield Road, Cambridge CB2 1EW, UK
[2]Bio21 Institute, University of Melbourne, Parkville, Vic. 3010, Australia
[3]Department of Chemistry, Imperial College, South Kensington, London SW7 2AZ, UK
[4]Cavendish Laboratory, Department of Physics, University of Cambridge, Madingley Road, Cambridge CB3 0HE, UK

ABSTRACT

This paper describes two aspects of research aimed at harnessing the triplet energy generated in electron-hole recombination in polymer electroluminescent devices. The purpose is to design solution-processible phosphorescent organometallic triplet emitters and to design high triplet energy polymer hosts that can transfer triplet energy to the phosphorescent guests. The method employed Suzuki cross coupling reactions to incorporate either phosphorescent cores or high energy triplet monomers covalently into polymer hosts to evaluate their optoelectronic properties. The results showed (i) efficient energy transfer from polyfluorene hosts to red phosphorescent guests and (ii) that pyridine and carbazole monomers could raise triplet energies of hosts. It is concluded that these approaches offer promise in the design of solution processible electrophosphorescent materials for red and green light emitting devices.

INTRODUCTION

The design and synthesis of organic light emitting materials and their evaluation in OLEDs has been a subject of intense research for the past two decades [1]. Many new light-emitting materials have been studied with a focus on improving device efficiencies and lifetimes. The major sources of emissive materials belong respectively to families of organometallic compounds, conjugated polymers and dendritic compounds. Following the pioneering work of King [2] and subsequently of the USC and Princeton groups [3] d^6-cyclometalated complexes have attracted considerable interest as phosphorescent emitters in OLEDs. A device incorporating the green emitter iridium (III) *tris*(2-phenylpyridinato-N,C) [$Ir(ppy)_3$] in the emissive layer was capable of very high efficiency owing to the fact that all the singlets and triplets generated by electron-hole recombination could be harnessed through phosphorescence. Other important contributions in the area of red Ir emitters have been reported by Tsuboyama *et al.* [4], and Tokito has described a high-efficiency white phosphorescent device using bis{2-[3,5-bis(trifluoromethyl)phenyl]-pyridinato-N,$C^{2'}$}iridium(III)] and red-emitting cyclometalated Ir complexes, where an external quantum efficiency of 12 % was reported [5]. Forrest and Thompson have fabricated a deep blue device by the combination of bis(4′,6′-difluorophenylpyridinato)tetrakis(1-pyrazolyl)borate with diphenyldi(o-tolyl)silane and *p*-bis(triphenylsilyl)benzene as high energy hosts [6]. The present work focuses on solution processible materials. As has already been demonstrated with singlet emissive devices, impressive results have been obtained with ink-jet deposition of conjugated polymers. Such an approach using solution processible electrophosphorescent materials in combination with conjugated polymer hosts, either as blends or covalently linked, was the focus of the present investigation. In the case of covalently linked electrophosphorescent polymers the organometallic emissive complex may be conjugatively linked to the polymer host through an aromatic or heteroaromatic ligand in conjugation (in the present work) with a polyfluorene host. Alternatively the organometallic complex may be tethered to the polyfluorene host by a linking chain. While this link provides no conjugation between the two materials it may have the advantage of spatial control in placing the organometallic complex on average within the optimum radius of the conjugated host for efficient Dexter transfer of the triplet excited state from the higher energy conjugated polymer host to the phosphorescent emitter. It is also possible that triplet state emission may be enhanced by direct electron hole recombination on the phosphorescent dopants as a result of charge trapping induced at the organometallic complex. At the outset of this work these questions had not been satisfactorily answered. A second aspect of the research was to design higher energy triplet hosts than polyfluorene for energy transfer to green phosphorescent hosts. During the course of this work some answers have emerged from other research groups. Several groups have reported phosphorescent PLEDs using poly(vinylcarbazole) (PVK) as the host polymer. Chen has studied Ir complexes with different triplet energies (from green to red) doped in poly(vinylcarbazole) (PVK) and polyfluorene hosts [7]. Devices fabricated with PVK as the polymer host exhibited a higher device efficiency than those using polyfluorene, which has poor host to guest energy transfer and triplet confinement as a result of the low triplet energy level. Many examples of solution processible electrophosphorescent emitters blended in various hosts have been described, but with the exception of Chen's work already mentioned above no significant results had been reported for polymeric covalently linked materials [8]. In this paper we report examples of red emissive iridium complexes covalently and conjugatively linked to polyfluorene hosts. We then describe approaches to the synthesis of potential high energy triplet host conjugated polymers for application in blue light emitting electrophosphorescent devices.

EXPERIMENTAL DETAILS

^{1}H and ^{13}C NMR spectra were recorded on Bruker DPX-400 (400MHz and 100MHz respectively) and Bruker DRX-500 (500 MHz and 125MHz respectively) instruments. IR spectra were recorded on a Perkin Elmer Spectrum RX I FT-IR spectrometer. GPC was preformed in $CHCl_3$ in a PL gel Mixed B column and with a differential refractometer, at a flow rate of 1 mL/min. UV-visible spectra were measured on a Hewlett-Packard 8452 diode array spectrophotometer. Photoluminescence spectra were recorded on an Aminco-Bowman Series 2 luminescence spectrometer. For photoluminescence measurements, the samples were held under vacuum in a continuous-flow helium cryostat. The temperature was controlled with an Oxford-Intelligent Temperature Controller (ITC 502). The samples were excited with the UV lines (355 and 365 nm) of an Argon ion continuous-wave laser. The emitted light was collected with an optical fiber and recorded by an Oriel spectrograph coupled to a cooled CCD array (Oriel Instaspec IV). Photoluminescence quantum yields were measured under a nitrogen atmosphere using the integrating sphere technique.

Light-emitting diodes were fabricated on indium-tin oxide (ITO) patterned glass substrates which were cleaned in an ultrasonic bath of acetone followed by isopropanol prior to oxygen plasma etching for ten minutes. A conventional photo-resist spin-coater was used to deposit first a 50 nm layer of filtered PEDOT:PSS [poly(3,4-ethylenedioxythiophene):poly(styrenesulfonic acid)] followed by a 70 nm layer of oligomer or polymer. The PEDOT:PSS layer was dried under nitrogen at 120 °C for 1 h to remove any residual water prior to spinning on the emitting layer. A 100 nm calcium cathode was evaporated onto the device through a shadow mask and capped with aluminum. Current-voltage characteristics were measured using a Keithley 230 V source and a 195 DMM and luminance using a silicon photodiode and Keithley 2000 multimeter. Electroluminescence spectra were taken with an optical fiber coupled to the Instaspec IV described above. All electrical measurements were carried out under vacuum. Synthesis of the polymeric organometallic complexes has been reported in an earlier publication [8].

DISCUSSION

Polymers based on polyfluorene and iridium emitters

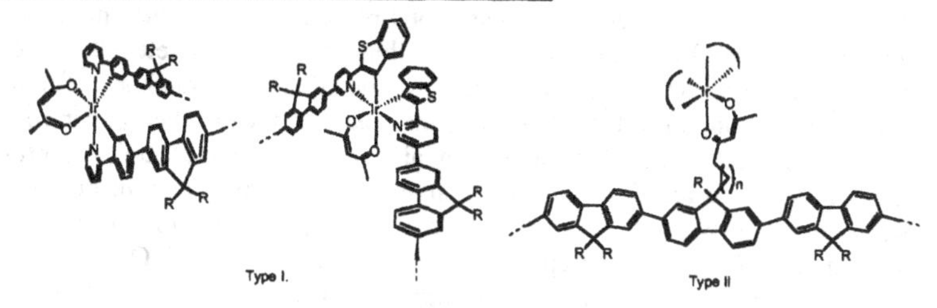

Figure 1. Type I and Type II phosphorescent copolymers.

Phosphorescent emitters based on iridium complexes were incorporated into the polymer chain either through conjugative linking through the ligands (Type I) or by tethering through a polymethylene linker attached to the 9-position of the fluorene (Type II) (Figure 1). The conjugatively linked materials were prepared by two methods. The first involved Suzuki statistical copolymerization of a fluorene-2,7-bisboronate with a molar equivalent of a mixture of the corresponding 2,7-dibromofluorene and up to 4 mol% of the dibromo-substituted iridium complex. The statistical incorporation of the iridium emitters by this method was less satisfactory

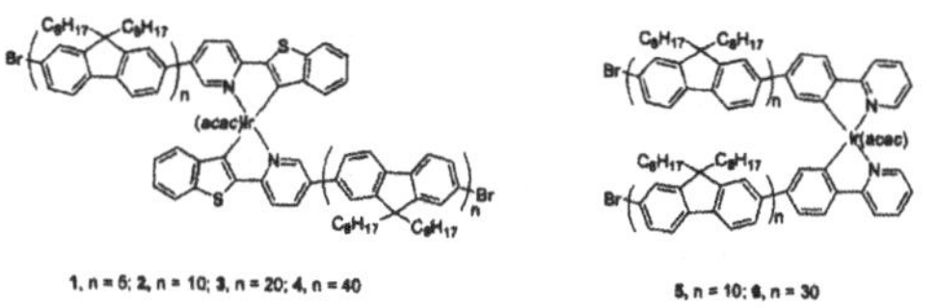

1, n = 5; 2, n = 10; 3, n = 20; 4, n = 40

5, n = 10; 6, n = 30

Figure 2. The iridium phosphorescent copolymers **1-6** used in the present study.

and less controlled than the alternative method. This employed an AB fluorene monomer (2-bromo-7-boronate) to chain extend the dibromo functionalized iridium complexes simultaneously in two directions. This resulted in well-defined polymeric materials with reproducible levels of incorporation of iridium complexes in the chain. The synthesis of these materials has already been reported [8]. The iridium complexes carried two 2-phenylpyridinato (*ppy*) or 2-(2′-benzo[*b*]thienyl)pyridinato (*btp*) ligands bearing suitable bromo-substituents for polymerization, and charge balance was complemented with a pentane-2,4-dionate (*acac*) ligand. Ouyr study has been limited to the use of *acac* ligands which may be incompatible with the requirements of a long lifetime electrophosphorescent device. Eventually a selection of alternative tris-cyclometalated complexes will have to be made to avoid the potentially labile *acac* ligands where device lifetimes are reportedly decreased in the presence of the hole transporting PEDOT/PSS [9].

A summary of the results that have recently been reported in full [8] now follows. The polymers that yielded the best results and the most information were the *btp* (red emitting) family (**1-4**) and the analogous *ppy* (yellow-green emitting) family (**5, 6**). Extension of the ligands *btp* and *ppy* by controlled Suzuki coupling reactions with fluorene monomers and formation of coordination complexes based on Ir(III) afforded materials that were shown to be readily spin-coated from solution. Photoluminescence studies showed that mixing of the triplet levels of the fluorene and cyclometalating ligand took place to give a hybrid mixed triplet state as the lowest energy level. Emission occurred from this state. The wavelength could be tuned from red to green by selection of the cyclometalating ligand and the number of fluorene repeat units. Triplet emission dominated the electroluminescence behavior, even under conditions of low iridium loadings, indicating that charge trapping at the metal centre was probably the dominant mechanism. The red devices showed significant improvements in device efficiency over polyfluorene on its own. This indicated that the design of the triplet energy of the host polyfluorene and the Ir(*btp*)$_2$(*acac*) acceptor complex had been optimized. On the other hand the green devices, while more efficient than mere blends of iridium-complex with a polyfluorene host, were not optimal. This is due to the fact that the triplet energy of the polyfluorene host was insufficiently above that of the Ir(*ppp*)$_2$(*acac*) emitter to guarantee complete energy transfer [8]. The new solution-processible triplet emitters may be optimized by the design of alternative iridium complexes and high energy triplet hosts. Approaches to the latter problem are described below.

High energy host polymers for energy transfer to blue-green emitters

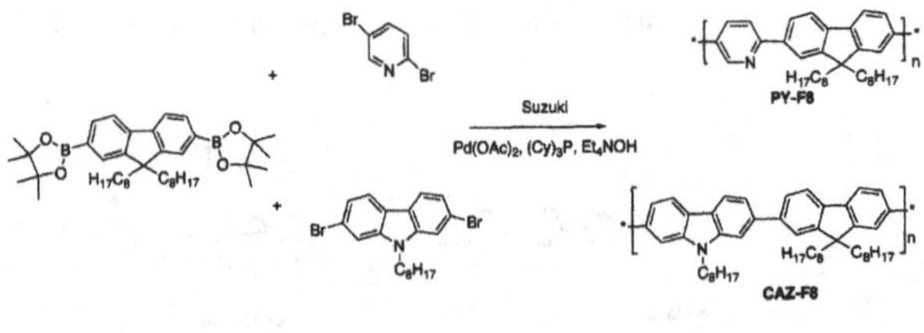

Figure 3. High triplet energy pyridyl (**PY-F8**) and carbazole (**CAZ-F8**) copolymers.

The goal of the present work was to raise the triplet energy of polyfluorene (2.1 eV) by preparing alternating copolymers of fluorene with monomers such as 4-hexylpyridine and 9-octylcarbazole [10] that have a high triplet energy (2.5 eV for polypyridine). This would provide the required high energy hosts for efficient energy transfer to blue phosphorescent organometallic emitters. The Suzuki copolymerization of 9,9-dioctylfluorene-2,7-bisboronate with 2,5-dibromo-4-hexylpyridine and 3,7-dibromo-9-octylcarbazole afforded respectively the copolymers **PY-F8** and **CAZ-F8** (Figure 3).

The molar mass M_n and photophysical data of the copolymers are listed in Table 1 and absorption and PL spectra of thin films are shown in Figure 4.

Polymer	M_n	Film absorption, λ_{max} (nm)	Film emission, λ_{em} (nm)	Optical band gap (eV)
PY-F8	17000	214, 355	407	3.12
CAZ-F8	6700	201, 239, 348	412	3.08

Table 1. Physical properties of the copolymers **PY-F8** and **CAZ-F8**.

The two copolymers, **PY-F8** and **CAZ-F8** exhibit a single absorption band with absorption maxima located at 355 nm and 348 nm respectively, which correspond to $\pi\pi\pi^*$ transitions of the aromatic rings. The shoulder peak observed at 320 nm of **CAZ-F8**, but not in **PY-F8**, originates from the ---* transitions based on the carbazole units. The absorption bands are more structured compared with pyridine-containing copolymers. Vibronic features 436 nm and 468 nm were observed for **CAZ-F8**. The PL spectra were recorded for the freshly synthesized copolymer films. The PL maxima of **PY-F8** (407 nm) and **CAZ-F8** (412 nm) are blue shifted compared with the polyfluorene (422 nm). This is due to the conjugation being interrupted by the octylpyridine or the carbazole units in the polymer chain. The vibronic structure of fluorene moieties at 426 nm is also observed. The Stokes shift of **PY-F8** is 52 nm, which is smaller than that of **CAZ-F8** (64 nm) (see Figure 4).

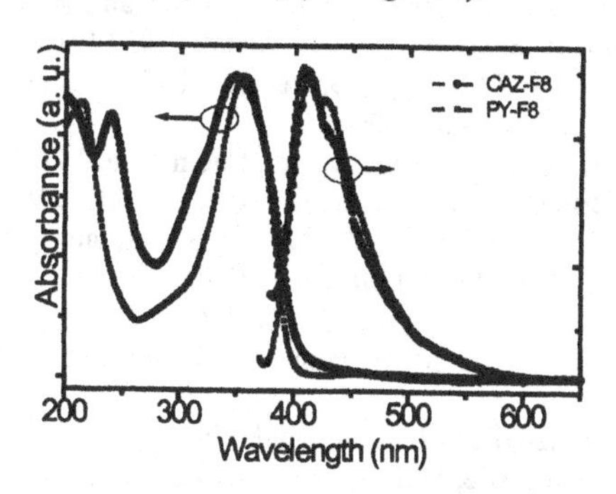

Figure 4. Normalized absorption and PL spectra of copolymers **PY-F8** and **CAZ-F8**.

Experiments are in hand to measure the triplet energies of the new polymers. However a reasonable estimate of the triplet energy may be made by referring to the extensive studies reported by Köhler and Beljonne that show it to be 0.7 eV below the corresponding singlet levels [11]. Preliminary results with blends of $Ir(ppy)_3$ imply that these materials have potential to transfer triplet energy to green phosphorescent hosts in electrophosphorescent devices.

CONCLUSIONS

In this paper we have described improved electroluminescent devices fabricated from iridium phosphorescent complexes conjugatively linked to polyfluorene hosts and the design of two potential high triplet energy alternating copolymers of fluorene with pyridine and carbazole repeat units respectively.

ACKNOWLEDGMENTS

We thank the Engineering and Physical Sciences Research Council (UK) for financial support and the provision of the Swansea National Mass Spectrometry Service. We thank the Royal Society (University Research Fellowship to A.K.), the European Commission ('LAMINATE', 'STEPLED'), the Croucher Foundation (Postdoctoral Fellowship to C.S.K.M.), the Commonwealth Scholarship Commission, the Cambridge Commonwealth Trust and the Ramsay Memorial Trust (scholarships to N.R.E.) for generous financial support. We acknowledge support from the Australian Research Council, CSIRO (Fellowship to S.E.W.) and the Victorian Endowment for Science Knowledge Initiative (Fellowship to A.B.H.).

REFERENCES

1. J. H. Burroughes, D. D. C. Bradley, A. R. Brown, R. N. Marks, K. Mackay, R. H. Friend, P. L. Burn and A. B. Holmes, Nature **347**, 539 (1990); A. Kraft, A. C. Grimsdale and A. B. Holmes, Angew. Chem. Int. Ed. Engl. **37**, 402 (1998); R. H. Friend, R. W. Gymer, A. B. Holmes, J. H. Burroughes, R. N. Marks, C. Taliani, D. D. C. Bradley, D. A. Dos Santos, J. L. Brédas, M. Lögdlund and W. R. Salaneck, Nature **397**, 121 (1999); J. L. Segura, Acta Polym. **49**, 319 (1998); M. T. Bernius, M. Inbasekaran, J. O'Brien and W. Wu, Adv. Mater. **12**, 1737 (2000); U. Mitschke and P. Bäuerle, J. Mater. Chem. **10**, 1471 (2000); U. Scherf and E. J. W. List, Adv. Mater. 2 **14**, 477 (2002); I. D. Rees, K. L. Robinson, A. B. Holmes, R. O'Dell and C. R. Towns, MRS Bull. **27**, 451 (2002).
2. K. A. King, P. J. Spellane and R. J. Watts, J. Am. Chem. Soc. **107**, 1431 (1985).
3. M. A. Baldo, D. F. O'Brien, Y. You, A. Shoustikov, S. Sibley, M. E. Thompson and S. R. Forrest, Nature **395**, 151 (1998); M. A. Baldo, S. Lamansky, P. E. Burrows, M. E. Thompson, S. R. Forrest, Appl. Phys. Lett. **75**, 4 (1999).
4. A. Tsuboyama, H. Iwawaki, M. Furugori, T. Mukaide, J. Kamatani, S. Igawa, T. Moriyama, S. Miura, T. Takiguchi, S. Okada, M. Hoshino and K. Ueno, J. Am. Chem. Soc. **125**, 12971 (2003).
5. S. Tokito, T. Iijima, T. Tsuzuki and F. Sato, Appl. Phys. Lett. **83**, 2459 (2003).
6. R. J. Holmes, B. W. D'Andrade, S. R. Forrest, X. Ren, J. Li and M. E. Thompson, *Appl. Phys. Lett.* **83**, 3818 (2003).
7. X. Chen, J.-L. Liao, Y. Liang, M. O. T. Ahmed, H.-E. Seng and S.-A. Chen, J. Am. Chem. Soc. **125**, 636 (2003).
8. A. J. Sandee, C. K. Williams, N. R. Evans, J. E. Davies, C. E. Boothby, A. Köhler, R. H. Friend and A. B. Holmes, J. Am. Chem. Soc. **126**, 7041 (2004).
9. A. van Dijken, A. Perro, E. A. Meulenkamp and K. Brunner, Org. Electr. **4**, 131 (2003).
10. A. van Dijken, J. J. A. M. Bastiaansen, N. M. M. Kiggen, B. M. W. Langeveld, C. Rothe, A. Monkman, I. Bach, P. Stossel and K. Brunner, J. Am. Chem. Soc. **126**, 7718 (2004).
11. A. Köhler and D. Beljonne, Adv. Funct. Mater. **14**, 11 (2004).

Mater. Res. Soc. Symp. Proc. Vol. 846 © 2005 Materials Research Society DD7.8

HIGHLY LUMINESCENT COMPOSITE FILMS FROM CORE-SHELL OXIDE NANOCRYSTALS

Valérie Buissette,[1] Mélanie Moreau,[1] Thierry Gacoin,[1] Thierry Le Mercier,[2] Jean-Pierre Boilot[1]
[1]Groupe de Chimie du Solide, Laboratoire de Physique de la Matière Condensée, CNRS UMR 7643, Ecole Polytechnique, 91128 Palaiseau cedex (France). [2] Centre de Recherches d'Aubervilliers, Rhodia, 52, rue de la Haie Coq, 93308, Aubervilliers (France).

ABSTRACT

Luminescent nanocrystals can find interesting applications for the elaboration of light emitting transparent materials. The work described here is based on the use of lanthanide doped vanadate (YVO_4:Eu) and phosphate (La,Ce,Tb)PO_4-0,7H_2O nanoparticles grown through aqueous colloidal synthesis, with average sizes below 10 nm. The well-dispersed colloids are transparent and respectively exhibit red and green luminescence under U.V. excitation with high luminescence yields (20 – 50 %). Improvement of luminescence properties of the nanocrystals is achieved through the elaboration of core/shell nanostructures, obtained after the growth at the surface of an amorphous silica shell or a crystalline lanthanum phosphate shell. Surface derivatization is further achieved through the controlled growth of an organically modified silica coating using a functionalized silane precursor. Concentrated sols are obtained, which are highly luminescent and well-dispersed. They can be spin-coated on various substrates, leading to perfectly transparent and highly luminescent thin films. Multi-layers films and heating treatments are performed, leading to optimized materials.

INTRODUCTION

Recently, the study of nanometric luminescent materials has become of great interest.[1] Indeed, microstructural characteristics of the phosphors play an important role in the efficiency of luminescent devices, and new developments as electroluminescent devices, integrated optics or biological labels[2] imply the design of new phosphors with controlled properties at the nanometric scale. The major part of the work has been focused on semiconductor nanocrystals.[3] However, a significant amount of research has been devoted to lanthanide doped nanocrystals,[4] but fewer work has focused on the synthesis of well-dispersed particles (yttrium vanadate,[5,6] lanthanum fluoride,[7] lanthanum phosphate[8] and yttrium oxide[9] nanoparticles doped with rare earth ions). The basic idea of the work presented here is to explore the potentialities of lanthanide (Ln) doped yttrium vanadate (YVO_4) and lanthanum phosphate ($LaPO_4$-xH_2O) synthesized through colloidal chemistry. This choice is explained by the excellent luminescence properties of the bulk material: red phosphor for YVO_4:Eu and green phosphor for (La,Ce,Tb)PO_4. In a first part, the synthesis process, the structural and luminescence properties of the two nanocrystalline systems are described. Then, we report on the synthesis of nanocomposite materials in which the nanocrystals are dispersed in a transparent sol-gel matrix.

EXPERIMENTAL SECTION

- <u>*Synthesis of the aqueous colloidal suspensions*</u>

The synthesis of $Y_{1-x}Eu_xVO_4$ colloids is based on the use of well-known citrate complexing agents, both to limit the growth of the particles and to ensure their stability.[5b] The whole process is carried out in water. A 0.1 $mol.L^{-1}$ solution of $(Y,Eu)(NO_3)_3$ (10 mL) is mixed with a 0.1 $mol.L^{-1}$ solution of sodium citrate (7.5 mL), followed by the addition of a 0.1 $mol.L^{-1}$ solution of Na_3VO_4 (7.5 mL), whose pH is 12.5. The clear and colorless resulting mixture is subsequently heated at 60°C for 30 minutes. Finally, the solution is cooled and dialyzed against water. The transparent colloidal suspensions can be concentrated up to 400 $g.L^{-1}$.

The aqueous colloidal synthesis of lanthanide phosphate nanoparticles $(La,Ce,Tb)PO_4\text{-}0.7H_2O$ is based on the use of a mixture of lanthanide salts and sodium tripolyphosphate TPP precursors in water.[10] Hydrolysis of TPP groups and subsequent growth of lanthanide phosphate particles are observed after aging at 90°C. The whole process is carried out in water. A 0.1 $mol.L^{-1}$ solution of $La(NO_3)_3$ (10 mL) is mixed with a 0.1 $mol.L^{-1}$ solution of sodium tripolyphosphate $Na_5P_3O_{10}$ (noted TPP) (10 mL). The clear and colorless mixture is subsequently aged at 90°C for 3h. The resulting lanthanum phosphate colloidal suspension is cooled and dialyzed for one day against deionized water to remove the excess of ions and phosphates species. This leads to a transparent colloidal solution. To increase the stability of concentrated solutions, a 0.1 M solution of sodium hexametaphosphate $(NaPO_3)_{12\text{-}13}.Na_2O$ (2 mL) is eventually added and the colloid is further dialyzed for 3 days.

- *Lanthanum phosphate coating of the particles*

The nanoparticles are coated with a crystalline lanthanum phosphate $LaPO_4\text{-}0.7H_2O$ shell. The principle is based on the lanthanum phosphate particles synthesis described above. To an aqueous colloidal suspension of either vanadate or phosphate nanoparticles (100 mL, [Ln] = 10 mM), a mixture of lanthanum chloride (5mL, [La] = 100mM) and tripolyphosphate sodium (10 mL, [TPP] = 100 mM) salts is added. The mixture is further aged at 90°C for 1 hour. A second addition of lanthanum tripolyphosphate in the same quantities is repeated, as well as the aging process. After 1 hour, the resulting suspension is dialyzed against water, providing a transparent core-shell colloidal suspension.

- *Silica coating of the particles*

The nanoparticles are coated by a functionalized alkoxide according to a process adapted from the work of Philipse in the case of boehmite nanoparticles.[11,12] In a first step, 50 mL of an aqueous solution of sodium silicate (3% SiO_2) is added to a colloidal solution of nanoparticles (10 $g.L^{-1}$). The resulting solution is left under stirring at ambient temperature for 18 hours, and then dialyzed against water for 48 hours. The colloid is then concentrated down to 100 mL and added dropwise into 300 mL of a solution of ethanol containing 5 equivalents of 3-(methacryloxylpropyl) trimethoxysilane (TPM) compared to the lanthanide. This mixture is then refluxed for 24 hours and transferred into 400 mL of n-propanol so that water can be further removed by azeotropic distillation. The grafted colloids are purified by ultra filtration against propanol.

STRUCTURAL CHARACTERIZATIONS OF THE PARTICLES

The YVO_4:Ln nanoparticles crystallize in the zircon type structure. The mean coherence length deduced from the X-rays diffraction (XRD) diagram is 8 nm, which is also the average size measured by transmission electron microscopy (TEM, figure 1a). Moreover, the hydrodynamic diameter measured by dynamic light scattering (DLS) is 9 nm with a standard deviation of 0.6. This implies that the nanoparticles are well-dispersed monocrystals. The excellent stability of the suspension is achieved through citrate complexing groups at the surface, leading to transparent colloids, up to 400 g/L.

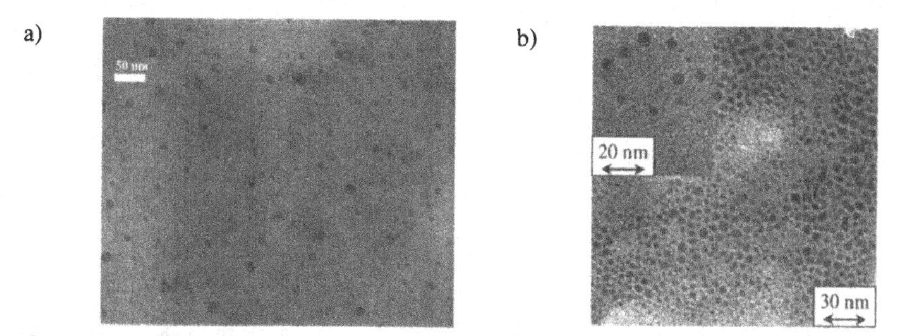

Figure 1 : TEM of (a) YVO_4:Eu and (b) (La,Ce, Tb)PO_4-0,7H_2O nanoparticles.

The phosphate nanoparticles crystallize in the rhabdophane type structure $LnPO_4$-xH_2O. The coherence length deduced from XRD is about 5 nm. The average size of the particles by TEM is 4 nm with a standard deviation of 1.3 (figure 1b). Moreover, the hydrodynamic diameter measured by DLS is 13 nm with a standard deviation of 0.6, which implies that the nanoparticles are well-dispersed monocrystals with an average size of about 5 nm. The stability of the suspension is achieved through polyphosphate complexing groups at the surface, leading to transparent colloids, up to 100 g/L.

Structural characterizations of the phosphate shell are difficult to perform. Optical characterizations through absorption measurements (4f5d absorption of Ce^{3+} in the U.V. region) show that the growing process of the shell is much faster in the presence of nanoparticles as germs than for homogeneous nucleation. Moreover, efficiency of Ce^{3+} to Tb^{3+} energy transfers, when a terbium phosphate shell is grown on cerium phosphate nanoparticles, shows that there is a good affinity between core and shell. As a consequence, the lanthanide phosphate phase is preferentially grown at the surface of the initial nanoparticles. The expected thickness of the shell is about 0.8 nm. In the case of phosphate nanoparticles, there is presumably an epitaxial growth of the shell, induced by the ripening process of the primary particles when aged in water.

Organo-silica coating of the nanoparticles is demonstrated by the stability of the grafted particles in organic media such as alcohols (propanol) and their instability in water. Thermogravimetric and elementary chemical analysis shows that the amount of silica in the suspension is about 2.6 equivalents compared to lanthanide ions (1eq. of silicate and 1.6 eq. of TPM) which corresponds to a shell about 5 nm thick at the surface. Subsequent heating process up to 600°C is possible, the size of the nanoparticles being unchanged.

LUMINESCENCE PROPERTIES OF THE PARTICLES

The transparent YVO_4:Eu colloidal suspensions exhibit strong red emission under U.V. excitation (figure 2a), the most intense peak is related to the europium $^5D_0$-$^7F_2$ transition at 617 nm. The emission mechanism involves absorption of energy through the vanadate groups of the host matrix, and further energy transfer from vanadate to europium luminescent ions. Their properties are similar to those of the bulk materials; the main differences are a lower luminescence yield, broader emission bands and a shift of the optimum content of Eu^{3+} ions. Spectroscopic differences are linked to the existence of surface sites for the emitting ions, as well as structural distortions induced by the small size of the particles.

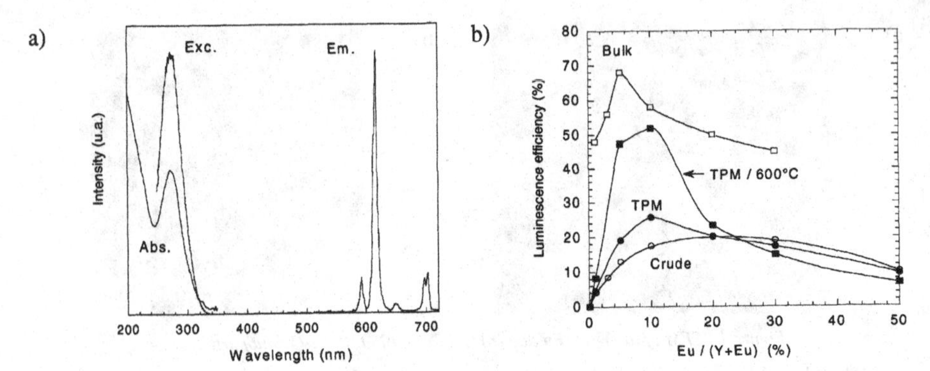

Figure 2 : (a) Absorption, excitation (λ_{em}=617 nm) and emission (λ_{exc}=280 nm) of YVO_4:$Eu_{10\%}$ colloid. (b) Luminescence efficiency as a function of Eu^{3+} content for YVO_4:Eu nanoparticles with various surface treatments and for the bulk material .Crude refers to the as synhetized particles, TPM to the grafted particles and TPM / 600°C to the grafted particles after heating treatment at 600°C.

The luminescence yield of the colloids is relatively high for aqueous suspensions (20 %) but is still lower than the bulk materials (70 %). The main quenching phenomenon is due to the presence of hydroxyls groups at the surface of the particles, which are efficient non-radiative recombination centers. Their elimination by deuteration induces an increase of the luminescence yield to 40 %. Others non-radiative pathways for excited states of the vanadate groups induce an absence of vanadate to vanadate energy transfers, which results in a poor luminescence efficiency at low europium content (figure 2b). The consequence is an apparent shift of the optimum concentration in europium doping ion in the nanocrystals (20 %) compared to the bulk (5 %). The passivation of the surface is realized through the growth of a silica shell in hydro alcoholic medium. In this case, the optimum luminescence yield reaches 25 % for an europium content of 10 %. Thus, the silica shell eliminates non-radiative centers for the vanadate groups, allowing energy transfers from vanadate to vanadate groups inside the particles, and thus high luminescence efficiency at low europium concentration. If the OH groups are also eliminated, either by transfer in deuterated water or by a subsequent heating process to 450°C, luminescence yields are about 50 % for 5-10 % europium doping.[5c]

The transparent (La,Ce,Tb)PO_4-xH_2O colloidal suspensions exhibit strong green emission under U.V. excitation (figure 3), the most intense peak is related to the terbium $^5D_4$-$^7F_5$ transition at 543 nm. The emission of Tb^{3+} and Ce^{3+} ions occurs after excitation in the 4f5d absorption bands of the cerium ions. Luminescence yields are high (50 %), which implies that the rhabdophane phase, which has not been described as a phosphor before, leads to efficient green nanophosphors. The partial quenching of the luminescence observed in aqueous colloids is essentially related to OH groups located at the surface and/or in the structure. After transfer in deuterated water, very efficient colloidal suspensions are obtained, with luminescence yields as high as 80 %. The spectroscopic characteristics of the nanoparticles are similar to those of the microcrystalline (La,Ce,Tb)PO_4 well-known phosphor. The major difference between rhabdophane nanoparticles and monazite bulk crystals is the large shift in the emission band of the cerium ions, due to the change of the crystalline phase.

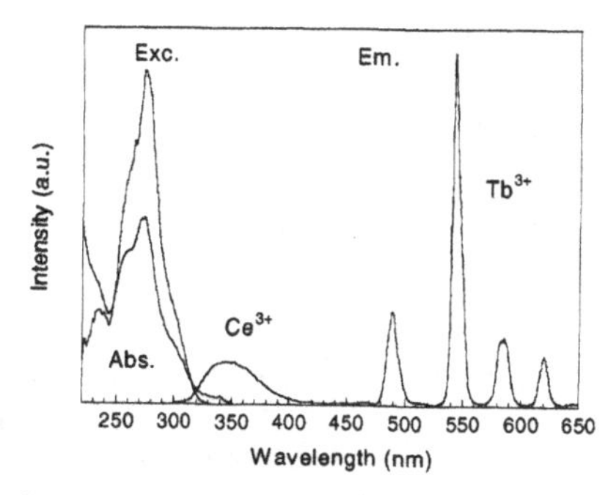

Figure 3 : Absorption, excitation (λ_{em} = 543 nm) and emission (λ_{exc} = 272 nm) spectra of (La,$Ce_{45\%}$,$Tb_{15\%}$)PO_4-xH_2O colloid in water.

As a consequence, energy transfers from Ce^{3+} to Ce^{3+} are absent in the nanoparticles, resulting in a modification of the Ce^{3+} and Tb^{3+} intensity ratio and of their optimum concentrations. The broadening of Tb^{3+} emission bands is linked to the existence of structural distortions induced by the limited coherence length of the particles.

The luminescence yields of the crude particles are very high, but they are altered when treated in oxidizing conditions, which is often the case in the realization of nanocomposite materials. For example, the coating of the crude particles by a silica shell in a basic medium reduced the luminescence yield by a factor of two, and a heating process to 450°C totally quenches their emission. These results are correlated to an increase of the Ce^{4+} content in the nanoparticles, from 0.5 % for the crude particles, to 5-10 % in oxidized particles. Ce^{4+} ions are well-known to efficiently quench the luminescence of Ce-Tb systems, for example in (La,Ce,Tb)PO_4 phosphors. Improvement of the luminescence properties is achieved through the elaboration of core/shell nanostructures. The coating of the particles by a lanthanum phosphate crystalline shell provides a good passivation of the surface towards the oxidation of the cerium ions. The luminescence yield of core-shell nanoparticles (La,Ce,Tb)PO_4-xH_2O / $LaPO_4$-xH_2O is unchanged after coating with a silica shell (50 %), and half of the luminescence remains after heating treatment to 200°C. Optimization of the passivation is being studied to improve the protection of the surface during heating processes.

DESIGN OF TRANSPARENT LUMINESCENT MATERIALS

Functionalization of the surface of the nanocrystals is essential to ensure their dispersion within a transparent dielectric host matrix. Surface derivatization is been achieved through the controlled growth of an organically modified silica shell using a functionalized silane precursor. Concentrated sols are obtained, which can be spin-coated on various substrates, leading to perfectly transparent and highly luminescent thin films. Concentrated TPM-coated YVO_4:Eu suspension leads to 50 nm thick transparent films. Multi-layers systems and heating treatments up to 600°C are possible, keeping intact the transparency of the film. The silica matrix provides a good dispersion of the particles. The heating process eliminates OH groups, which results in the improvement of the luminescence by a factor of 2.5 (fig. 4). The luminescence of multi-layers films increases exponentially with the number of deposited layers to an optimum thickness of about 1 µm. This is explained by the increase of the number of absorbed photons in the film, which can be described by a Beer-Lambert exponential law.

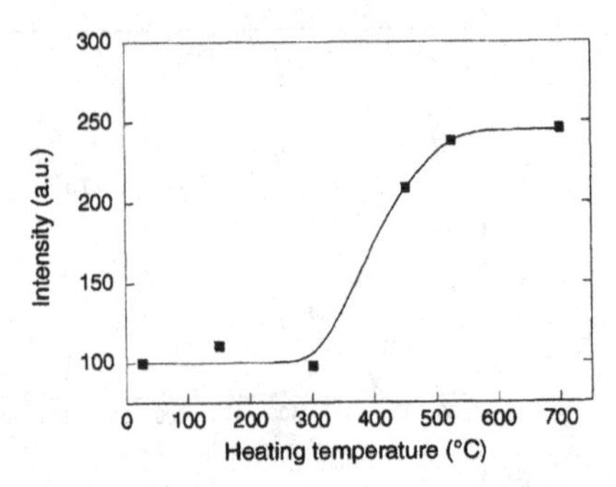

Figure 4 : Luminescence of thin film of TPM-grafted YVO_4:Eu nanoparticles for successive heating treatments

In the same way, (La,Ce,Tb)PO_4-xH_2O / LaPO_4-xH_2O core-shell nanoparticles can be further functionnalized by TPM. The organo-modified sols can be concentrated, and thin films thus obtained are highly luminescent in the green. Subsequent heating treatments up to 200°C slightly alter the luminescence efficiency. Improvements of the luminescence properties are obtained when thicker phosphate protecting shells are used.

CONCLUSION

Colloidal synthesis of vanadate (YVO_4:Eu) and phosphate ((La,Ce,Tb)PO_4-xH_2O) have been described, leading to well-dispersed nanocrystalline particles with sizes 5-8 nm. Both systems lead to efficient nanophosphors. Structural distortions induced by the small size are responsible for the broadening of the spectra but have minor effects on the luminescence efficiency. The luminescence yields are limited by surface quenching, essentially by the hydroxyls groups. Passivation of the surface is achieved in the case of YVO_4:Eu nanoparticles by the coating of a silica shell, and for (La,Ce,Tb)PO_4-xH_2O by a crystallized LaPO_4-xH_2O shell. The resulting core-shell nanoparticles are highly luminescent, well-dispersed, thus being suitable for applications in transparent luminescent devices.

[1] Jüstel, T.; Nikol, H.; Ronda, C. *Ang. Chem.*, **1998**, *37*, 3084.
[2] Chan, W.C.; Nie, S.M., *Science*, **1998**, *281*, 2013.
[3] Peng, X., Schlamp, M.C.; Kadavanich, A.; Alivisatos, A.P. *J. Am. Chem. Soc.*, **1997**, *119*, 7019.
[4] Tissue, B.M.; *Chem. Mat.*, **1998**, *10*, 2837.
[5] (a) Huignard, A.; Buissette, V.; Laurent, G.; Gacoin, T.; Boilot, J-P. *Chem. Mat.*, **2002**, *14*, 2264-2269. (b) Huignard, A.; Buissette, V.; Franville, A-C.; Gacoin, T.; Boilot, J-P. *J. Phys. Chem. B*, **2003**, *107*, 6754-6759.
[6] Riwotzki, K.; Haase, M. *J. Phys. Chem. B.*, **1998**, *102*, 10129.
[7] Stouwdam, J.W.; Van Veggel, F.C.J.M. *Nano Lett.*, **2002**, 2,7, 733-737.
[8] Riwotzki, K.; Meyssamy, H.; Schnablegger, H.; Kornowski, A.; Haase, M. *Angew.Chem Int.Ed.*, **2001**, *40*, 3, 573.
[9] Bazzi, R.; Flores-Gonzales, M.A.; Louis, C.; Lebbou, K.; Dujardin, C.; Brenier, A.; Zhang, W.; Tillement, O.; Bernstein, E.; Perriat, P. *J.Lumin*, **2003**, *102-103*, 445-450.
[10] V.Buissette, M.Moreau, T.Gacoin, J-P.Boilot, J-Y.Chane-Ching, T. Le Mercier, *Chem.Mat*, **2004**, 16, 3767.
[11] A.P. Philipse, A. Nechifor, C. Pathmamanoharan *Langmuir* 10, 4451 (1991).
[12] A. Huignard, T. Gacoin, F. Chaput, J.-P. Boilot, P. Aschehoug, B. Viana *Mat. Res. Soc. Pro.* vol. 667 (2001).

Mater. Res. Soc. Symp. Proc. Vol. 846 © 2005 Materials Research Society DD7.11

Influence of Halides on the Luminescence of Oxide/Anthracene/Polymer Nanocomposites

Dorothée V. Szabó, Heike Reuter, Sabine Schlabach, Christoph Lellig, Dieter Vollath[1]
Forschungszentrum Karlsruhe GmbH, Institute for Materials Research III,
D- 76021 Karlsruhe
[1]NanoConsulting,
D-76297 Stutensee

ABSTRACT

Nanocomposites made of an oxide core of a wide band gap insulator, a lumophore monolayer of anthracene and an outer protecting layer of PMMA are studied regarding their luminescence properties and the influence of halides stemming either from the precursor used for synthesis or from the lumophore itself. Halide-free nanocomposites exhibit luminescence spectra resembling to that of anthracene with some significant differences concerning the intensity ratio and an additional peak at 420 nm. Nanocomposites made from chlorides show excimer-like spectra with broad maxima. In microanalysis residual chlorine can be detected. Chlorine-free oxide kernels, coated with 9,10 dichloroanthracene exhibit luminescence spectra resembling to a superposition of the pure lumophores 9 chloro- and 9,10 dichloroanthracene. It can be shown that the origin of the halide strongly influences, but does not quench the luminescence spectra of the powders. Suspensions of the chlorine containing nanocomposites in ethanol exhibit modified anthracene like spectra. This is a strong indication for dechlorination by proton-transfer in ethanol. Suspensions of the same material in water lead to spectra showing a superposition of excimer spectrum and modified anthracene spectrum. Here a partial dechlorination occurs.

INTRODUCTION

Luminescence in oxide/mPMMA nanocomposites [1,2], surface modified oxides [3] and polymer-dielectric nanocomposites [4] is a well-described phenomenon. As in oxide/mPMMA nanocomposites the luminescence emerges from the bonding between mPMMA and oxide, containing carbonyl groups responsible for luminescence [5], the variability in emission wavelength is very limited. Therefore, a completely new concept of nanoparticles for luminescence applications was developed: three-layered nanocomposites [6,7]. These particles consist of an oxide kernel, a monolayer of an organic lumophore, and a polymer layer for protection outside.

Such multilayer nanocomposites are synthesized by a gas phase process, the Karlsruhe Microwave Plasma Process [8]. This process is characterized by low reaction temperatures, resulting in narrow particle size distribution of the nanoparticles and the possibility of in-situ coating of particles. Usually water-free chlorides, carbonyls, or metal-organics are well suited precursors for the synthesis of ceramic nanoparticles. As lumophores commercially available materials such as anthracene, pyrene, or perylene can be used. The protecting polymer coating generally is made from MMA (methyl methacrylic acid) which polymerizes under the UV radiation of the plasma.

Due to interactions between ceramic kernel, lumophore interlayer and polymer coating, these materials exhibit luminescence properties, differing from those of the pure lumophore. This is very promising for applications as "security markers". Due to the materials concept, nanocomposites with combined properties (e.g. superparamagnetism plus luminescence) can be realized, too. At a first glance these materials seem to be less toxic than quantum dots. This opens application potential in biology and diagnostics. Besides a broad variability in colors, particles according to this concept are easily dispersible in many different liquids like ethanol or water. Suspendability is an important property regarding applications in liquid systems. Especially dispersion in water would be impossible without polymer layer, as the lumophores usually are hydrophobic.

Chlorides are well suited precursors for the microwave plasma synthesis. On the other side it is well known that elements of the halogen group are quenchers of fluorescence for anthracene [9]. Influence of these elements has never been investigated regarding luminescence of oxide/anthracene/PMMA nanocomposites. Thus, in this study wide band insulator oxide nanoparticles are prepared from halide and butoxide precursors, respectively, coated with anthracene or its chlorine containing derivatives and a protecting PMMA layer. The influence of the precursor used, equivalent to residual halides and their origin, on luminescence spectra is investigated using powders and diluted suspensions.

EXPERIMENTAL

The nanocomposites were synthesized by the Karlsruhe Microwave Plasma Process, a gas phase process. For this study wide band insulators, such as crystalline ZrO_2 and HfO_2 or amorphous Al_2O_3 were used as ceramic kernels. Zirconium chloride, $ZrCl_4$, hafnium chloride, $HfCl_4$, aluminum chloride, $AlCl_3$, zirconium bromide, $ZrBr_4$, aluminum bromide, $AlBr_3$, and aluminum iodide, AlJ_3, all in their water free state, or metalorganic, halide free Zr-t-butoxide, $Zr(OC(CH_3)_3)_4$, Hf-t-butoxide, $Hf(OC(CH_3)_3)_4$, and Al-s-butoxide, $Al(OCH(CH_3)C_2H_5)_3$ were precursors for the synthesis of the oxide core. The precursors were evaporated outside the reaction chamber and introduced with argon carrier gas into the microwave plasma zone. As reaction gas a mixture of Ar/20 vol% O_2 was used. In the plasma all compounds are partially ionized and dissociated and the reaction to nanoparticles occurs. The lumophores (anthracene, or its chloride containing derivates 9 chloroanthracene and 9,10 dichloroanthracene) were introduced behind the plasma zone as a vapor, too, and condensed on the oxide nanoparticles. The final protecting polymer coating was made from evaporated MMA (methyl methacrylic acid) which condensed on the nanoparticles and polymerized under the UV radiation of the plasma. The nanocomposites were investigated by x-ray diffraction (Philips X'Pert, Cu-K_α) to determine the particle size of the oxide and to determine residual lumophore in the powder. Furthermore the powders were analyzed by EDX in an analytical TEM (Philips Tecnai F20-ST). Applying this analysis, the focus was on residual halide content. The influence of halides, originated either from the precursor or being part of the lumophore, on luminescence of powders and diluted suspensions was studied in detail. Therefore emission spectra of powders and suspensions were acquired with a Fluorolog FL3-22 (Jobin Yvon), excited at 325nm with a xenon lamp, and a bandpass of 4 nm on the emission and excitation side, respectively. Luminescence spectra were corrected for instrument response.

RESULTS AND DISCUSSION

The analysis of X-ray diffraction data showed small oxide nanocrystals of 3 to 4 nm for ZrO_2 and HfO_2, respectively, and amorphous Al_2O_3. An excess of anthracene or its derivates could not be detected in any case. This means that the particles are coated uniformly by the lumophore. Energy dispersive x-ray microanalysis revealed that nanocomposites made from chloride precursors generally contain 1 to 2 at% chlorine. Whereas the residual bromine content of nanocomposites made from bromide precursors was between 0.1 at%. In case of iodide precursor, residual iodine was not detectable. Therefore, one has to assume that chlorine is adsorbed on the nanoparticles, whereas bromine may be weakly, and iodine is not adsorbed. Obviously the adsorption ability decreases with increasing atomic weight of the adsorbent. For nanocomposites made from the butoxide, coated with 9,10 dichloroanthracene, a residual chlorine content of 0.5 at% is determined.

Luminescence spectra of the halide free nanocomposite powders are very similar. They resemble molecule spectra of anthracene, exhibiting a slight shoulder at 405 to 406 nm, an additional peak appearing as strong maximum at 420 nm, and peaks at 444 to 445 nm, and 471 to 474 nm. The most remarkable difference to the pure anthracene spectrum is the significant difference in the intensity ratio, and the additional peak with highest intensity at 420 nm. This is shown in figure 1. These spectra are different from emission spectra of anthracene, adsorbed on SiO_2 and SiO_2-TiO_2 surfaces, as described by Worrall et al. [10]. The spectra also differ from anthracene-labeled PMMA [11] so that a solution of anthracene in the coating of PMMA can be excluded.

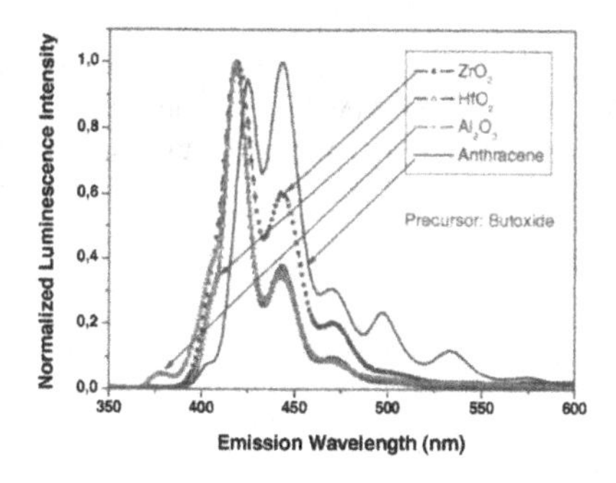

Figure 1: Normalized luminescence emission spectra of different nanocomposite powders type oxide/anthracene/PMMA made from butoxide precursors. For comparison the molecule spectrum of pure anthracene is shown.

Figure 2: Normalized luminescence emission spectra of different nanocomposite powders type oxide/anthracene/PMMA made from chloride precursors.

Nanocomposite powders made from chloride precursors totally differ from the spectra described before. As already mentioned they contain residual chlorine. The ZrO_2 and HfO_2 nanocomposites, respectively crystalline materials, exhibit excimer like spectra with a broad maximum at 545 to 550 nm. For the Al_2O_3/anthracene/PMMA one observes an excimer like spectrum with a broad maximum around 575 to 580 nm. This is shown in figure 2 for ZrO_2, HfO_2 and Al_2O_3 based materials. The maxima of emission differ from those known for oxide/mPMMA nanocomposites at around 400 to 410 nm [1] and also for pure ZrO_2 at 470 nm [12]. These spectra further differ significantly from the spectra of 9 chloroanthracene and 9,10 dichloroanthracene. Therefore, a formation of chloroanthracenes during particle synthesis from chlorine of the precursor can be excluded. It is more probable that the chlorine is adsorbed on the nanoparticles. Halide-free ZrO_2 or HfO_2 kernels, coated with 9,10 dichloroanthracene

(DCA) exhibit luminescence spectra resembling to a superposition of 9 chloro- and 9,10 dichloroanthracene. Obviously in the lumophore a partial loss of chlorine occurs during synthesis. The spectra differ from those of the nanocomposites made from chloride precursors. This is shown in figure 3 for ZrO_2-nanocomposites.

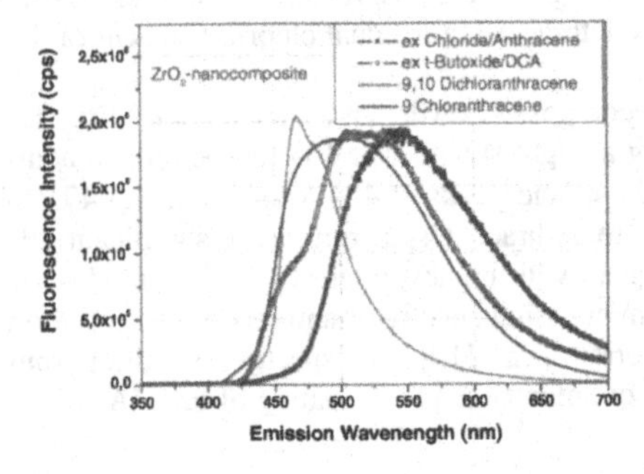

Figure 3: Luminescence emission spectra of chlorine containing nanocomposites of ZrO_2.

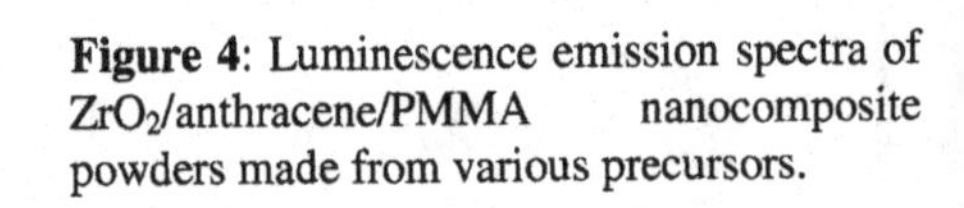

Figure 4: Luminescence emission spectra of ZrO_2/anthracene/PMMA nanocomposite powders made from various precursors.

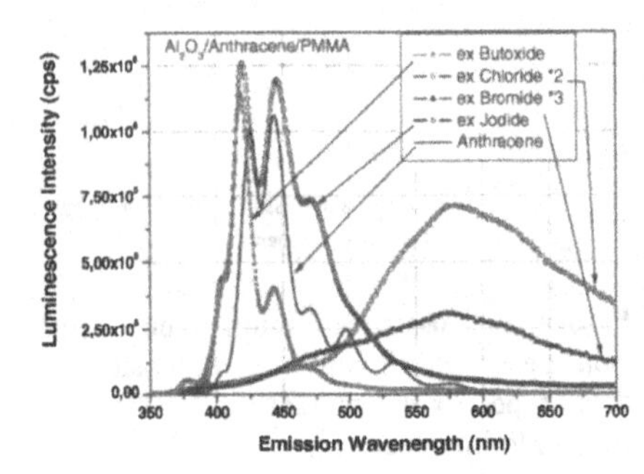

Figure 5: Luminescence emission spectra of Al_2O_3/anthracene/PMMA nanocomposite powders made from various precursors.

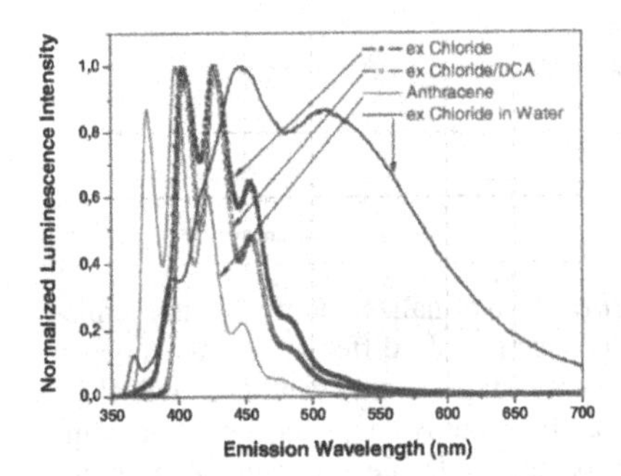

Figure 6: Normalized luminescence emission spectra of diluted suspensions of ZrO_2-nanocomposites in ethanol or water.

Comparing one type of nanocomposite, made from different precursors, it can be observed that with increasing atomic number of the halide the material behaves different. Figure 4 shows an excimer-like spectrum for ZrO_2/anthracene/PMMA nanocomposites made from chloride, whereas for the material made from bromide an anthracene-like molecule spectrum with reduced intensity is observed. A total quenching of luminescence due to halides is not observed. The bromine obviously is not adsorbed to the ceramic kernel so that the material behaves more than a halide-free material. The Al_2O_3 nanocomposite made from bromide shows a luminescence spectrum similar to the one made of chloride: an excimer like spectrum with a broad maximum around 575 nm and less intensity. The luminescence intensity decreases with increasing atomic number of the halide. Such a behavior is observed e.g. for halogen benzenes [5]. In contrast, the Al_2O_3 nanocomposite made from the iodide exhibits a molecule like spectrum with high intensity, but different from the spectrum of the nanocomposite made from the butoxide. This is shown in figure 5. As already mentioned, by EDX microanalysis no iodine was detected.

Table I: Emission maxima of anthracene bands measured in oxide/lumophore/PMMA nanocomposite powders and suspensions. Due to the similarity of the characteristics for ZrO_2 and HfO_2-nanocomposites only the results for ZrO_2 nanocomposites are presented. In the sample description PMMA is not mentioned. (S) denotes a shoulder; (M) denotes the maximum.

Sample	λ_1 (nm)	λ_2 (nm)	λ_3 (nm)	λ_4 (nm)	λ_5 (nm)	λ_6 (nm)	λ_7 (nm)
Pure anthracene		403 (S)	425	443 (M)	470	497	533
ZrO_2/anthracene ex butoxide		405 (S)	419 (M).	443	472	508	
ZrO_2/anthracene ex Cl							545-550 (M)
ZrO_2/anthracene ex Br		403 (S)	419 (M)	444	470		
Al_2O_3/anthracene ex butoxide	377	404 (S)	418 (M)	443	473		
Al_2O_3/anthracene ex Cl; ex Br							575-580 (M)
Al_2O_3/anthracene ex J		404 (S)	420 (M)	445	471		
Anthracene EtOH	377	398 (M)	421	447	476		
ZrO_2/anthracene ex butoxide, EtOH		404 (S)	427 (M).	454	483		
ZrO_2/anthracene ex Cl, EtOH	388 (S)	403	427 (M)	453	483		
ZrO_2/DCA ex Cl, EtOH		404	427 (M)	455	484		
ZrO_2/anthracene ex butoxide, H_2O	381	402	419 (M)				
ZrO_2/anthracene ex Cl, H_2O	366	395	416 (S)	447 (M)		511	
Al_2O_3/anthracene ex J, H_2O	381	402	419 (M)	445	474		

Diluted suspensions in ethanol (EtOH) of all chlorine containing nanocomposites exhibit modified anthracene like spectra, but with different intensity ratios. This is a strong indication for dechlorination by chlorine-hydrogen exchange [13] in ethanol. In these cases the origin of the chlorine, either precursor or lumophore, seems to be less important. Nanocomposites made from chloride as well as chlorine-free nanocomposites containing chloroanthracene exhibit very similar spectra. Suspensions of this material in water lead to spectra showing a superposition of excimer spectrum and modified anthracene spectrum, indicating a partial dechlorination by chlorine-hydrogen exchange. The results are presented in figure 6. Similar spectra are observed in other solvents like 2-propanole or cyclohexane. It is not clear yet where the chlorine from the precursor is located. This question has to be answered by XPS analysis. The results of luminescence measurements of powders and in suspensions are summarized in table I.

SUMMARY AND CONCLUSIONS

The concept of ceramic/lumophore/polymer nanoparticles leads to a large variety of tunable luminescence effects with the possibility to design materials that can be exploited technically. Application potential is seen in the fields of biology and diagnostics, as well as in the field of "security markers" for protection of trademarks. Such particles can be produced with the Karlsruhe Microwave Plasma Process. The selection of precursors has to be done very carefully since halides in the system exhibit a strong influence on luminescence. It was shown that the origin of the halide strongly affects the luminescence spectra of the powders, but a quenching of luminescence due to halides was not observed. Nanocomposites made from chlorides show excimer-like spectra with broad maxima, whereas halide-free nanocomposites exhibit luminescence spectra resembling to that of anthracene with some significant differences concerning the intensity ratio and the position of the maxima. Suspensions of chlorine containing nanocomposites in ethanol and water exhibit modified anthracene like spectra indicating a dechlorination. The dechlorination is partial in the case of water suspension. The location of the chlorine of the precursor is not clear yet.

REFERENCES

1. D. Vollath, I. Lamparth, D.V. Szabó, in "*Nanophase and Nanostructured Materials IV*" edited by S. Komarneni, J.C. Parker, R.A. Vaia, G.Q. Lu, J.-I. Matsushita, (Mater. Res. Soc. Proc. **703**, Pittsburgh, PA, 2002) V7.8.1-V7.8.6.
2. D. Vollath, D.V. Szabó, S. Schlabach, J. Nanoparticle Res. **6**, 181-191 (2004).
3. W. Dong, C. Zhu, *J. Phys. Chem. Solids* **64**, 265-271 (2003).
4. S. Musikhin, L. Bakueva, E.H. Sargabt, A. Shik, *J. Appl. Phys.* **91**, 6679-6683 (2002).
5. T. Förster, in "Fluoreszenz organischer Verbindungen", Vandenheock & Ruprecht, Göttingen (Germany), p. 97-100 (1951).
6. D. Vollath, I. Lamparth, D.V. Szabó, *Berg und Hüttenmännische Monatshefte* (*BHM*), **147**, 350-358 (2002).
7. D. Vollath, I. Lamparth, F. Wacker, German Patent Application DE 10203907.0 (2002).
8. D. Vollath, D. V. Szabó, in "*Innovative Processing of Films and Nanocrystalline Powders*" edited by K.-L. Choy, Imperials College Press, p. 210-251 (2002).
9. J.R. Lakowicz, in "*Principles of Fluorescence Spectroscopy*", Kluewer Academic / Plenum Publishers, New York, p. 238 (1999).
10. D.R. Worrall, S.L. Williams, A. Eremenko, N. Smirnova, O. Yakimenko, G. Starukh, *Colloids and Surfaces* A **230**, 45-55 (2004).
11. C.S. Kim, S. M. Oh, S. Kim, C.G. *Cho, Macromol. Rapid Commun.* **19**, 191-196 (1998).
12. Q. Li, D. Ai, X. Dai, J. Wang, *Powder Technology* **137**, 34-40 (2003).
13. K. Hamanoue, T. Nakayama, K. Ikenaga, K. Ibuki, *J. Phys. Chem.* **96**, 10297-10302 (1992).

Mater. Res. Soc. Symp. Proc. Vol. 846 © 2005 Materials Research Society DD7.13

Ambipolar injection in a submicron channel light-emitting tetracene transistor with distinct source and drain contacts

J. Reynaert*, D. Cheyns*, D. Janssen, V.I. Arkhipov, G. Borghs, J. Genoe, P. Heremans
IMEC, Kapeldreef 75, B-3001 Leuven, Belgium
*also with ESAT, Katholieke Universiteit Leuven, Leuven, Belgium

ABSTRACT

Over the last decade, organic semiconductor thin film transistors have been the focus of many research groups because of their potential application in low-cost integrated circuits. Recently, an organic light-emitting field-effect transistor (OLEFET) was reported. In an OLEFET structure, optimal injection of both holes and electrons into the light-emitting layer are required for maximum quantum efficiency, whereas the gate serves as a controlling electrode. In this work, we achieved an OLFET structure with interdigitated hole-injecting Au and electron-injecting Ca contacts within a submicrometer channel length. Both contacts are bottom contacts to the upper-lying tetracene organic semiconductor. The study of IV-characteristics and light emission from these devices shined light on the underlying physics of the OLEFETs.

INTRODUCTION

Since the first organic light-emitting transistor (OLEFET) has been reported [1], several small molecules [2, 3] or polymers [4, 5] have been studied with respect to their application in OLEFET structures. In this work, we have achieved a transistor structure featuring a submicrometer channel length with interdigitated coplanar hole-injecting Au and electron-injecting Ca contacts. A semiconducting tetracene thin film was deposited on top of the contacts. Tetracene is known as a hole-conducting organic material with a reported mobility [6] of the order of 0.1 $cm^2/Vsec$. Fabrication of transistor structures with asymmetric Au and Mg contacts but with much larger channel distances (min. 140 μm) has been reported recently [2]. However, to our best knowledge, the present paper is the first report on an organic transistor structure with a submicrometer channel and distinct contact metals.

EXPERIMENTAL RESULTS

By the use of a shadowmask technology, an interdigital structure could be made with alternating Au and Ca fingers, with a distance between two adjacent fingers varying from 5 micrometer down to hundreds of nanometers. These fingers serve as coplanar source and drain contacts, with the organic semiconductor thin film being deposited on top of these contacts. The gate contact is foreseen at the Si substrate backside, which is separated from the transistor channel by a 100 nm silicon oxide dielectric. The complete thin film transistor structure is shown in Fig.1. A detailed description of the device fabrication will be given elsewhere.

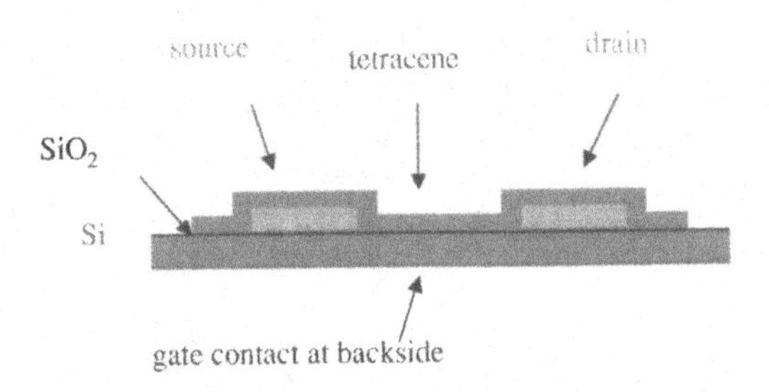

Figure 1: Thin film transistor structure

Fig. 2a shows a top view of a transistor fabricated with the developed shadow mask technology for a Au source and drain contact. The drain-source current I_{ds} versus drain-source voltage V_{ds} is plotted in Fig. 3 for various source-gate voltages V_{gs} and reveals a typical hole-channel organic transistor behavior. The light output from this device was recorded during measurement (Fig. 2b). The amount of light was very limited and remains restricted to the border line of the Au drain contact.

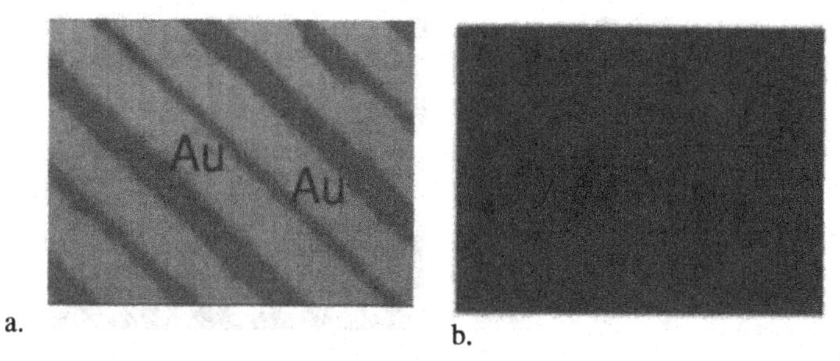

Figure 2: a. Top view of the tetracene thin film transistor measured in Fig. 3. The smaller contact interdistance is 4.6 μm, the larger 13.7 μm, b. Recorded light-emission from the structure in Fig. 2a.

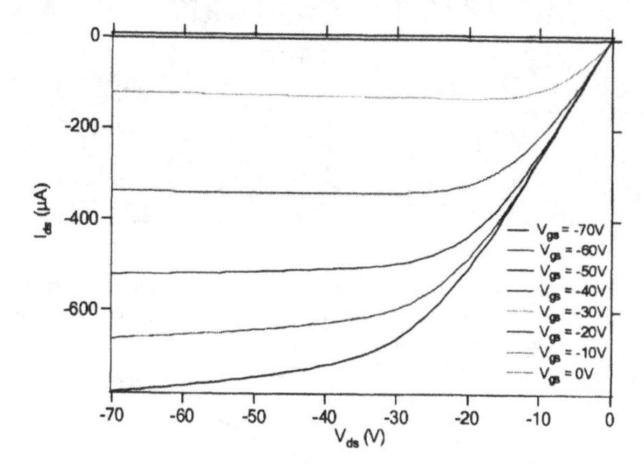

Figure 3: I_{ds} vs. V_{ds} –characteristics for a tetracene thin film transistor with effective channel length L_{eff}=3.46 μm and W=7560 μm.

In Fig. 4a, the top view is given of a thin film transistor with hole-injecting Au contact and electron-injecting Ca contact with an effective channel length of 4μm. The I_{ds}-V_{ds} transistor characteristics for this structure are shown in Fig. 5 for various values of V_{gs}-bias. The transistor behavior is not conform with a hole-channel organic transistor, nor is there any sign of ambipolar transistor behavior [7] due to an electron channel formation through improved electron injection from the Ca contact. One can see from the recorded light output at V_{gs}=-40V, -50V and –60V that the light-emission is increased for higher V_{gs}-bias, given in Fig. 4b, c and d. The location of the hole-electron recombination zone, visualized by the light-emission in the transistor channel, does not vary with applied V_{gs}-bias. This is again an indication that no electron channel is formed.

In Fig. 6a, the top view is given of a thin film transistor with hole-injecting Au contact and electron-injecting Ca contact, with a submicrometer channel length of 0.7 μm. The I_{ds}-V_{ds} transistor characteristics for this structure are shown in Fig. 7 for various values of V_{gs}-bias.

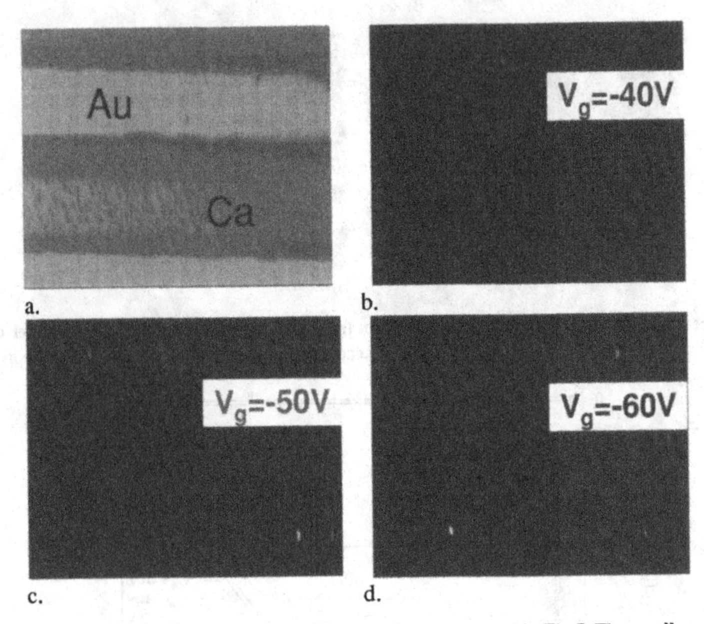

Figure 4: a. Top view of the tetracene thin film transistor measured in Fig.5. The smaller contact interdistance is 5.4 μm, the larger 14.8 μm, b, c, d: Recorded light-emission from the structure in Fig. 4a for V_{gs} at –40V, -50V and –60V respectively.

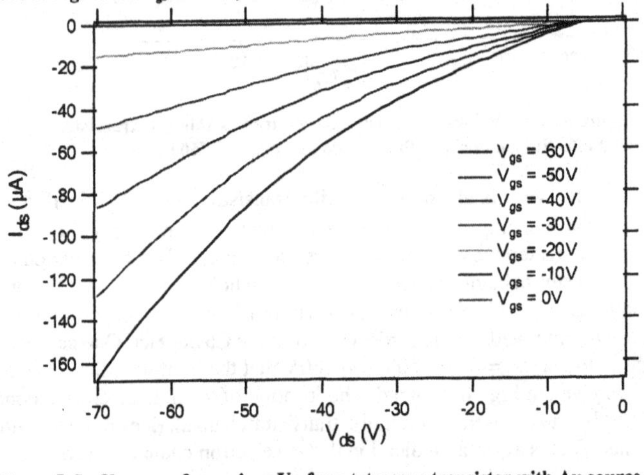

Figure 5: I_{ds}-V_{ds} curve for various V_{gs} for a tetracene transistor with Au source and Ca drain contact. The effective channel length and width are L_{eff}=4 μm and W=7560 μm.

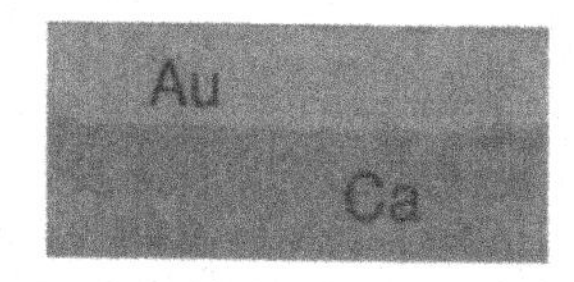

a.

Figure 6: a. Top view of the tetracene thin film transistor measured in Fig.7. The contact interdistance is 0.7 μm, b. Recorded light-emission from the structure in Fig. 6a for V_{gs} varying from 0V to –70V.

From the transistor characteristics in Fig. 7 we see a small V_{gs}-dependence of I_{ds} at intermediate V_{ds}-bias, which totally disappears at higher V_{ds}-bias. This is typical for short channel transistors [8-10]. Also the recorded light-emission from Fig. 6b reveals no V_{gs}-dependence.

The transistor characteristics and recorded light-outputs, shown above, will be interpreted after having studied the carrier injection from Au and Ca into tetracene, presented in the next section.

V_g=0V
V_g=-10V
V_g=-20V
V_g=-30V
V_g=-40V
V_g=-50V
V_g=-60V
V_g=-70V

b.

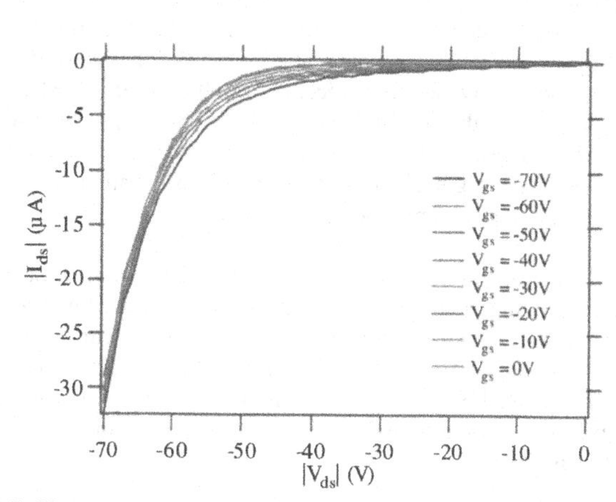

Figure 7: I_{ds}-V_{ds} curve for various V_{gs} for a tetracene transistor with Au source and Ca drain contact. The channel length and width are L=0.7 μm and W=5670 μm.

DISCUSSION

In order to study the carrier injection from Au and Ca contacts in tetracene, we made use of tetracene crystals. The current injection in tetracene crystal is found to be contact-limited [11]. Several contact combinations were made on a set of two tetracene crystals, as shown in Fig.8a and b. The fact of having several contact combinations on the same tetracene crystals has as an advantage that differences in crystal quality are avoided when comparing current-voltage (IV) characteristics with different contact combinations [11].

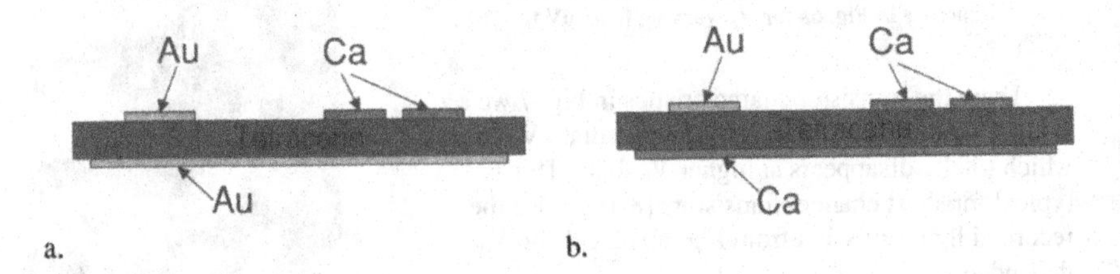

Figure 8: a,b: Two sets of tetracene crystals were used for IV characterization. The crystal thickness for the devices represented in *a* and b is 9.5μm and 13.5 μm respectively.

When tetracene is sandwiched between two Au contacts, one expect to inject holes from the highest biased Au contact due to the high work function of Au. The holes are ejected at the other Au contact. The crystal device structure as well as the carrier transport are represented in Fig 9a. When the hole-ejecting contact is replaced by Ca, as shown in Fig. 9b, the current decreases by orders of magnitude, indicating that holes experience a barrier for being ejected at the Ca contact. In the last experiment, we also replaced the remaining Au contact by Ca. Due to the low work function of the Ca contact, the observed current is due to only electrons injected from the lower biased contact (see Fig. 9c). This current is equal to the current measured in Fig. 9b, from which we can conclude that the observed hole-ejection barrier does not prevent electrons from being injected at the Ca contact. The origin of this hole-ejection barrier can be found in an interface reaction between tetracene and calcium.

The occurrence of a barrier for hole-ejection from tetracene towards a Ca contact, paves the way to understanding of the behavior of the tetracene thin film transistor with Ca drain contact. The I_{ds}-current represented in Fig. 4 is an electron current as holes can not reach the Ca drain contact. Given electrons are the minority carriers in the transistor channel, which is still populated by holes, the I_{ds}-current should be proportional to the emitted light intensity. An increased light output for increasing V_{gs} and V_{ds} indeed gives a first indication that this is the case.

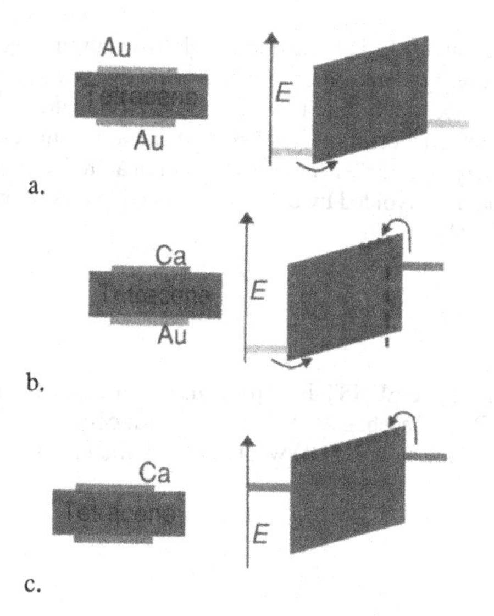

Figure 9a, b, c: Different sandwich contact combinations with Au and Ca contacts on a tetracene crystal (left side), energy band diagram including carrier injection from Au and Ca contacts (right side).

The hole ejection barrier and electron injection at the Ca contact are illustrated in further details in Fig. 10. Under a negative gate bias, holes are accumulated in the transistor channel, though the altered interface at the calcium contacts prevents them of being ejected. Injection of electrons from the calcium contact is favored by the V_{ds}-field and by the space charge of holes accumulated in the transistor channel. Under such circumstances, injected electrons recombine with holes close to the drain contact.

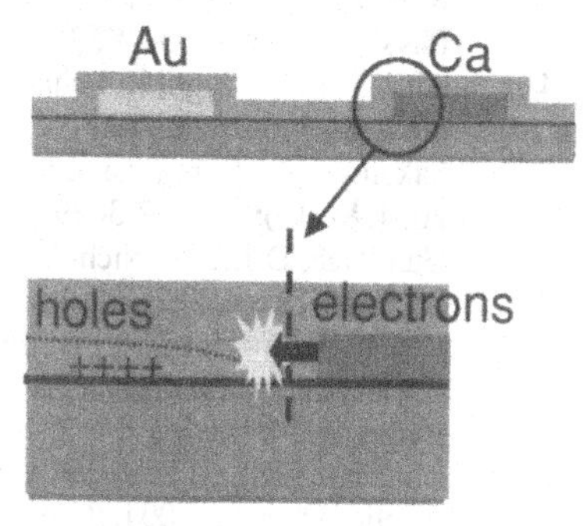

Figure 10: Hole ejection and electron injection with subsequent light emission at the Ca drain contact

CONCLUSION

We report on the study of submicron-channel tetracene OLEFETs bordered by source and drain contacts of distinct metals, favoring hole- and electron-injection respectively. In case of submicron channel length, short-channel effects dominate the transistor behavior, while for longer channels I_{ds} is limited to the electron current injected from the Ca contact. The study of current-voltage characteristics of sandwich contacted tetracene crystals indicates the occurrence of a barrier for hole ejection from tetracene into the calcium contact, while electrons can still be injected from calcium. Our results indicate that the use of calcium can create a barrier for hole ejection in a predominantly hole-conducting device. This could be avoided by using a less reactive contact metal with still a low workfunction, such as Mg.

ACKNOWLEDGEMENTS

This work is funded within the EU-IST-FET program under project IST-33057 (ILO). The authors want to thank R. Schmechel, M.A. Loi and M. Muccini for interesting scientific discussions. One of the authors (JR) wants to thank the Flemish IWT-fund for his scholarship.

REFERENCES

1. Hepp, A., H. Heil, W. Weise, M. Ahles, R. Schmechel, and H. von Seggern, Physical Review Letters, 2003. **91**(15): p. 157406.
2. Rost, C., D.J. Gundlach, S. Karg, and W. Riess, Journal of Applied Physics, 2004. **95**(10): p. 5782-5787.
3. Rost, C., S. Karg, W. Riess, M.A. Loi, M. Murgia, and M. Muccini, Applied Physics Letters, 2004. **85**(9): p. 1613-1615.
4. Ahles, M., A. Hepp, R. Schmechel, and H. von Seggern, Applied Physics Letters, 2004. **84**(3): p. 428-430.
5. Sakanoue, T., E. Fujiwara, R. Yamada, and H. Tada, Applied Physics Letters, 2004. **84**(16): p. 3037-3039.
6. Gundlach, D.J., J.A. Nichols, L. Zhou, and T.N. Jackson, Applied Physics Letters, 2002. **80**(16): p. 2925-2927.
7. Geens, W., *Study of the potential energy conversion efficiency of organic solar cells based on donor/acceptor heterojunctions*. 2002, University of Antwerp, Belgium.
8. Chabinyc, M.L., J.P. Lu, R.A. Street, Y.L. Wu, P. Liu, and B.S. Ong, Journal of Applied Physics, 2004. **96**(4): p. 2063-2070.
9. Austin, M.D. and S.Y. Chou, Applied Physics Letters, 2002. **81**(23): p. 4431-4433.
10. Sze, S.M., *Physics of Semiconductor Devices*. 2nd ed. 1981, New York: Wiley.
11. Reynaert, J., V.I. Arkhipov, G. Borghs, and P. Heremans, Applied Physics Letters, 2004. **85**(4): p. 603-605.

Nanocomposite Optical Materials

Mater. Res. Soc. Symp. Proc. Vol. 846 © 2005 Materials Research Society

Self-assembled oligonucleotide semiconductor conjugated to GaN nanostructures for biophotonic applications

A. Neogi[1]; *J. Li*[1]; *A. Sarkar*[2]; *P. B. Neogi*[3]; *B. Gorman*[1], *T. Golding*[1]; *H. Morkoc*[4]

[1]Department of Physics, University of North Texas, Denton, TX, USA.
[2]Michigan Molecular Institute, Midlands, MI, USA.
[3] Department of Biology, University of North Texas, Denton, TX, USA.
[4] Department of Electrical Eng.,Virginia Commonwealth University, Richmond, VA, USA.

ABSTRACT

We investigate the optical properties of a new class of wide-bandgap semiconductor based biomaterial system. We have synthesized a guanosine derivative with a strong dipole moment, which self-assemble in ~ 50 –100 nm confined pits to form a ribbon like semiconductor structure (SAGC). SAGC were successfully self-assembled on GaN/AlN QD matrix and the luminescence from GaN QDs can be resonantly transferred to the SAGC molecules resulting in a significant enhancement in emission from the guanine molecules. We also propose the design of ultraviolet-visible photonic bandgap structures based on hybrid SAGC-GaN photonic crystal.

INTRODUCTION

In recent years, there have been significant interests in using novel solid-state nanomaterials for biological and medical applications. The unique physical properties of nanoscale solids (dots or wires) in conjunction with the remarkable recognition capabilities of biomolecules could lead to miniature biological electronics and optical devices including probes and sensors. Such devices may exhibit advantages over existing technology not only in size but also in performance.

DNA or oligonucleotide has a fundamental role in biological processes. The combination of molecular biology (for engineering DNA with the desired functional and/or self-assembling properties) and nanotechnology (for device fabrication) thus becomes the tool to realize a new class of nanophotonic elements. DNA has been one of the most investigated class of biomolecules, leading to a somewhat controversial description of its electrical properties, and, hence, of its potential for electronic applications. Depending on the interconnection mechanism (chemical bonding of the DNA on a metal by a selected sequence of oligonucleotide[1], mechanical contact with a gold interdigitated patterns[2] or single DNA molecule immobilized in a metal contact[3]), the DNA molecules have been found to be conductive, non-conductive or rectifying. Recently self-assembled DNA bases on graphite for adenine[4] and for guanine have been achieved. In particular, the self-organization of organic molecules on flat surfaces gives structures with a high degree of order, thereby opening a wide range of applications in electronic[5] and optical devices[6]. The spontaneous self-assembly of small molecules from solution directly onto solid surfaces has been used to design two-dimensional organized structures [5,7,8].

Among the conjugated bases forming DNA, the guanine represents a versatile molecule that, depending on the environment, can undergo different self-assembly pathways. In the presence of cations, it is notorious for its propensity to associate into planar tetrameric nanostructures, namely G-quartets (G4), which are a thermodynamically stable architecture consisting of a cyclic array of four guanine joint by hydrogen bondings[14, 15]. In these structures, each base behaves equally both as donor and acceptor in four H bonds with its neighbors. By varying the ratio of the concentration of cation vs. molecule, one can build up supramolecular structures composed of stacks of guanine quartets; this approach allows the development of octamers, quadruplexes or even polymeric species (Fig. 1c) [16, 17]. In this regard, by simply casting a guanine solution containing ~ 10^{-2} M cations like K^+ or Na^+, Henderson and coworkers have grown dry tubular nanostructures, known as G wires, lying flat at a surface[18]. By tuning the chemical composition of a guanosine derivative, it is possible to form G4 also in absence of a templating metal cation[19]. Alternatively, in the absence of cations, guanine derivatives have usually been found to self-assemble into hydrogen-bonded networks [2, 5]. This material system has been utilized to design molecular electronic devices such as MOSFETs or photodetectors. Maruccio et al[10] have demonstrated a field effect transistor based on a deoxyguanosine derivative (a DNA base). This three-terminal field effect nanodevice was fabricated starting from a deoxyguanosine derivative ($dG(C_{10})_2$). Guanosine has been chosen because of its peculiar sequence of H-bond donor or acceptor groups, and because it has the lowest oxidation potential among the DNA bases, which favors self-assembly and carrier transport, respectively. Such a guanosine supramolecular assembly has the form of long ribbons (see Fig. 1b), with a strong intrinsic dipole moment along the ribbon axis that causes current rectification in transport experiments.

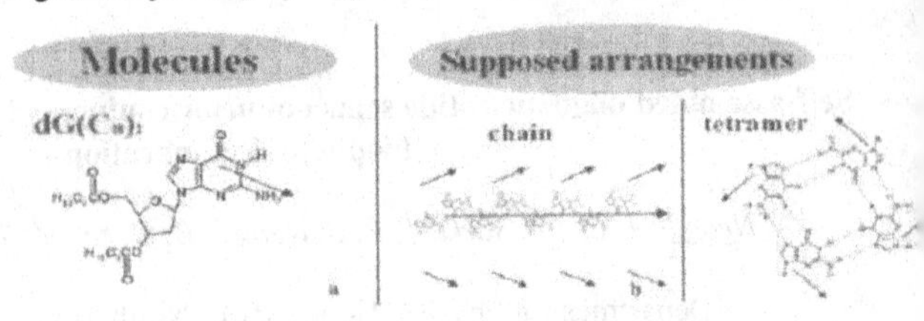

Fig. 1. (a) Scheme of the DG molecule. The arrow indicates the intrinsic dipole direction. (b) Scheme of the ribbon-like supramolecular structures formed by the DGs by self-assembling in both chlorinated solvents and in the solid state. The long arrow indicates the large total dipole of the ribbon. The short arrows indicate the dipoles of the individual molecules forming the ribbon. (c) Tetramer form of guanosine

Due to the strong intrinsic dipole moment of these self-assembled deoxyguanosine (SADG) molecule we investigate the feasibility of using it for photonic or optoelectronic applications. The unique physical properties of nanoscale solids (dots or wires) in conjunction with the remarkable recognition capabilities of guanosine molecule could lead to miniature biological electronics and optical devices including probes and sensors. Such devices may exhibit advantages over existing technology not only in size but also in performance. In this work we investigate two different classes of photonic devices, one based on the luminescence properties of the material, which can be used as optical sensor and the other based on reflection or transmission properties due to the existence of a photonic bandgap based on modified guanosine crystals.

EXPERIMENT

- **Enhancement of fluorescence in SAGC molecules conjugated to GaN quantum dots:**

For utilizing SADG molecules in photonic applications based on fluorescence properties, it ought to be conjugated to inorganic nanoparticles similar to the studies on semiconductor quantum dots conjugated with biomolecules[20]. GaN semiconductor nanostructures provide the ideal inorganic material system for the fabrication of photonic devices due to its transparency in the ultraviolet-visible wavelength regime. Despite the fact that guanosine derivatives such as 8-oxo-deoxyguanosinemolecules has been successfully self-assembled on solid substrates such as mica, sapphire and even on semiconducting substrates viz. SiO_2, the effect of self-assembly on highly polar semiconductor surface such as GaN has not been investigated. GaN is a highly polar semiconductor with a large strain induced polarization field developed due to the noncentrosymmetric nature of the crystalline symmetry and the absence of naturally available substrate for the epitaxial growth of these materials[21, 22]. The surface of GaN layers can be tailored to have by positive or negative polarity via Ga or N termination of the cap layer. The existence of surface polarization charges at the surface and interface results in domain formation of electronic charges as observed by electron-force microscopy[7]. The formation of domains with a specific polarity can be utilized for the enhancement of the self-assembly behavior of guanosine molecules by suitable modification of the side chains. The primary choice of conjugating deoxyguanosine to GaN interface is due to the similarity in their optical properties[8-10]. In analogy to solid-state physics, these biomolecular systems have been termed as self-assembled guanosine crystals (SAGCs)[8]. The mobility of charge carriers for electronic transport in the deoxyguanosine film depends on the ordering of the molecules.

Under appropriate conditions, the modified guanosine molecules self-assemble to form a ribbon like structure. Guanine also has a strong dipole moment of the order of 7 D which provides a polarity to each self-assembled guanosine conjugate supramolecular structures. SAGCs were obtained from a solution of deoxyguanoscine derivative in chloroform. The optimum self-assembly on confined nitride nanostructure were observed by depositing a 1 μl drop of a 3.5×10^{-2} M solution of $dG(C10)_2$ in chloroform. Ideal self-assembly of SAGCs have been observed in 50-100 nm confined spaces. The self-assembly process resulting in the formation of an organic semiconductor occurs only at specific concentration of the solution and great care has to be taken in order to control the solid-state assembly of the molecules. The self-assembled deoxyguanosine films behave like wide gap semiconductors, with energy gap in the range of 3 – 3.5 eV and electron effective mass $m_e > 2\ m_o$(similar to GaN)[9,10].

Our group has been involved in the fabrication and characterization of GaN nanostructures and QDs for various optoelectronic applications . GaN QDs in AlN confined layer structures have been fabricated by molecular beam epitaxy with very high optical quality. The structural dot density is about 3×10^{10} cm^2, though it has been observed from near-field optical spectroscopy that the fraction of optically active dot is relatively much lower. Despite the huge density of dislocations in group-III nitrides hetero-epitaxially grown on silicon or sapphire, intense PL emission is observed. Due to the very large built-in electric field (piezoelectric and spontaneous polarization), which occurs in wurtzite nitrides (~ 5 MV/cm for AlN/GaN system), a giant quantum confined Stark effect occurs. Using various combinations of substrates, buffer layers and growth conditions, we have been able to tailor the size of the dots due to the modification of the strain between the core GaN layer and the AlN layer. The strain induced polarization effects lead to emission in the wide UV-visible electromagnetic spectral range covering from 300 to 600 nm. This characteristics offer a wide spectral bandwidth to GaN QD based molecular beacons. Compared to bulk GaN the quantum efficiency of photoluminescence (PL) or fluorescence in GaN QD is increased by two-three orders of magnitude. The emission is confined to a narrow band centered at a wavelength characteristic of the dot size and built in strain.

The SAGC semiconductor energy gap of onset (3.4 eV) demonstrated by current-voltage measurement[8] demonstrates the band-like description of this particular biomolecular material at the optimum concentration. The absorption onset depends on the concentration of the SAGC molecules. This energy range is resonant GaN/AlN QD emission energy, facilitating energy transfer of radiative component of the electron-hole recombination energy from the GaN semiconducting layer to the biomolecular layer through the tailored Ga polarized surface interaction. Fig.2 depicts the PL spectra of SAGC coated GaN QDs, probed by optical parametric amplifier at excitation wavelength ~ 295 nm (4.2 eV). Exciton recombination from the confined states in GaN QDs results in emission at ~ 3.7 eV. SAGCs, which normally does not fluoresce, is observed to emit at 3.3 eV. This occurs along with a reduction in the PL emission from GaN QD layers.

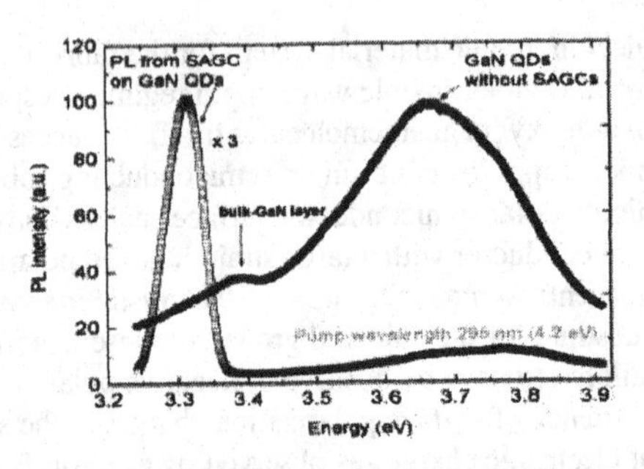

Figure 2. Resonant energy transfer from GaN QDs to SAGC showing enhanced luminescence from molecular structures

- **UV-Visible Photonic Bandgap Structures based on hybrid organic-inorganic semiconductors**

For the control of UV or visible light using photonic crystals, the fabrication of convenient structures poses two problems.

- Materials - to avoid light absorption by the photonic crystal, wide gap semiconductors, as for example GaN, ZnO must be used. Unfortunately, these materials have weak dielectric constants, which reduces the photonic band gap width.
- Scale - Photonic crystals must also have a period smaller than the wavelength of light to be controlled. This leads to the second problem, due to the technological challenge involved in the nanoscale lithography and etching of material structure.

Photonic crystals fabricated using organic or polymeric materials usually have bandgap in the infrared regime due to lower contrast in its dielectric constant. To overcome the low dielectric contrast in organic semiconductors, the use of organic semiconductors coupled with high dielectric constant inorganic compounds has been proposed to achieve UV-visible photonic crystal. Thus, *the future of using hybrid organic-inorganic material system depends on the development of materials with larger inorganic content and higher refractive index.* Among inorganic materials available for the fabrication of UV-visible photonic bandgap structures, nitride based semiconductors are an essential compound due to its transparency in the visible wavelength regime. The recent success in the nanofabrication of GaN based PCs has opened up the possibilities of developing new material system for bio-conjugation using hybrid organic molecules than can be coupled to GaN material system for the development of biomolecular or biophotonic sensors.

One of the main drawbacks of the nitride system is its relatively low dielectric constant (ε =6.45 at 400nm) in the ultraviolet regime, which is further reduced in the infrared regime and results in relatively smaller spatial dimensions. As a result, efforts have been directed towards optimizing the design of the GaN based structure with larger periodic dimensions, which can be achievable by conventional etching or semiconductor processing techniques. *Using SAGC semiconductors that has a lower dielectric constant integrated with relatively high contrast GaN semiconductors is an effective way to design photonic crystals for the UV-Visible regime.*

The MIT Photonic Bands (MPB) package is being used to solve this expression for the definite-frequency eigenstates for all of the band structures in this work[11]. We have modeled various geometries of GaN-SAGC material system using triangular, square and graphite lattice structures with cylindrical holes. The band gap of square structure opens up at r/a over 0.4, where r is the radius of holes, and 'a' is the period. And the maximum ratio of gap to mid-gap wavelength is about 18.08% for air hole. The lattice and band structures are shown in fig. 3 for triangular structure and fig. 4 for graphite structure. The graphite structure was first discussed by Cassagne et. al.[12] The notation is similar to that used by Lommer.[13] The structure and gap parameters are in table 1 for triangular structure and in table 2 for graphite information.

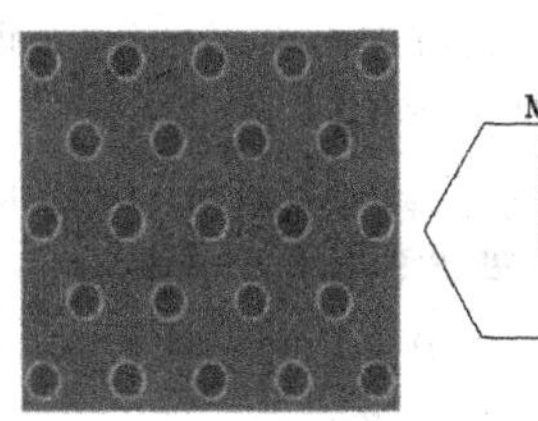

Fig.3 a) Two dimensional triangular lattice structure with cylindrical holes and the first Brillouin zone for the fabrication of self-assembled olegonucleotide conjugated GaN photonic crystals.

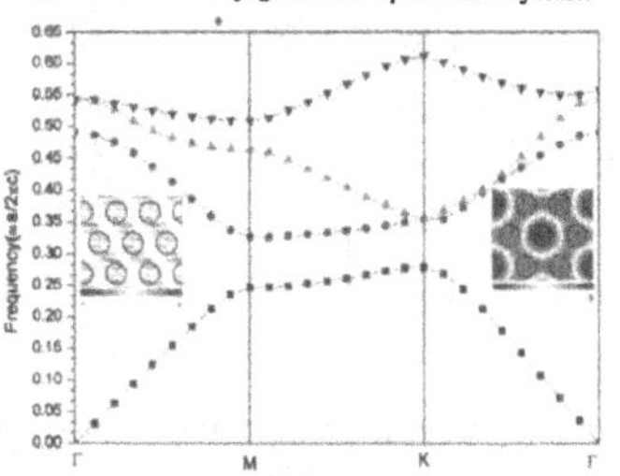

Fig 3 b) Photonic bandstructure designed for gap in the UV-visible wavelength range (386-451 nm) with r=0.3a and magnetic field distribution of TE mode at M(0, 0.5, 0) (left); at K(1/3, 1/3, 0) (right) for the 1st or fundamental band.

It is observed from the tables 1 and 2 that the band gap can be tuned by changing the cylindrical-hole diameter and lattice period to confine the guanosine chain within 50-100 nm. Triangular structure has only TE mode band gap. The gap-midgap ratio of TM mode in triangular structure with air holes reaches its maximum 33.80% at r=0.413a. Graphite structure has both TM and TE modes and usually the gap of TM mode is wider than that of TE mode. The gap-midgap ratios of TM and TE modes reach 27.98% and 17.02% at r=0.5a and r=0.428a, respectively. Both the TE and TM mode bandgaps may be formed at different positions and bandwidths. But the gap of the triangular structure is wider than that of graphite structure for same filling factor, which is the ratio of the area of holes in one Bravais lattice to that of one Bravais lattice. So it is easier to get a triangular structure by using focused ion-beam milling technique.

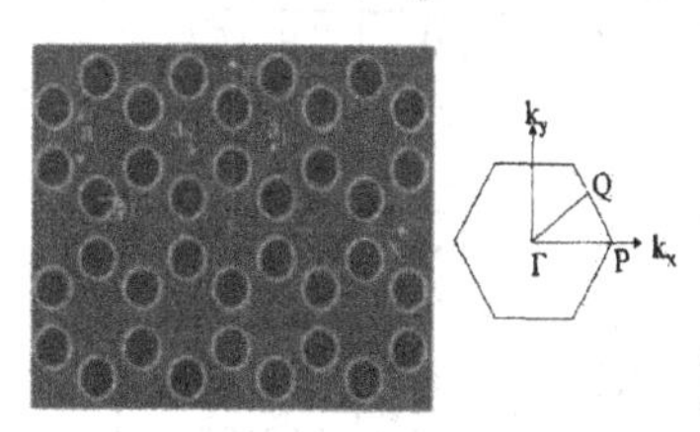

Fig 4. a) Two dimensional graphite lattice structure with cylindrical holes and the first Brillouin zone for the fabrication of self-assembled olegonucleotide conjugated GaN photonic crystals.

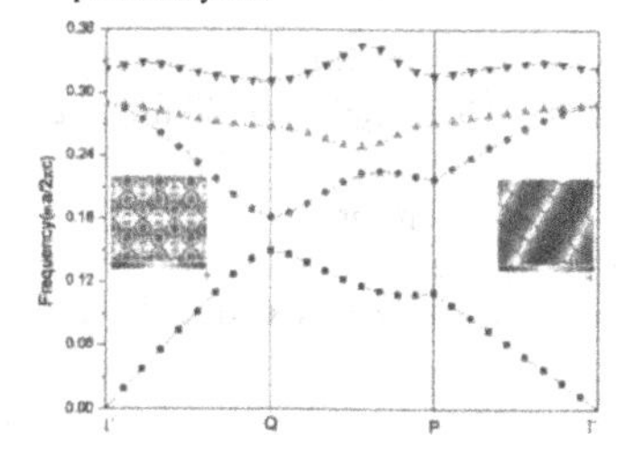

Fig. 4 b) Photonic bandstructure designed for gap in the UV-visible wavelength range (520-623 nm) with r=0.45a and electric field distribution of TM mode at Q(0, 0.5, 0) (left); at K(1/3, 1/3, 0) (right) for the 1st or fundamental band.

Embedding SAGC within the GaN air holes reduces the gap because of the smaller dielectric contrast. In our calculation the refractive index of SAGC is 1.19 measured using variable angle spectroscopic ellipsometer. The gap-midgap ratio of triangular GaN/SAGC structure reaches its maximum 23.78% at r=0.394a. But that of graphite GaN/SAGC structure reach the maximum 27.98% and 17.02% for TM and TE modes, respectively, with the same radii as that of GaN/Air structure.

Efforts are therefore directed towards the

achievement of longer guanosine chain so that wider gap or the gap in visible or infrared regime can be achieved.

Table 1: Proposed GaN-SAGCs Photonic crystal structure with bandgap in the UV-visible region for TE mode in triangular structure

r/a	a(nm)	r(nm)	Filling Factor	Type	Wavelength (nm) From	Wavelength (nm) To	Gap-Midgap Ratio
0.25	115	29	22.67%	GaN/Air	377	424	11.70%
				GaN/SAGC	392	425	8.06%
0.3	126	38	32.65%	GaN/Air	363	448	20.98%
				GaN/SAGC	386	451	15.54%
0.35	139	49	44.44%	GaN/Air	349	467	28.90%
				GaN/SAGC	381	472	21.30%
0.4	158	63	58.04%	GaN/Air	343	482	33.71%
				GaN/SAGC	388	492	23.63%

Table 2: Proposed GaN-SAGCs Photonic crystal structure with bandgap in the UV-visible region for TM mode in graphite structure

r/a	a(nm)	r(nm)	Filling Factor	Type	Wavelength (nm) From	Wavelength (nm) To	Gap-Midgap Ratio
0.3	58	17	21.77%	GaN/Air	384	420	8.91%
				GaN/SAGC	388	421	8.15%
0.35	60	21	29.63%	GaN/Air	376	425	12.18%
				GaN/SAGC	382	427	10.07%
0.4	63	25	38.69%	GaN/Air	368	434	16.28%
				GaN/SAGC	377	437	14.66%
0.45	68	31	48.97%	GaN/Air	362	450	21.66%
				GaN/SAGC	375	455	19.25%
0.5	73	37	60.46%	GaN/Air	348	462	27.98%
				GaN/SAGC	366	468	24.49%

The proposed material system can also be broaden to encompass the infrared wavelength regime extending from 1.3 μm to 1.55 μm for the TE mode with SAGC length (r) ~ 90 nm. Our estimation also shows that TM active PBGs in the infrared regime can be achieved if the ordering of the guanine molecules can be extended to ~ 165 nm. We therefore have demonstrated the enhancement of luminescence from SAGCs conjugated to GaN nanostructures. Based on the self-assembly on GaN nanostructures, we have designed GaN-SAGC based hybrid photonic crystal structures for bandgap in the UV-IR range.

1. E. Braun, Y. Eichen, U. Sivan, G. Ben-Yoseph, *Nature*, 391 ,775-780(1998).
2. Y. Okahata, T. Kobayashi, K. Tanaka, and M. Shimomura, *J. Am. Chem. Soc.*, 120, 6165-6166(1998).
3. D.Porath, A.Bezryadin, *Nature*, 403, 635-637(2000).
4. T. Uchihasi, T. Okada, Y. Sugawara, K. Yokoyama and S. Morita, *S. Phys. Rev. B*, 60, 8309-8313 (1999)
5. S.J. Sowerby, M.Edelwirth, and W. M. Heckl, W.M., *J. Phys. Chem. B* 102, 5914-5922(1998).

6. L. A. Bumm, J. J. Arnold, M. T. Cygan, T. D. Dunbar, T. P. Burgin, L. Jones II, D. L. Allara, J. M. Tour, and P. S. Weiss. *Science* 271, 1705-1707(1996).
7. R. H. Friend, R. W. Gymer, A. B. Holmes, J. H. Burroughes, R. N. Marks, C. Taliani, D. D. C. Bradley, D. A. Dos santos, J. L. Brédas, M. Lögdlund and W. R. Salaneck, *Nature* , 397, 121-128(1999).
8. M.S. Vollmer, F. Effenberg, R. Stecher, B. Gompf, W. Eisenmenger, *Chem. Eur. J*, 5,96-101(1999).
9. P.Samori,V. Francke,K. Mullen,J.P. Rabe, *Chem. Eur. J*, 5, 2312-2317(1999).
10. G. Maruccio, P. Visconti, V. Arima, S. D'Amico, A. Biasco, Eliana D'Amone, R. Cingolani, R. Rinaldi, S. Masiero, T. Giorgi, and G. Gottarelli, Nano Lett. 3, 479 (2003).
11. S. Johnson, and J. Joannopoulos, *Opt. Express*, 8, 173-190(2001).
12. D. Cassagne, C. Jouanin, D. Bertho, *Phy. Rev. B*, 52, R2217 (1995)
13. W. M. Lomer, *Proc. R. Soc. London Ser. A*, 227, 330 (1955)
14. J.M.Williamson, *Curr. Opin. Struct. Biol.* 3, 357 (1993)
15. G. Gottarelli, S. Masiero , E. Mezzina, S. Pieraccini ,J.P. Rabe, P. Samor'ý P and G. P. Spada, *Chem.Eur.J.*,6,3242-3248(2000).
16. G. Gottarelli, S. Masiero, E. Mezzina, G. P. Spada, P. Mariani,M. Recantini, *Helv.Chim. Acta*, 81, 2078-2091 (1998)
17. S. L. Forman, J.C. Fettinger, S. Pieraccini,G. Gottarelli, J.T. Davis, *J. Am. Chem. Soc*, 122(17); 4060-4067(2000).
18. T.C. Marsh, J Vesenka, and E Henderson, *Nucl. Acids. Res.*, 23, 696- 700(1995).
19. J.L. Sessler, M. Sathiosatham, K. Doerr,V. Lynch, K.A. Abboud, *Angew. Chem. Int. Ed.*, 39, 1300-1303 (2000).
20. D. Gerion, W. J. Parak, S. C. Williams, D. Zanchet, C. M. Micheel, and A. P.Alivisatos, *J. Am. Chem. Soc.* 124, 7070 (2002)
21. A. Neogi, et al,, IEEE Transactions on Nanotechnology, **2**, 10, 2003
22. A. Neogi et al, Applied Physics Letters, (2005)[In Press].

Mater. Res. Soc. Symp. Proc. Vol. 846 © 2005 Materials Research Society DD8.3

Supernatant Controlled Synthesis of Monodispersed Zinc Sulfide Spheres and Multimers

Yanning Song and Chekesha M. Liddell
Department of Materials Science and Engineering
Cornell University, Ithaca, NY 14853

ABSTRACT

Uniform zinc sulfide spheres and multimers ranging in size from ~ 90 nm to 1.0 micron were produced in large quantities by adding varying amount of supernatant in a secondary nucleation process. The particle morphology was investigated using scanning and transmission electron microscopy. Smaller particles exhibited increased surface roughness in the nitrate system. Characterization by x-ray and electron diffraction showed that the particles were built up from nanocrystallites. The relationship between the particle size, porosity, and refractive index was studied by modeling UV-Vis spectra using the Mie scattering method. Monodispersed zinc sulfide spheres and multimers in the size range from 100 nm to 600 nm can be used as high refractive index building blocks for photonic crystals with band gaps covering the entire visible spectrum as well as portions of the near-IR and UV regions.

INTRODUCTION

Photonic crystals are materials with a periodic refractive index (or dielectric constant) that creates an optical band gap [1]. They have enabled new ways to control light by preventing its propagating through the structure due to Bragg reflection, in the same way as semiconductors manipulate electrons [2].

The self-assembly of monodispersed colloidal particles [3] has been demonstrated as an easy route to the formation of three-dimensionally ordered structures with submicron periodicity. The structures prepared by these methods, have generally led to close packed structures such as face-center cubic (FCC). Theoretical studies show that due to symmetry-induced degeneracy, only a pseudogap exists no matter how large the refractive index contrast in the crystal [4]. Larger and more stable band gaps can be generated by using non-spherical objects as the building blocks to form lattices with reduced symmetry [5]. However, only a few methods (i.e., ion irradiation [6] and geometrical confinement [7]) are available for the synthesis of non-spherical inorganic colloids with monodispersed size and well-defined shape.

Zinc sulfide is a semiconductor with direct band gap of 3.54 eV and refractive index of 2.43 (488 nm), appropriate for photonic applications including photonic crystals. Monodispersed zinc sulfide particles have been prepared by several methods. For example, Wilhelmy and Matijevic [8] used the thermal decomposition of thioacetimide (TAA) to prepare micron sized spherical particles by aging the reaction mixture for several hours using a two step procedure. Celikkaya and Akinc [9] found that the particle size and size distribution is affected considerably by the chemical nature of the supporting anions and by the sulfide ion generation rate. Liddell et al. determined that non-spherical zinc sulfide multimers in the size range 600 nm to 3 microns could be synthesized by precipitation induced clustering of preformed monodispersed spheres in acidic aqueous solutions of metal salt and thioacetamide. The cluster formation process was directed by varying the reactant concentration, temperature and time of reaction [10-12]. The present work presents a method for enhanced control of colloid size and degree of clustering in the size regime below 600 nm through a secondary reaction process in which the amount of primary supernatant was systematically varied.

EXPERIMENTAL

Synthesis

Colloidal particles of ZnS were prepared by the thermal decomposition of thioacetamide (TAA, CH_3CSNH_2) in acidic aqueous solutions of zinc salt [8, 9]. The particles were precipitated via a two-stage process from solutions with fixed composition— 0.0153 M $Zn(NO_3)_2$, 0.00168 M $Mn(NO_3)_2$, and 0.143 M TAA. Small amounts of manganese seem to promote a higher degree of monodispersity in the ZnS precipitate, however the effect has not been investigated in detail here. TAA was added to the cation solutions at room temperature and the solutions were acidified with 0.800 ml of 15.8 M HNO_3 in order to control the release of sulfide ions. The reactions were carried out in 200 ml Erlenmeyer flasks sealed with laboratory film. Reaction stages proceeded as follows:

STAGE 1. Primary seeds were produced by immersing the reaction vessels in a constant temperature water bath for 5 hours at 29 °C. The seed solutions were placed in a second bath at 85 °C to grow colloidal particles. The reactions were terminated after 20–24 min by quenching the solutions to below 10 °C in an ice bath. The reaction product was centrifuged at 4000 rpm for 14 minutes.

STAGE 2. The supernatant from the Stage 1 reaction was pipetted from the centrifuge tube and added to fresh seed solutions, prepared as described above (5 h, 29 °C). Stage 2 reactions were terminated after 19-24 min at 85 °C by quenching the solutions to below 10 °C in an ice bath. The onset of precipitation (a noticeably cloudy suspension) in samples containing supernatant typically occurred earlier than was observed for samples without such an addition. The ZnS particles were separated from the mother liquid by centrifuging and were redispersed and washed three times with water. The particles were collected by filtration on 0.1 μm cellulose ester membranes.

Sample identification and conditions for those discussed in this study are listed in table I.

Characterization

Dry ZnS powder was resuspended in water, ultrasonically dispersed and pipetted onto a bare aluminum SEM mount. The particles were allowed to settle for 30 minutes before removing the

Table I. Preparation conditions for zinc sulfide samples

Sample	Reaction Time at 85°C Stage 1 (min)	Supernatant Added (ml)	Reaction Time at 85°C Stage 2 (min)
24xxxx	24	-	-
240524	24	5	24
246024	24	60	24
224019	22	40	19
226019	22	60	19

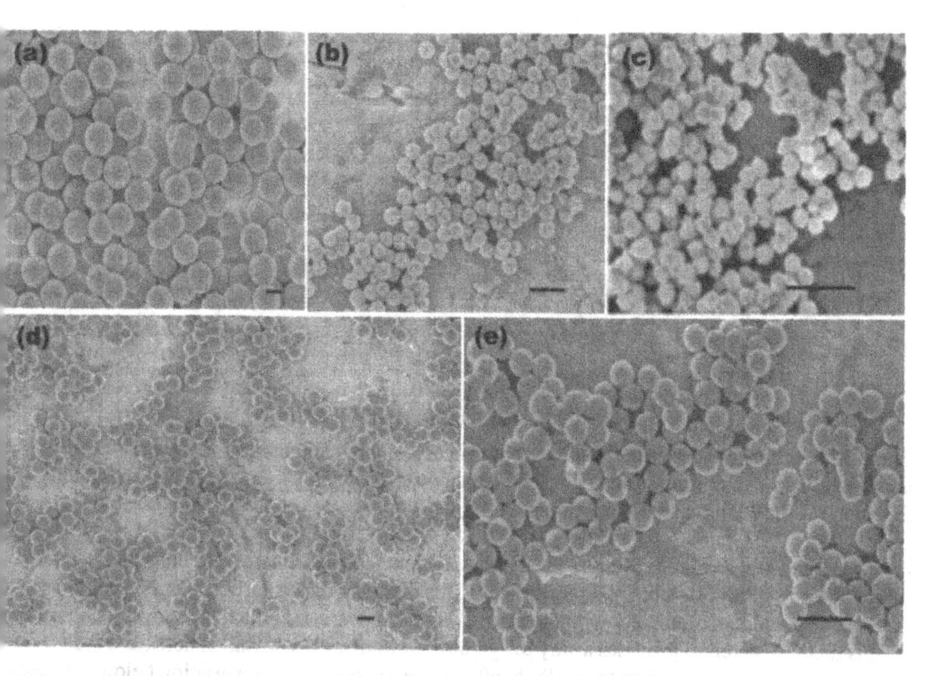

Figure 1. SEM micrographs of (a) Sample 24xxxx (b) Sample 240524 (c) Sample 246024 (d) Sample 224019 (e) Sample 226019. Scale bars represent 500 nm.

excess liquid and drying the mount in a flow of nitrogen gas. SEM images were obtained at 1-5 kV using a LEO 1550 Scanning Electron Microscope. The average particle size, standard deviation and coefficient of variation (CV) were determined from the SEM images, typically by measuring 50 to 100 particles. For TEM imaging, the particles were deposited from suspension onto 300-mesh copper grids with carbon stabilized formvar support film. The particles were examined using a JEOL 1200EX Transmission Electron Microscope at an operating voltage of 120 kV. X-ray diffraction patterns were collected from dry ZnS samples using Cu Kα radiation on a Scintag diffractometer equipped with an intrinsic germanium detector. A Shimadzu UV-3101PC UV/Vis/Near-IR Spectrophotometer was used to record absorption/extinction spectra (wavelength scan range from 200 to 1100 nm) for ZnS particles that had been resuspended and sonicated immediately prior to measurement.

DISCUSSION

The scanning electron microscope (SEM) image in figure 1(a) shows that the ZnS powder synthesized in the absence of supernatant consists of spheres having diameter 0.912 ± 0.055 μm as well as dimers and other clusters. As described in prior work, clusters are formed via the coagulation of spheres during synthesis from homogeneous solution at elevated temperature [10].

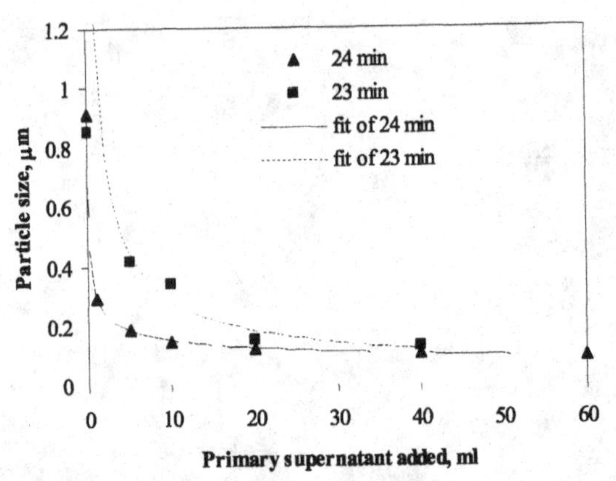

Figure 2. Relationship between the amount of primary supernatant added and the particle diameter. Curve fitting is empirically based on the power function $y = k\,x^{a}$ (a is approximately -2/3 and -1/3 for 23 min and 24 min respectively).

When supernatant was introduced into the synthesis, up to 0.5 g of spheres and clusters with smaller average sizes were obtained. Specifically, the diameter of the particles was found to vary inversely and monotonically with the amount of the supernatant added, as shown in figure 2.

This method therefore provides a way to synthesize larger yields of ZnS spheres and well defined clusters in the size range from 100 to 600 nm, with coefficient of variation below 6%. A similar method based on the addition of primary supernatant has been employed in the synthesis of ZnO colloidal spheres [13]. As discussed previously, there is a greater potential for obtaining full photonic band gaps for assemblies of non-spherical high refractive index building blocks.

Particles obtained from the same supernatant at shorter reaction times were found to be larger than those obtained at longer reaction times. For example, the particle sizes for samples containing 40 ml of supernatant and aged at 19 min, 23 min and 24 min were 0.515 ± 0.021 μm (sample 244019), 0.128 ± 0.007 μm (sample 244023), and 0.103 ± 0.006 μm (sample 244024), respectively. This may be an indication that larger particles are sacrificed for the growth of smaller particles. Further experiments are required to investigate this possibility and the origin of the effect.

The phase purity of the ZnS particles was studied by x-ray diffraction, as shown in figure 3(a). The diffraction peaks were indexed to the planes of sphalerite (ICSD 05-0566), cubic ZnS. As displayed in Figure 3(b), transmission electron microscopy confirmed that the particles are polycrystalline aggregates of nanocrystals. The crystallites were estimated from x-ray diffraction data to be 15 nm in size based on the Scherrer formula. Nanocrystal aggregation generally leads to high porosity and thus low refractive index colloids.

Figure 4 displays the analysis of the ZnS particle porosity and the resulting refractive index. Experimental UV-vis spectra for aqueous colloidal ZnS suspensions were simulated utilizing the program *Mie Simulator* by Bernhard Michel [14]. For modeling purposes, particle sizes were fixed to the average values determined by SEM and only the zinc sulfide filling fraction (f_{ZnS}) was varied to obtain the simulated spectra.

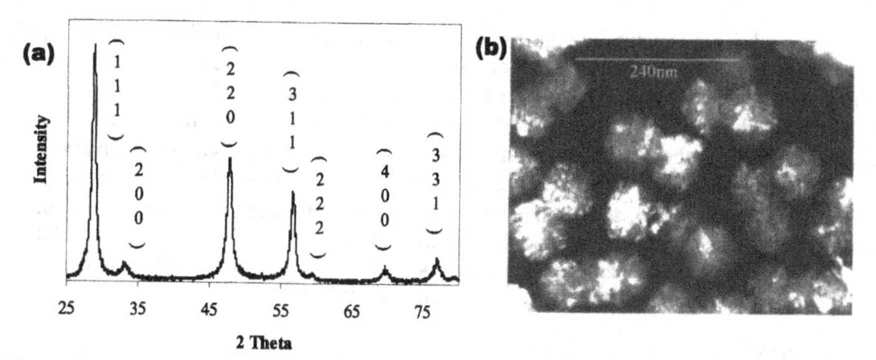

Figure 3. (a) XRD pattern and (b) TEM micrograph for sample 246024. Peaks are well indexed to sphalerite ZnS and polycrystalline substructure is apparent.

The filling fraction provided the best fit between experimental and simulated spectra and was used to estimate the porosity of colloid according to the relation, porosity = $(1-f_{ZnS})*100\%$. The average effective refractive index of the particles in suspension was evaluated as

$$n_{eff} = f_{ZnS}\, n_{ZnS} + (1-f_{ZnS})n_{water}$$

where n_{ZnS} was obtained from the literature.

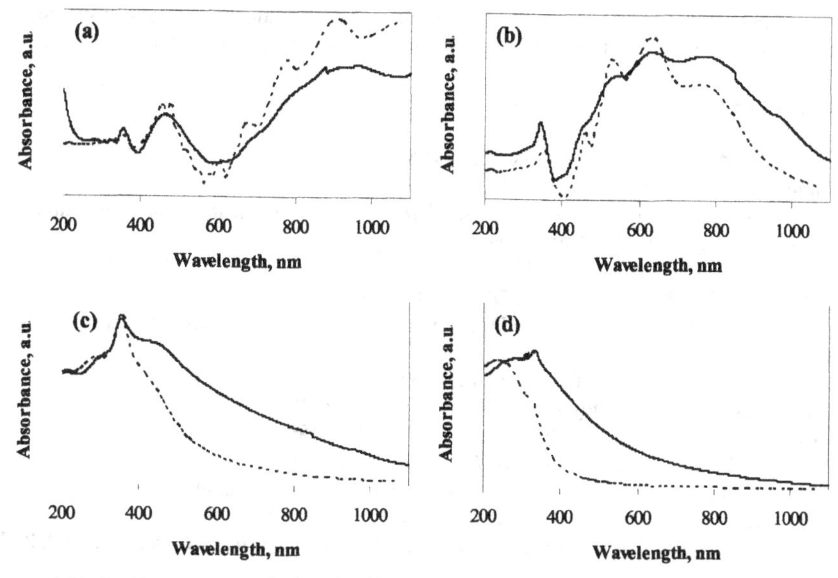

Figure 4. Extinction spectra calculated using Mie theory (dotted line) superimposed on experimental UV/VIS/Near-IR spectra (solid line). (a) Sample 24xxxx, f=0.74; (b) Sample 224019, f=0.78; (c) Sample 240524, f=0.70; (d) Sample 246024, f=0.73.

The observed ripple structure in Figure 4(a) and 4(b) indicates a narrow size distribution [15]. The corresponding samples exhibit these short range oscillations, due to the internal resonance or breathing modes of the spheres. The ripple structure is superimposed on a long range wave structure. When the particle size decreased from 0.912 ± 0.055 μm (figure 4 (a)) to 0.511 ± 0.021 μm (figure 4(b)), the wave structure shifted to the left. For sample 240524 with particle size 0.185 ± 0.011 μm (figure 4(c)), only a shoulder at around 500 nm was observed. When the particle size was even smaller, as for sample 246024 (0.095 ± 0.006 μm), the peaks moved out of the visible range. The average filling fraction of ZnS for samples with particle size from ~100 nm to ~900 nm, corresponds to an average particle porosity of 27 %. The effective refractive index of these particles ranges from 2.21 at 413 nm to 2.06 at 709 nm.

CONCLUSION

In summary, uniform zinc sulfide colloidal clusters with particle sizes from ~90 nm to 1.0 micron have been prepared. The synthesis involves the addition of varying amounts of supernatant in a two-stage process. This method allowed the controllable production of ZnS clusters in large quantities. The particles exhibited morphology and optical properties that may be suitable for photonic crystals applications. Although the colloids were aggregates of nanocrystals, they still exhibited a high refractive index.

ACKNOWLEDGEMENTS

The authors thank Khalilah Bey (senior thesis) for her contribution to the colloid synthesis. We also acknowledge the Cornell Center for Materials Research for use of shared facilities.

REFERENCES

1. E. Yablonovitch, *Phys. Rev. Lett.* **58,** 2059 (1987).
2. S. John, *Phys. Rev. Lett.* **58,** 2486 (1987).
3. *Colloids and Colloid Assemblies: Synthesis, Modification, Organization and Utilization of Colloid Particles*, Editor: F. Caruso, Wiley-VCH (2004)
4. J. W. Haus, *J. Mod. Opt.*, **41**, 195 (1994)
5. K. Busch and S. John, *Phys. Rev. E*, **58,** 3896 (1998).
6. E. Snoeks, A. van Blaaderen, T. van Dillen, C. M. van Kats, K. Velikov, M. L. Brogersma and A. Polman, *Adv. Mater.*, **13**, 1511 (2001).
7. Y. Lu, Y. Yin and Y. Xia, *Adv. Mater.*, **13**, 415 (2001).
8. D. M. Wilhelmy and E. Matijevic, *J. Chem. Soc.* **80**, 563 (1984).
9. A. Celikkaya and M. Akinc, *J Am. Ceram. Soc.* **73**, 245 (1990).
10. C. M. Liddell and C. J. Summers, *Adv. Mater.* **15**, 1715 (2003).
11. C. M. Liddell and C. J. Summers, A. M. Gokhale, *Mater. Char.*, **50**, 69 (2003).
12. C. M. Liddell and C. J. Summers, *J. Colloid Inter. Sci.* **274**, 103 (2004)
13. E. W. Seelig, B. Tang, A. Yamilov, H. Cao and R.P.H. Chang, *Mater. Chem. Phys.* **80,** 257 (2003).
14. M. Bernhard, *MieCalc*, (2002). Online. Internet. 1 Nov. 2004. Available: http://www.unternehmen.com/Bernhard-Michel/MieCalc/eindex.html.
15. S. M. Scholz, R. Vacassy, J. Dutta and H. Hofmann, *J. Appl. Phys.* **83**, 7860 (1998).

Mater. Res. Soc. Symp. Proc. Vol. 846 © 2005 Materials Research Society DD8.7

Organic/Inorganic Hybrid Silicate Materials for Optical Applications; Highly Fluorinated Hybrid Glasses Doped with (Erbium-ions/CdSe nanoparticles) for Laser Amplifier Material

Kyung M. Choi and John A. Rogers

Bell Laboratories, Lucent Technologies, Murray Hill, New Jersey, U. S. A.

A new family of organic/inorganic hybrid silicate materials, bridged polysilsesquioxanes, was designed and synthesized through a molecular-level mixing technique. Since hybrid materials in the molecular-composite level, whose domain sizes are in the nanometer-scale, and whose constituents often lose individual identities and thus create new properties, we obtained a set of improved properties from those organically modified glasses. By modifying the Si-O-Si polymeric network, in this study, we produced controllable, porous hybrid glasses for facile and uniform doping of various ions, metals or semiconductor particles. By taking advantage of void volume created in those molecularly modified silicate systems, novel optical materials with designed properties can thus be achieved. Via a chemical strategy, we designed hexylene- or fluoroalkylene-bridged hybrid glasses doped with both Er^{+3} ions and CdSe nano-particles for the development of new laser amplifier materials. In photoluminescence experiments, a significant enhancement in fluorescence intensity at 1540 nm has been obtained from the fluoroalkylene-bridged glass. The presence of CdSe nano-particles, by virtue of their low phonon energy, also appears to significantly influence the nature of the surrounding environment of Er^{+3} ions in those modified silicate systems, resulting in the increased fluorescence intensity.

Introduction

The development of new photonic materials has been actively pursued for decades. Unique or efficient optical device materials have been sought for various information processing devices and technologies such as optical fibers, waveguides, optical displays, optical switching devices, laser devices, optical lenses, laser amplifiers, and holographic devices. However, there are many inherent limitations in modifying or creating material structures with the requisite properties for these applications in single component materials. The use of multi-component material systems such as the organic/inorganic combination employed in this work, allows us to go beyond such limitations and create new optical materials, which hitherto has not been possible.

In general, organic polymers offer easy processibility while inorganic materials offer superior thermal stability and compatibility with common inorganic substrates. For this reason, organic/inorganic hybrid silicates have been widely investigated as a new class of optical device materials with a reliability of properties that are not found in conventional silicate materials since the development of hybrid materials to optimize materials' properties became critical to bring new advances.

Since this paper has to cover up full of our presentation at MRS 2004 Fall meeting, some data presented here may be duplicated with those that we already submitted for other publication.

Organic/inorganic hybrid materials can range, from physical mixtures of inorganic oxides and organic compounds including blends, composites, and nanocomposites, to molecular composites that utilize formal chemical linkages between the organic and inorganic domains. Since the resulting properties of these hybrid materials are influenced by each of the components and domain sizes that are assembled, hybrid materials in the molecular composite state (domain sizes in the nanometer regime) often provide improved properties, which would not be expected from individual organic or inorganic components.

For example, hybrids are often microscopically homogeneous materials with a completely uniform distribution of organic and inorganic moieties in a domain size at the molecular level and thus the resulting silicates could be made essentially free of phase separation problems.

In this work, we employed the molecular-level mixing technique to produce useful optical device materials based on organic/inorganic hybrid silicate materials. These optical materials based on hybrid systems can be achieved by inserting functional organic domains between inorganic oxide groups.[1-6] The molecular modification of silicate structures has been shown to produce useful optical glasses with controllable porous structures (Figure 1).[5,6]

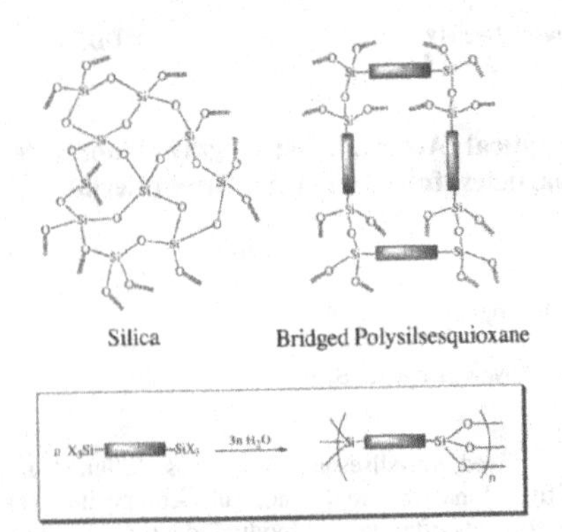

Figure 1. Structural differences between conventional silica and the organically modified hybrid silica; the shaded rectangles correspond to the variable organic fragments.

Rare earth ion-doped glasses are of great interest for a wide range of photonic applications.[7,8] Er^{3+} ion-doped silica in particular has been widely studied for laser amplifier or planar waveguide applications.[9,10] In previous works, scientists have been working on the development of inorganic silica doped with Er^{3+} ion to achieve high NIR efficiency and low phonon energy of the matrix to shorten the pumping distance, and to obtain proper gain and life time.[7-11]

In this work, we have designed and synthesized fluorine-modified silica based on polysilsesquioxane. The size of the inserted organic spaces was varied in an attempt to control the pore size of the hybrid silica. New alkylene- or fluoroalkylene-bridged hybrid glasses have been prepared for use as the basis for Er^{3+} ion as well as CdSe nanoparticles doping.

There has been much research and development effort in doping Er^{3+} ions homogeneously into the normal glass matrix. Work has been reported for Er^{3+} doping in other types of matrices, such as lithium niobate or chalcogenide.[12,13] In this work, we demonstrated that highly fluorinated hybrid silica, polysilsesquioxane, can be a promising candidate matrix for Er^{3+} ion-doped laser amplifier applications because of the low phonon energy of fluorinated silica matrix.

In order to carry out this approach, three different sol-gel processable monomers have been prepared with their structures based of normal tetraethoxysilane (TEOS)-based silica, alkylene-bridged and fluoroalkylene-bridged polysilsesquioxanes. Those sol-gel monomers, are then polymerized to produced condensed xerogels. In Figure 2, the chemical structures of modified silicate materials are shown. In fluoroalkylene-bridged xerogel, fluorine has been inserted into inorganic oxides to play the role of resulting the phonon energy of the hybrid glass matrix for improved laser performance as a result of decreasing the extent of electron-phonon interactions.

Erbium isopropoxide was used as the source for Er^{3+} ions in this work; erbium isopropoxide was doped into sol-gel mixtures prepared from the aforementioned three sol-gel processable monomers. We have also incorporated CdSe nanoparticles into these systems because of the low phonon energy associated with CdSe (a phonon energy of nano-sized CdSe: 200 cm^{-1}). After the sol-gel mixtures containing the monomers and erbium isopropoxide were prepared, the CdSe nanoparticles were then added into the mixtures. Incorporating the CdSe nano-particles into the silicate matrix contributes to further reducing the phonon energy of the matrix. The process of doping CdSe particles was performed based on the previous work, which shows that a strong enhancement of 1540 nm fluorescence can be achieved in highly concentrated semiconductor nanoparticle environments.

In order to produce efficient laser amplifier materials, we need to consider several key parameters. For example, the materials' homogeneity is important in photonic materials, due to light scattering issue. The dispersion of the rare earth ions in sol-gel glasses was recently demonstrated in a fluorescence line narrowing study.[11] We have examined the homogeneity of three silicate systems doped with Er^{3+} ions or CdSe nanoparticles.

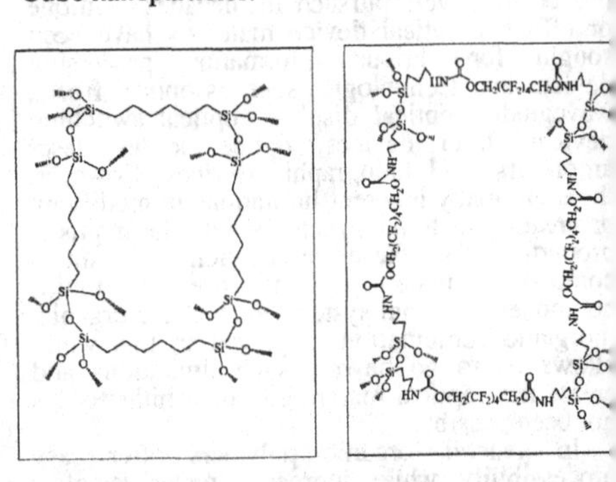

Figure 2. Chemical structures of different silicate systems; (left) hexylene- and (right) fluoroalkylene-bridged polysilsesquioxane.

In this study, we demonstrated that the TEOS-based silicate mixtures show a substantial degree of undesirable phase separation after doping with either Er^{3+} ions or CdSe nanoparticles while fluoroalkylene-based modified silicate matrix doesn't show any significant phase separations. Additionally, the inorganic silica material containing the erbium ions has been heated to remove OH-groups in order to avoid the quenching effect due to the presence of high-phonon energy OH-groups (3000-3500 cm^{-1}). During this heating process, the silica may capture some impurities, which would serve as light scattering centers. However, sol-gel process in this study was performed at a low temperature (< 200 °C). For this reason, the hybrid silicate materials would be free of light scattering problem arising from residual impurities.

Results and Discussions

Chemical Homogeneity: In the development of optical materials, the chemical homogeneity is essential due to light scattering issue. We take advantages from hybrid silicate matrices, which often result in homogeneous distribution of dopants by controlling pore structures.

We demonstrated the homogeneity of three different silicate systems doped with Er^{+3} ions or CdSe nano-particles (Figure 3). Ethanol was used for the sol-gel mixtures. A visual inspection was performed for this study.

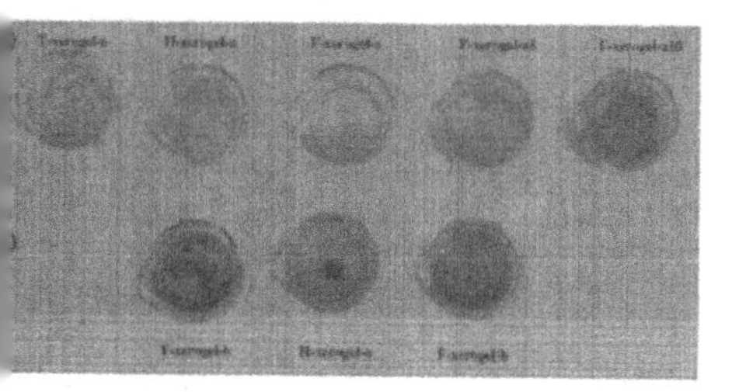

Figure 3. Visual inspection of mixing homogeneity (a) Er^{+3} ion-doped and (b) biphase of (Er^{+3}/CdSe)-doped sol-gel mixtures (The mixing composition is listed in Table 1); xerogels-a5, and -a10 denote different erbium concentrations, five and ten times higher erbium-ion concentrations than that of xerogel-a.

In the investigation shown in Figure 3, we found that the TEOS-based silica results in a substantial degree of undesirable phase separation after doping.

In Figure 3, erbium isopropoxide reveals a pink color. As demonstrated in the mixing test, the T-xerogel-a shows a significant phase separation after the doping of erbium-ion source, which shows that the TEOS-based sol-gel mixture has a rather limited solubility of Er^{+3} ions (T-xerogel-a). In contrast, modified sol-gel mixtures accommodate and homogeneously distribute the Er^{+3} source without any phase separation.

We then doped CdSe nano-particles into the three different sol-gel mixtures since those nanocrystalline particles are prepared for easy incorporation into a variety of matrices without significant phase separation. The incorporation of such nano-particles into the hybrid silicate systems is expected to influence the optical behavior of optical materials, for example, photoluminescence intensity, in this case.

In the mixing test, the CdSe nano-particles segregated and didn't get incorporated well into the TEOS-based mixture (T-xerogel-b; orange color indicates CdSe nano-particles, which agglomerated and phase separated within the TEOS-based mixture. In contrast, for the case of hybrid silicate systems, it shows less phase separation in the same concentration; we observed some orange-colored clusters toward the middle of the container.

In fluoroalkylene-bridged case, which is denoted by F-xerogel-b (Figure 3), the CdSe nano-particles were completely incorporated without any phase separation. This result demonstrates that the fluoroalkelene sol-gel system has the capability of uniformly incorporating those dopants without significant phase separation. The sol-gel mixtures were then polymerized to produce the condensed xerogels.

Spectroscopic Characterizations: The chemical characterizations for those xerogels can be also determined by spectroscopic techniques including solid state nuclear magnetic resonance, infrarad, and Raman spectroscopes.

We employed solid state NMR technique to determine the degree of condensation. ^{29}Si solid state NMR was used to identify the formation of the "Si-O-Si" linkage in three silicate systems. Single pulse magic angle spinning (SP/MAS) NMR technique was employed for the calculation of the degree of condensation.

We obtained the ^{29}Si solid state SP/MAS solid state NMR spectra and curve fitting lines of both normal silica and the modified silicate systems. From the ^{29}Si SP/MAS solid state NMR spectra for T-xerogel, hexylene- and flouroalkylene-bridged systems, the degrees of condensations were calculated to be 78.8, 79.1, and 91.1 %, respectively.

In the fluoroalkylene-bridged system, the higher degree of condensation, disappearance of T^1 peak, and enhanced T^3 peak indicates a lower level of OH-group content as well as high hydrophobicity raised from the strategic fluorine-insertion. Therefore, such modified hybrid silica may provide erbium-ions with a lower-hydroxyl environment than purely inorganic silicate xerogels.

We also carried out photoluminescence experiments. The results of photoluminescence study are shown in Figures 4-6. We determined the effect of different silicate matrices on fluorescence arising from erbium ions at 1540 nm wavelength using an Ar^+ ion laser pumped at 488 nm wavelength.

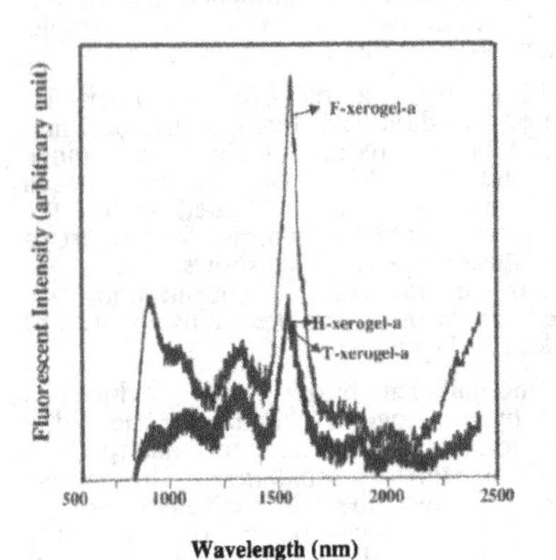

Figure 4. Different silicate matrix effect; comparative fluorescence intensities of Er^{+3}ions doped into three silicate matrices using a low power of 1.5 W/cm^{-2}.

A comparative fluorescence intensities of erbium ions doped into three different silicate matrices is shown in Figure 4 (T-xerogel-a, H-xerogel-a, F-xerogel-a). In Figure 4, fluorescence intensity of erbium-ions increased significantly more in the fluoroalkylene-bridged silicate matrix (the top curve) than the other two silicate systems, which were overlapped in the two bottom curves. We believe that the enhanced fluorescence intensity of the fluoroalkylene-based silica is attributed to mainly its low phonon energy in the fluorinated matrix combined with highly uniform dispersion of erbium ions. It shows a silicate matrix effect, which the advantage of the modified silicate matrices. This approach permits the easy molecular-level mixing technique without encountering the problem of phase separation, which is often difficult to suppress in many other inorganic single components.

Further investigations in fluorescence measurements have also been carried to examine the effort of other parameters. The erbium-ion concentration has been found to significantly affect the fluorescence intensity (Figure 5).

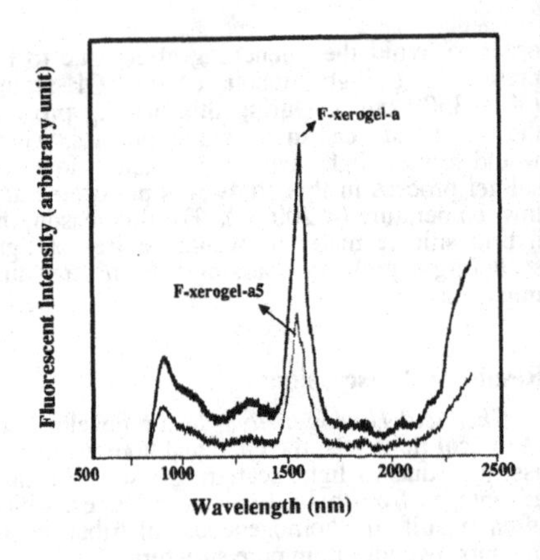

Figure 5. Erbium concentration effect; comparative fluorescence intensities of different Er^{+3} ion-concentrations doped into fluoroalkylene-bridged silicate matrices (F-xerogels-a1 and -a5) using a power of 3.0 W/cm^{-2}.

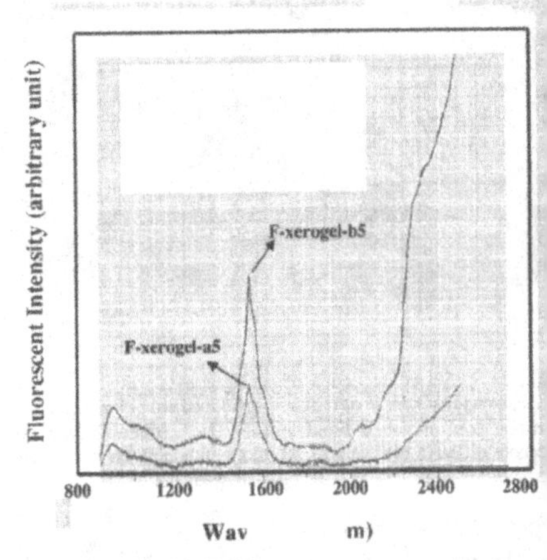

Figure 6. CdSe effect; comparative fluorescence intensities of fluoroalkylene-bridged silica doped with Er^{+3} ions and doped with both Er^{+3}/CdSe (F-xerogels-a5 and -b5) using a high power of 3 W/cm^{-2}.

For the study of erbium-ion concentration effect, a couple of fluoroalkylene-bridged silicate glasses doped with two different levels of erbium concentrations were prepared; F-xerogel-a (1.49 atomic % of erbium by XPS study) and F-xerogel-a5 (2.98 atomic % of erbium by XPS analysis).

As shown in Figure 5, the result reveals that the fluorescence intensity increases (the upper curve) when the erbium doping level was low (1.49 atomic %) due to "self-quenching effect".

We then examined CdSe effects in fluorescence intensity using the fluorinated xerogel (Figure 6). It is anticipated that the presence of CdSe in the silicate matrix affects the fluorescence intensity because of its low phonon energy of CdSe (200 cm^{-1}).

As shown in Figure 6, the fluorescence intensity of F-xerogel-a5 is dramatically increased in F-xerogel-b5 when additionally doped with CdSe nano-particles. This result reveals that the nature of the surrounding environment of the erbium-ions significantly influences the photoluminescence intensity. From the experiments shown in Figures 4-6, we conclude that the photoluminescence intensity is affected from the presence of both fluorine and CdSe nano-particles.

By taking advantage of the unique structural features and uniform doping capability in fluorinated silica, we introduce here that the structural characteristics of silicate matrices are very important to improve performance of such optical devices like laser amplifiers due to the limitations in normal inorganic silica.

Table 1. Various Combinations of Silicate Matrices and Dopants.

Xerogels	Silicate matrices	Dopants
T-Xerogel	TEOS	None
H-Xerogel	Hexylene-	None
F-Xerogel	Fluoroalkylene-	None
T-Xerogel-a	TEOS	Erbium ions
T-Xerogel-b	TEOS	Erbium ions/CdSe
H-Xerogel-a	Hexylene-	Erbium ions
H-Xerogel-b	Hexylene-	Erbium ions/CdSe
F-Xerogels-a, -a5, and -a10	Fluoroalkylene-	Erbium ions
F-Xerogels-b, -b5, and -b10	Fluoroalkylene-	Erbium ions/CdSe

References

1. (a) P. A. Agaskar, V. W. Day, W. G. Klemperer, *J. Am. Chem. Soc.* **109,** 5554 (1987). (b) H. Schmidt, "*Sol-Gel Science and Technology*"; World Scientific: Singapore, p. 432 (1989). (c) B. M. Novak, *Adv. Mater.* **5,** 422 (1993). (d) Y. Haruvy, S. E. Webber, *Chem. Mater.* **3,** 501 (1991).

2. (a) R. J. P. Corriu, J. J. E. Moreau, P. Thepot, M. W. C. Man, C. Chorro, J. P. Lere-Porte, J. L. Sauvajol, *Chem. Mater.* **6,** 640 (1994). (b) R. H. Baney, M. Itoh, A. Sakakibara, I. Suzuki, *Chem. Rev.* **95,** 1409 (1995).

3. (a) K. M. Choi, K. J. Shea, *Chem. Mater.* **5,** 1067 (1993). (b) K. M. Choi, K. J. Shea, K. J. *J. Am. Chem. Soc.* **116,** 9052 (1994).

4. (a) K.M. Choi, K.J. Shea, K. J. *J. Phys. Chem.* **98,** 3207 (1994). (b)K. M. Choi, K. J. Shea, K. J. *J. Phys. Chem.* **99,** 4720 (1995).

5. (a)K. J. Shea, D. A. Loy, *Chem. Rev.* **95,** 1431 (1995). (b) K. J. Shea, D. A. Loy, O. W. Webster, *J. Am. Chem. Soc.* **114,** 6700 (1992). (c) K. J. Shea, D. A. Loy, *MRS Bulletin* **26,** 368 (2001).

6.K.M.Choi,K.J.Shea, "*Photonic Polymer SystemsFundamentals, Methods, and Applications*" edited by D.L. Wise et al., World Scientific Publishing Co. Pte. Ltd. , **49** (1998).

7. P. Urquhart, *IEE Proceedings* **135,** 385 (1988).

8. (a) D. J. DiGiovanni, "*Optical Waveguide Materials*" edited by M. M. Broer;, G. J. Sigel, R. T. Kersten, H. Kawazoe, Mater. Res. Soc. Proc. **244,** 135 (1992). (b) J. L. Wagener, P. F. Wysocki, M. J. F. Digonnet, H. J. Shaw, D. J. DiGiovanni, *Proceedings of SPIE,* **1789,** 80 (1993). (c) D. J. DiGiovanni, *Ceramic Transactions* **75,** 73 (1997).

9. V. P. Gapontsev, S. M. Matitsin, A. A. Isineev, V. B. Kravchenko, *Optics and Laser Tech.* **14,** 189 (1982).

10. B. J. Ainslie, *J. Lightwave Tech.* **9,** 220 (1991).

11. M. J. Lochhead, K. L. Bray, *Chem Mater.* **7,** 572 (1995).

12. H. Lin, S. Jiang, J. Wu, F. Song, N. Peyghambarian, E. Y. B. Pun, *J. Phys. D.; Applied Physics* **36,** 812 (2003).

13. W. Sohler, H. Suche, *Optical Engineering* **66,** 127 (2000).

Mater. Res. Soc. Symp. Proc. Vol. 846 © 2005 Materials Research Society DD8.9

Microscopic Theory of Surface-Enhanced Raman Scattering in Noble-Metal Nanoparticles

Vitaliy N. Pustovit and Tigran V. Shahbazyan
Department of Physics, Jackson State University, Jackson MS 39217

ABSTRACT

We study the role of a strong electron confinement on the surface-enhanced Raman scattering from molecules adsorbed on small noble-metal nanoparticles. We describe a novel enhancement mechanism which originates from the different effect that confining potential has on s-band and d-band electrons. We demonstrate that the interplay between finite-size and screening efects in the nanoparticle surface layer leads to an enhancement of the surface plasmon local field acting on a molecule located in a close proximity to the metal surface. Our calculations, based on time-dependent local density approximation, show that the additional enhancement of the Raman signal is especially strong for small nanometer-sized nanoparticles.

INTRODUCTION

An recent interest in single-molecule surface-enhanced Raman scattering (SERS) stems from the discovery of enormously high (up to 10^{15}) enhancement of Raman spectra from certain (e.g., Rhodamine 6G) molecules fixed at the nanoparticle surfaces in gold and silver colloids [1–3]. Major SERS mechanisms include electromagnetic (EM) enhancement by surface plasmon (SP) local field near the metal surface [4–8] and chemical enhancement due to dynamical charge transfer between a nanoparticle and a molecule [9–12]. Although the origin of this phenomenon has not completely been elucidated so far, the EM enhancement was demonstrated to play the dominant role, especially in dimer systems when the molecule is located in the gap between two closely spaced nanoparticles [10–13].

An accurate determination of the SERS signal intensity for molecules located in a close proximity to the nanoparticle surface is a non-trivial issue. The classical approach, used in EM enhancement calculations [4–8], is adequate when nanoparticle-molecule or interparticle distances are not very small. For small distances, the quantum-mechanical effects in the electron density distribution can no longer be neglected. These effects are especially important in noble-metal particles where the SP local field is strongly affected by highly-polarizable (core) d-electrons. In the bulk part of the nanoparticle, the (conduction) s-electrons are strongly screened by the localized d-electrons. However, near the nanoparticle boundary, the two electron species have different density profiles. Namely, delocalized s-electrons spill over the classical boundary [14], thus increasing the effective nanoparticle radius, while d-electron density profile mostly retains its classical shape. The incomplete embedding of the conduction electrons in the core electron background [15–18] leads to a reduced screening of the s-electron Coulomb potential in the nanoparticle surface layer. The latter has recently been observed as an enhancement of the electron-electron scattering rate in silver nanoparticles [19, 20].

Here we study the role of electron confinement on SERS from molecule adsorbed on the surface of small Ag particles. To this end, we develop a microscopic theory for SERS in noble-metal particle, based on the quantum extension of two-region model [15–17, 21], which describes the role of the surface-layer phenomenologically while treating conduction electrons quantum-mechanically within time-dependent density functional theory. We find that the reduction of screening near the surface leads to an additional enhancement of the Raman signal from a molecule located in a close proximity to the nanoparticle. In particular, we address the dependence of SERS on nanoparticle size and show that the interplay of finite-size and screening effects is especially strong for small nanometer-sized particles.

QUANTUM TWO-REGION MODEL

We consider SERS from a molecule adsorbed on the surface of Ag spherical particle with radius R. For $R \ll \lambda$, the frequency-dependent potential is determined from Poisson equation,

$$\Phi(\omega, \mathbf{r}) = \phi_0(\mathbf{r}) + e^2 \int d^3r' \frac{\delta N(\omega, \mathbf{r}')}{|\mathbf{r} - \mathbf{r}'|}, \tag{1}$$

where $\phi_0(\mathbf{r}) = -e\mathbf{E}_i \cdot \mathbf{r}$ is potential of the incident light with electric field amplitude $\mathbf{E}_i = E_i \mathbf{z}$ along the z-axis, and $\delta N(\omega, \mathbf{r})$ is the induced density (hereafter we suppress frequency dependence). There are four contributions to $\delta N(\mathbf{r})$ originating from valence s-electrons, $\delta N_s(\mathbf{r})$, core d-electrons, $\delta N_d(\mathbf{r})$, dielectric medium, $\delta N_m(\mathbf{r})$, and the molecule, $\delta N_0(\mathbf{r})$. The density profile of delocalized s-electrons is not fully inbedded in the background of localized d-electrons but extends over that of localized d-electrons by $\Delta \sim 1-3$ Å [15–17]. Due to localized nature of the d-electron wave-functions, we can adopt a phenomenological description by assuming a uniform bulk-like d-electron ground-state density n_d in the region confined by $R_d < R$ with a sharp step-like edge. Then the induced charge density, $e\,\delta N_d(\mathbf{r}) = -\nabla \cdot \mathbf{P}_d(\mathbf{r})$, is expressed via electric polarization vector vanishing outside of the region $r < R_d$,

$$\mathbf{P}_d(\mathbf{r}) = \frac{\epsilon_d - 1}{4\pi}\,\theta(R_d - r)\mathbf{E}(\mathbf{r}) = -\frac{\epsilon_d - 1}{4\pi e}\,\theta(R_d - r)\nabla\Phi(\mathbf{r}), \tag{2}$$

where $\epsilon_d(\omega)$ is the core dielectric function which can be taken from experiment. Similarly, dielectric medium contribution, which is nonzero for $r > R$, is given by

$$e\,\delta N_m(\mathbf{r}) = -\nabla \cdot \mathbf{P}_m(\mathbf{r}), \quad e\,\mathbf{P}_m(\mathbf{r}) = -\frac{\epsilon_m - 1}{4\pi}\,\theta(r - R)\nabla\Phi(\mathbf{r}), \tag{3}$$

where ϵ_m is medium dielectric constant. We represent the molecule by a point dipole with dipole moment $\mathbf{p}_0 = \boldsymbol{\alpha}_0\mathbf{E}$, where $\boldsymbol{\alpha}_0$ is the molecule polarizability tensor, so that

$$e\,\delta N_0(\mathbf{r}) = -\nabla \cdot \mathbf{P}_0(\mathbf{r}), \quad e\,\mathbf{P}_0(\mathbf{r}) = \delta(\mathbf{r} - \mathbf{r}_0)\,\mathbf{p}_0 = -\delta(\mathbf{r} - \mathbf{r}_0)\,\boldsymbol{\alpha}_0\nabla\Phi(\mathbf{r}_0), \tag{4}$$

where $\mathbf{r}_0$ is the vector pointing at the molecule location (we chose origin at the sphere center).

Using Eqs. (2-4), the potential $\Phi(\mathbf{r})$ in Eq. (1) can be expressed in terms of only induced s-electron density, δN_s. Substituting the above expressions into $\delta N = \delta N_s + \delta N_d + \delta N_m + \delta N_0$ in the rhs of Eq. (1) and integrating by parts, we obtain,

$$\begin{aligned}
\epsilon(r)\Phi(\mathbf{r}) \;=\;& \phi_0(\mathbf{r}) + e^2 \int d^3r' \frac{\delta N_s(\mathbf{r}')}{|\mathbf{r} - \mathbf{r}'|} + \frac{\epsilon_d - 1}{4\pi} \int d^3r' \nabla' \frac{1}{|\mathbf{r} - \mathbf{r}'|} \cdot \nabla'\theta(R_d - r)\Phi(\mathbf{r}') \\
&+ \frac{\epsilon_m - 1}{4\pi} \int d^3r' \nabla' \frac{1}{|\mathbf{r} - \mathbf{r}'|} \cdot \nabla'\theta(r - R)\Phi(\mathbf{r}') - \nabla_0 \frac{1}{|\mathbf{r} - \mathbf{r}_0|} \cdot \boldsymbol{\alpha}_0 \nabla_0 \Phi(\mathbf{r}_0),
\end{aligned} \tag{5}$$

where $\epsilon(r) = \epsilon_d$, 1, and ϵ_m in the intervals $r < R_d$, $R_d < r < R$, and $r > R$, respectively. Since the source term has the form $\phi(\mathbf{r}_0) = \phi(r_0)\cos\theta = -eE_i r\cos\theta$, we expand Φ and δN_s in terms of spherical harmonics and, keeping only the dipole term ($L = 1$), obtain,

$$\begin{aligned}\epsilon(r)\Phi(r) &= \phi_0(r) + e^2\int dr' r'^2 B(r,r')\delta N_s(r') - \frac{\epsilon_d - 1}{4\pi}R_d^2\partial_{R_d}B(r,R_d)\Phi(R_d) \\ &+ \frac{\epsilon_m - 1}{4\pi}R^2\partial_R B(r,R)\Phi(R) - \nabla_0[B(r,r_0)\cos\theta_0]\cdot\boldsymbol{\alpha}_0\nabla_0[\Phi(r_0)\cos\theta_0], \quad (6)\end{aligned}$$

where θ_0 is the angle between molecule position and incident light direction (z-axis), and

$$B(r.r') = \frac{4\pi}{3}\left[\frac{r'}{r^2}\theta(r-r') + \frac{r}{r'^2}\theta(r'-r)\right] \quad (7)$$

is the dipole term of the radial component of the Coulomb potential.

The second terms in rhs of Eq. (6) is the s-electrons contribution to total induced potential, while the third and fourth terms originate from the scattering due to change of dielectric function at $r = R_d$ and $r = R$, respectively. The last term represents the potential of the molecular dipole. The latter depends on the molecule orientation with respect to the nanoparticle surface. In the following we assume averaging over random orientations and replace the polarizability tensor by isotropic α_0.

The values of Φ at the boundaries and at the molecule position can be found by setting $r = R_d, R, r_0$ in Eq. (6). In doing so, the total potential is expressed in terms of only s-electron induced density δN_s. Within TDLDA formalism, the latter can be related back to the potential via

$$\delta N_s(\mathbf{r}) = \int d^3r'\Pi_s(\mathbf{r},\mathbf{r}')\Big[\Phi(\mathbf{r}') + V'_x[n(r')]\delta N_s(\mathbf{r}')\Big], \quad (8)$$

where $\Pi_s(\mathbf{r},\mathbf{r}')$ is the polarization operator for noniteracting s-electrons, $V'_x[n(r')]$ is the (functional) derivative of the exchange-correlation potential (in the endependent-particle approximation) and $n(r)$ is electron ground-state density. The latter is calculated in a standard way using Kohn-Sham equations for jelium model, and then is used as imput in the evaluation of Π_s and V'_x. Eqs. (6) and (8) determine the self-consistent potential in the presence of molecule, nanoparticle, and dielectric medium.

ENHANCEMENT OF RAMAN SIGNAL

In the conventional SERS picture, the enhancement of Raman signal from the molecule comes from two sources: far-field of the radiating dipole of the molecule in the local nanopar-tricle field, and the secondary scattered field of this dipole by the nanoparticle. Accordingly, we present total potential as a sum $\Phi = \phi + \phi^R$, where ϕ^R is the potential of the radiating dipole. Since the molecular polarizability is very small, in the following we will restrict our-selves by the lowest order in α_0, i.e., ϕ is the potential in the absence of molecule and ϕ_R determines the Raman signal in the first order in α_0. Inclusion of higher orders leads to the renormalization of molecular and nanoparticle polarizabilities due to image charges; these effects are not considered here. In the same manner, the induced s-electron density can be decomposed into two contributions, $\delta N_s = \delta n + \delta n^R$, originating from the electric field of incident light and that of the radiating dipole.

Keeping only zero-order terms in Eq. (6), we have

$$\epsilon(r)\phi(r) = \bar{\phi}(r) - \frac{\epsilon_d - 1}{3}\beta(r/R_d)\phi(R_d) + \frac{\epsilon_m - 1}{3}\beta(r/R)\phi(R), \tag{9}$$

where

$$\bar{\phi}(r) = \phi_0(r) + e^2 \int dr' r'^2 B(r, r')\delta n_s(r'), \tag{10}$$

and $\beta(r/R) = \frac{3}{4\pi}R^2\partial_R B(r, R)$ is given by

$$\beta(x) = x^{-2}\theta(x-1) - 2x\theta(1-x). \tag{11}$$

The boundary values of ϕ can be obtained by matching $\phi(r)$ at $r = R_d, R$,

$$\begin{aligned} (\epsilon_d + 2)\phi(R_d) + 2a(\epsilon_m - 1)\phi(R) &= 3\bar{\phi}(R_d) \\ (\epsilon_d - 1)a^2\phi(R_d) + (2\epsilon_m + 1)\phi(R) &= 3\bar{\phi}(R) \end{aligned} \tag{12}$$

where $a = R_d/R$. Substituting $\phi(R_d)$ and $\phi(R)$ back into Eq. (9), we arrive at

$$\epsilon(r)\phi(r) = \bar{\phi}(r) - \beta(r/R_d)\frac{\lambda_d}{\eta}\left[\bar{\phi}(R_d) - 2a\lambda_m\bar{\phi}(R)\right] + \beta(r/R)\frac{\lambda_m}{\eta}\left[\bar{\phi}(R) - a^2\lambda_d\bar{\phi}(R_d)\right], \tag{13}$$

where

$$a = R_d/R, \quad \lambda_d = \frac{\epsilon_d - 1}{\epsilon_d + 2}, \quad \lambda_m = \frac{\epsilon_m - 1}{2\epsilon_m + 1}, \quad \eta = 1 - 2a^3\lambda_d\lambda_m \tag{14}$$

It is convenient to separate out δn_s-independent contribution by writing

$$\phi = \varphi_0 + \delta\varphi_0 + \delta\varphi_s, \tag{15}$$

where $\varphi_0 = \phi_0/\epsilon(r) = -eE_i r/\epsilon(r)$,

$$\delta\varphi_0(r) = \frac{1}{\epsilon(r)}\Big[-\beta(r/R_d)\phi_0(R_d)\,\lambda_d(1 - 2\lambda_m)/\eta + \beta(r/R)\phi_0(R)\,\lambda_m(1 - a^3\lambda_d)/\eta\Big], \tag{16}$$

and

$$\delta\varphi_s(r) = \int dr' r'^2 A(r, r')\delta n_s(r'), \tag{17}$$

with

$$\begin{aligned} A(r, r') = \frac{e^2}{\epsilon(r)}\Big[B(r, r') \; &- \; \beta(r/R_d)\Big[B(R_d, r') - 2a\lambda_m B(R, r')\Big]\lambda_d/\eta \\ &+ \; \beta(r/R)\Big[B(R, r') - a^2\lambda_d B(R_d, r')\Big]\lambda_m/\eta\Big]. \end{aligned} \tag{18}$$

Note that $\delta\varphi_s(r)$ as well as the total potential $\phi(r)$ are continuous at $r = R_d, R$.

Turning to Eq. (8), we use decompositions $\Phi = \phi + \phi^R$ and $\delta N_s = \delta n_s + \delta n_s^R$ to obtain decoupled equations for quantities of zero and first orders in α_0. Then, expanding both parts in spherical harmonics and keeping only the dipole ($L = 1$) terms, we obtain

$$\delta n_s(r) = \int dr' r'^2 \Pi_s(r,r') \Big[\phi(r') + V_x'(r')\delta n_s(r')\Big]. \tag{19}$$

Using Eqs. (15) and (17) then leads to a closed equation for δn_s,

$$\begin{aligned}\delta n_s(r) &= \int dr' r'^2 \Pi_s(r,r') \Big[\varphi_0(r') + \delta\varphi_0(r')\Big] \\ &+ \int dr' r'^2 \Pi_s(r,r') \Bigg[\int dr'' r''^2 A(r', r'')\delta n_s(r'') + V_x'(r')\delta n_s(r')\Bigg],\end{aligned} \tag{20}$$

with $A(r,r')$ given by Eq. (18). The effect of d-electrons and dielectric medium is thus encoded in the functions $A(r,r')$ and $\delta\varphi_0(r)$, which reduce to $B(r,r')$ and 0, respectively, for $\epsilon_d = \epsilon_m = 1$.

Turning to the first order in α_0, the equation for ϕ^R takes the form,

$$\epsilon(r)\phi^R(r) = \bar{\phi}^R(r) - \frac{\epsilon_d - 1}{3}\beta(r/R_d)\phi^R(R_d) + \frac{\epsilon_m - 1}{3}\beta(r/R)\phi^R(R), \tag{21}$$

with

$$\bar{\phi}^R(r) = \phi_0^R(r) + e^2 \int dr' r'^2 B(r,r')\delta n_s^R(r'), \tag{22}$$

where

$$\phi_0^R(r) = -\alpha_0 \nabla_0 [B(r, r_0)\cos\theta_0] \cdot \nabla_0[\phi(r_0)\cos\theta_0], \tag{23}$$

is the potential of the molecular dipole in the presence of local field and $\delta n_s^R(r)$ is the induced charge of s-electrons due to molecular potential. The frequency dependence of the Raman field $\phi^R(\omega_s, r)$ is determined by the Stokes shift $\omega_s = \omega - \omega_0$, where ω_0 is the vibrational frequency of the molecule as determined by α_0. The last two terms in Eq. (21) describe potential of molecular dipole scattered from the nanoparticle boundaries at R_d and R. For simplicity, we only consider the case when the molecule is located at the z-axis ($\theta_0 = 0$) so that molecular dipole potential is given by

$$\phi_0^R(r) = -\frac{4\pi\alpha_0}{3r_0^2}\,\beta(r/r_0)\,\frac{\partial\phi(r_0,\omega)}{\partial r_0} = -\chi(\omega)\,eE_i\Big[r\theta(r_0 - r) - \frac{r_0^3}{2r^2}\,\theta(r - r_0)\Big], \tag{24}$$

where

$$\chi(\omega) = \frac{8\pi\alpha_0}{3eE_i r_0^3}\frac{\partial\phi(r_0)}{\partial r_0} \tag{25}$$

Marching ϕ^R at $r = R_d$ and $r = R_d$, we obtain

$$\phi^R = \varphi_0^R + \delta\varphi_0^R + \delta\varphi_s^R = \chi(\omega)\Big[\tilde{\varphi}_0 + \delta\varphi_0 + \delta\tilde{\varphi}_s\Big], \tag{26}$$

where

$$\tilde{\varphi}_0(r) = -\frac{eE_i}{\epsilon(r)}\Big[r\theta(r_0 - r) - \frac{r_0^3}{2r^2}\,\theta(r - r_0)\Big], \tag{27}$$

$\delta\varphi_0$ is given by Eq. (16) with ω_s instead of ω and

$$\delta\tilde{\varphi}_s(r) = \int dr' r'^2 A(r, r')\delta\tilde{n}_s(r'), \tag{28}$$

where $A(r, r')$ is given by Eq. (18), with ω_s instead of ω, and $\delta\tilde{n}_s(r) = \delta n_s^R(r)/\chi(\omega)$ satisfies Eq. (20) but with $\tilde{\varphi}_0(r)$ instead of $\varphi_0(r)$ (and ω_s instead of ω). In the following we consider the case when the molecule is located at the distance of several angstroms from the nanoparticle classical boundary, so the overlap between the molecular orbitals and the s-electron wave function is small. Then we have

$$\int dr' r'^2 \Pi_s(r, r')\tilde{\varphi}_0(r') \simeq \int dr' r'^2 \Pi_s(r, r')\varphi_0(r'), \tag{29}$$

leading to $\delta\tilde{n}_s(r) \simeq \delta n_s(r)$ and, correspondingly, $\delta\tilde{\varphi}_s(r) \simeq \delta\varphi_s(r)$. The Raman signal is determined by the far-field asymptotics of $\phi^R(r)$. From Eqs. (16-18), we find for $r \gg R$

$$\delta\varphi_0(r) = \frac{eE_i}{\epsilon_m r^2}\,\alpha_d, \quad \delta\varphi_s(r) = \frac{eE_i}{\epsilon_m r^2}\,\alpha_s, \tag{30}$$

with

$$\begin{aligned}
\alpha_d(\omega_s) &= R^3\Big[1 - (1 + \lambda_m)(1 - a^3\lambda_d)/\eta\Big], \\
\alpha_s(\omega_s) &= \frac{4\pi}{3eE_0}\Bigg[\int_0^\infty dr' r'^3 \delta n_s(r') - \Big[1 - (1 + \lambda_m)(1 - \lambda_d)/\eta\Big]\int_0^R dr' r'^3 \delta n_s(r') \\
&- \Big[1 - (1 + \lambda_m)(1 - a^3\lambda_d)/\eta\Big]R^3 \int_R^\infty dr' \delta n_s(r') \\
&+ \Big[(1 + \lambda_m)\lambda_d/\eta\Big]\int_{R_d}^R dr'(r'^3 - R_d^3)\delta n_s(r')\Bigg],
\end{aligned} \tag{31}$$

and using Eq. (27) we obtain for the far field

$$\phi^R(r) = \frac{eE_i r_0^3}{2\epsilon_m r^2}\,\chi\,\Big[1 + 2(\alpha_d + \alpha_s)/r_0^3\Big]. \tag{32}$$

Turning to χ, we note that for small molecule-nanopatricle overlap, the local potential at the molecule location can be evaluated using the far-field expressions Eqs. (30). We then obtain

$$\frac{\partial\phi_0}{\partial r_0} = -\frac{eE_i}{\epsilon_m}\Big[1 + 2(\alpha_d + \alpha_s)/r_0^3\Big], \tag{33}$$

and, substituting into Eq. (32), we finally arrive at

$$\phi^R(r) = -\frac{4\pi\alpha_0 eE_i}{3\epsilon_m r^2}\Big[1 + 2g(\omega)\Big]\Big[1 + 2g(\omega_s)\Big], \tag{34}$$

with

$$g = \frac{\alpha}{r_0^3}, \quad \alpha = \alpha_d + \alpha_s. \tag{35}$$

The above expression generalizes the well-known classical result [4, 5] to the case of noble-metal particle with different distributions of d-electron and s-electron densities. The surface-enhanced Raman field retains the same functional dependence on the nanoparticle polarizability, however the latter contains all the information about the surface layer effect.

Finally, the enhancement factor is given by the ratio of Raman to incident field intensities,

$$A(\omega, \omega_s) = \left|1 + 2g(\omega) + 2g(\omega_s) + 4g(\omega)g(\omega_s)\right|^2. \tag{36}$$

DISCUSSION

For large nanoparticle with $R \sim 100$ nm, the classical EM theory provides an enhancement of the Raman signal as large as 10^6-10^7 [6]. In reality, the EM enhancement is inhibited by various factors. In noble-metal particles, the interband transition between d-electron and s-electron bands reduce the SP oscillator strength leading to a weakening of the local fields. For nanoparticle radius below 15 nm, finite-size effects become important. The SP resonance damping comes from the electron scattering at the surface leading to the size-dependent SP resonance width $\gamma_s \simeq v_F/R$. At the resonance frequency, the size-dependence of SERS is quite strong. Indeed, if molecular vibrational frequencies are smaller that the SP width, the enhancement factor decreases as $A \propto R^4$ for small nanoparticles, resulting in several orders of magnitude drop in the Raman signal.

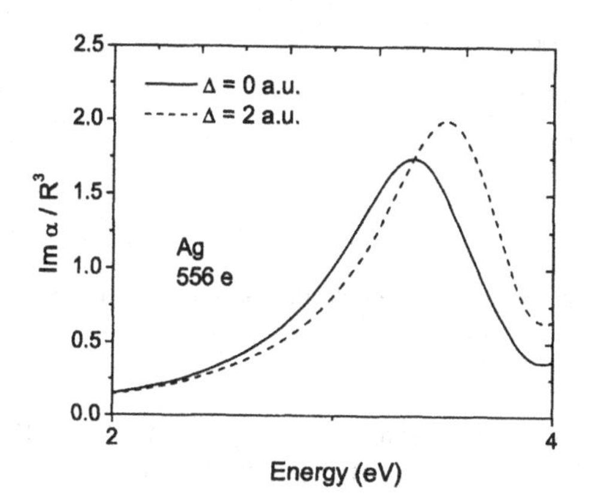

Figure 1: Calculated nanoparticle polarizability for different surface layer thicknesses

Our main observation is that, in small nanoparticle, the local field enhancement due to reduced screening in the surface layer can provide an additional enhancement of the Raman signal. Although the thickness of the surface layer (0.1-0.3 nm) is small as compared to oveall nanoparticle size [15–17], such an enhancement can be condiderable for a molecule located in a close proximity to the surface. In Fig. 1 we show the calculated polarizability with and without surface layer. In the presence of the surface layer, the SP energy experiences a blueshift [15, 16] due an effective decrease in the d-electron dielectric function in the nanoparticle. At the same time, an increase in the peak amplitude, which accompanies the blueshift, indicates a stronger local field at resonance energy acting on a molecule in a close proximity to nanoparticle surface.

In Fig. 2 we show the results of our numerical calculations of SERS with and without surface layer thicknesses, Δ for different nanoparticle sizes. Although the overall magnitude of the enhancemet increases with Δ, a more important effect is its size-dependence. For finite Δ, the enhancement factor descreases more slowly that in the absence of the surface layer: as nanoparticle size decreases, the signals strength ratio *increases*. The reason is that, as the nanoparticle becomes smaller, the fraction of the surface layer increases, and so does the contribution of the unscreened local field into SERS.

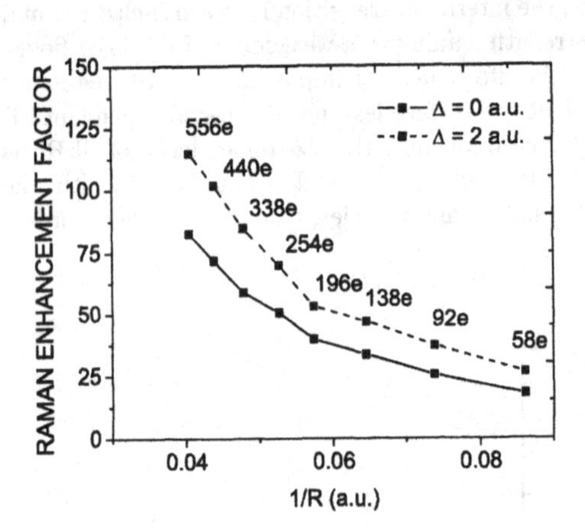

Figure 2: Size-dependence of enhancement factor for different surface layer thicknesses.

ACKNOWLEDGMENTS

This work was supported by NSF under grants DMR-0305557 and NUE-0407108, by NIH under grant 5 SO6 GM008047-31, and by ARL under grant DAAD19-01-2-0014.

REFERENCES

1. S. Nie and S. R. Emory, Science **275**, 1102 (1997).
2. K. Kneipp *et al.,* Rev. Lett. 78, 1667 (1997).
3. K. Kneipp *et al.,* Chem. Rev. **99**, 2957 (1999).
4. M. Kerker, D.-S. Wang, and H. Chew, Appl. Optics **19**, 4159 (1980).

5. J. Gersten and A. Nitzan, J. Chem. Phys. **73**, 3023 (1980).
6. G. S. Schatz and R. P. Van Duyne, in *Handbook of Vibrational Spectroscopy*, edited by J. M. Chalmers and P. R. Griffiths (Wiley, 2002) p. 1.
7. H. Xu *et al.*, Phys. Rev. Lett. **83**, 4357 (1999).
8. H. Xu *et al.*, Phys. Rev. B **62**, 4318 (2000).
9. A. Otto *et al.*, J. Phys. Cond. Matter **4**, 1143 (1992).
10. M. Michaels, M. Nirmal, and L. E. Brus, J. Am. Chem. Soc. **121**, 9932 (1999).
11. A. M. Michaels, J. Jiang, and L. E. Brus, J. Phys. Chem. B **104**, 11965 (2000).
12. Otto, Phys. Phys. Stat. Sol. (a) 4, 1455 (2000).
13. W. E. Doering and S. Nie, J. Phys. Chem. B **106**, 311 (2002).
14. W. Ekardt, Phys. Rev. B **31**, 6360 (1985).
15. A. Liebsch, Phys. Rev. **48**, 11317 (1993).
16. V. V. Kresin, Phys. Rev. **51**, 1844 (1995).
17. A. Liebsch and W. L. Schaich, Phys. Rev. **52**, 14219 (1995).
18. J. Lermé *et al.*, Phys. Rev. Lett. **80**, 5105 (1998).
19. C. Voisin *et al.*, Phys. Rev. Lett. **85**, 2200 (2000).
20. C. Lopez-Bastidas, J. A. Maytorena, and A. Liebsch, Phys. Rev. **65**, 035417 (2001).
21. A. A. Lushnikov, V. V. Maksimenko, and A. J. Simonov, Z. Physik B **27**, 321 (1977).

Mater. Res. Soc. Symp. Proc. Vol. 846 © 2005 Materials Research Society DD8.10

Photoelectrochemical Behaviors of Pt/TiO_2 Nanocomposite thin films Electrodes Prepared by PLD/Sputtering Combined System

Takeshi Sasaki*, William T. Nichols, Jong-Won Yoon and Naoto Koshizaki
Nanoarchitectonics Research Center,
National Institute of Advanced Industrial Science and Technology (AIST),
Tsukuba Central 5, 1-1-1 Higashi, Tsukuba 305-8565, Japan

ABSTRACT

Sputtering of TiO_2 and pulse laser deposition (PLD) of Pt were performed simultaneously to build the nanocomposite with homogenous dispersion of Pt in TiO_2 matrix. The films exhibited a pronounced decrease in the optical band gap energy with increasing Pt content indicating the formation of new energy states in the optical gap of TiO_2 matrix. As-deposited nanocomposite films were amorphous and crystallized to Pt metal and rutile type of TiO_2 after post-annealing at 600 °C. Pt nanoparticle size in the annealed nanocomposite films increased from 6 to 9 nm with the repetition rate of the laser. The photoelectrochemical measurements of the Pt/TiO_2 nanocomposite films in aqueous 0.1 *M* Na_2SO_4 solution indicate that a dramatic increase in anodic photocurrent at 1.0 V vs. Ag/AgCl at the wavelength region of visible light occurred in annealed films. Pt nanoparticles in the matrix of TiO_2 can play an important role in the optical and photoelectrochemical proprieties of Pt/TiO_2 nanocomposite thin films.

INTRODUCTION

Titanium dioxide (TiO_2) is a promising photoactive material commonly used for photocatalytic and photoelectrode applications. These applications are somewhat limited, however, because TiO_2 has a large bandgap and negligible response in the visible spectrum. To overcome this difficulty, TiO_2 can be doped with nanoparticles of noble metals such as Au, Ag and Pt [1]. Platinum dispersions in TiO_2 are particularly interesting because of its potential as a photoelectrode for decomposing water [2,3]. In these cases the Pt nanoparticles are generally deposited on the surface of TiO_2 powders. These Pt particles trap photoexcited electrons in the conduction band of TiO_2, which is followed by the evolution of hydrogen in aqueous solution.

A supported thin film nanocomposite Pt/TiO_2 film is more stable and in many applications, more practical than a powder, while still maintaining a large surface to volume ratio. Numerous synthesis techniques have been successfully used to prepare functional Pt/ TiO_2 nanocomposites, such as sol-gel processing, magnetron sputtering and PLD [4-8]. Magnetron sputtering has the advantage of uniform deposition in large areas and good control of the TiO_2 films. Unfortunately, it is difficult to produce narrow size distributions of nanoparticles such as metals and oxides because of unstable Ar plasma under the preparation conditions. PLD, on the other hand, is capable of producing excellent size control but at limited uniform film growth. We report the preparation and characterization of Pt/TiO_2 nanocomposites using a combination of magnetron sputtering for TiO_2 films and PLD for Pt nanoparticles. The photoelectrochemical properties of obtained nanocomposite thin films were also demonstrated.

EXPERIMENTAL

Pt/TiO_2 nanocomposite films were deposited on quartz glass and indium tin oxide (ITO) coated glass substrates by a simultaneous deposition of Pt and TiO_2 using PLD/sputtering combined system as shown in Fig. 1. The simultaneous deposition allows to simplify the deposition procedure rather than the alternate deposition. TiO_2 was deposited by RF magnetron sputtering under pressure of 0.53 Pa in argon with 80 W of r.f. power at room temperature. Laser deposition of platinum was carried out with various repetition rates between 0.1 - 0.4 Hz corresponding to 6 - 24 pulses per minute. A constant laser energy of 200 mJ per pulse was used and focused to the same 2 mm focal spot. Total time of deposition for each sample was 2 hours. All samples were annealed in air at 600 °C for one hour. Phase identification for the nanocomposite films was conducted with a grazing incidence x-ray diffractometer (XRD) using CuKα radiation. The Pt/Ti atomic ratios in the nanocomposite films were estimated by x-ray fluorescence spectroscopy (XRF). The optical transmittance of the nanocomposite films was measured using a spectrometer. For the electrochemical measurements, a conventional three-electrode system was used: nanocomposite films on ITO glass substrate, Pt and Ag/AgCl electrodes were used as working, counter and reference electrodes, respectively. The potential of the working electrode in 0.1 M Na_2SO_4 aqueous solution was controlled using a potentiostat. Photocurrents were measured under the irradiation of the light through a monochromator and an electronic shutter from a 500 W xenon lamp.

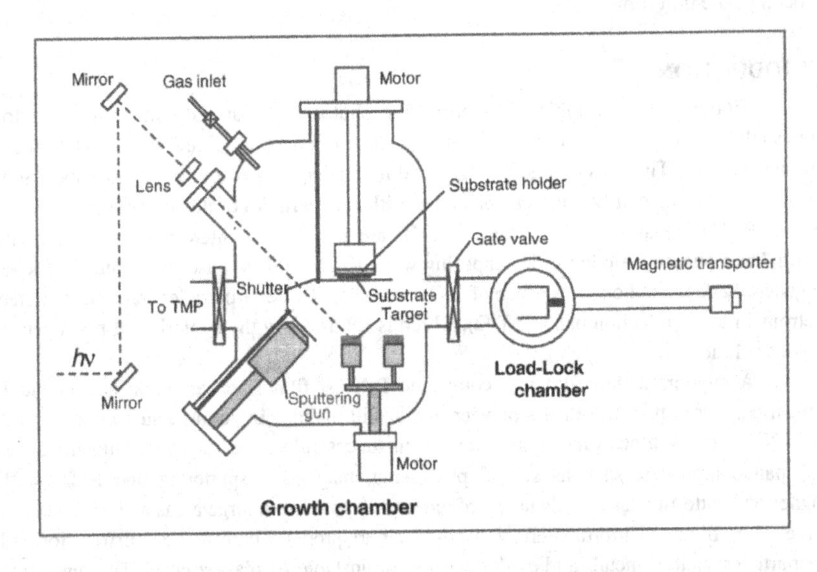

Fig. 1. Schematic diagram of the PLD/Sputtering combined system.

RESULTS AND DISCUSSION

The UV-vis spectra demonstrate a dramatic increase in absorbance compared with a magnetron sputtered pure TiO_2 film, as shown in Fig. 2. The ripples in the spectra in the visible region result from interference of light reflected from the film interfaces. The optical band gap (ΔE) of the obtained films was estimated from the UV-vis spectra as shown in Fig. 3. The procedure was followed as reported by Abass et al [9]. For the magnetron sputtered

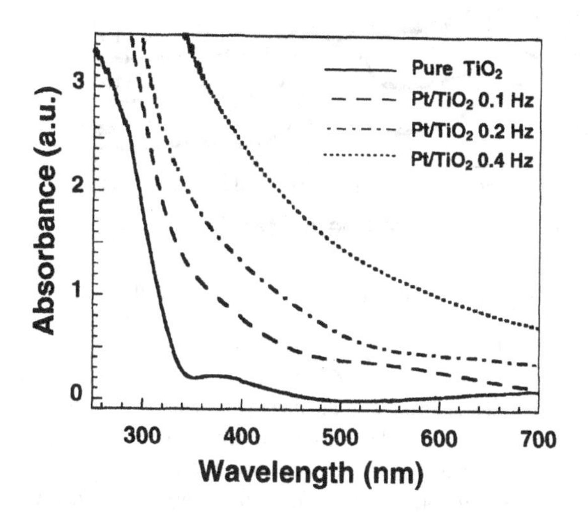

Fig. 2. The UV-VIS spectra of as deposited' Pt/TiO_2 nanocomposite thin films prepared at different laser repetition rates compared to a pure TiO_2 film

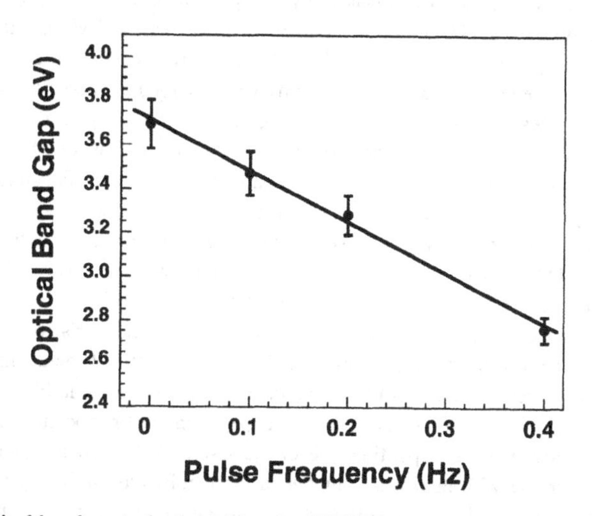

Fig. 3. Optical band gap calculated from the UV-VIS spectra.

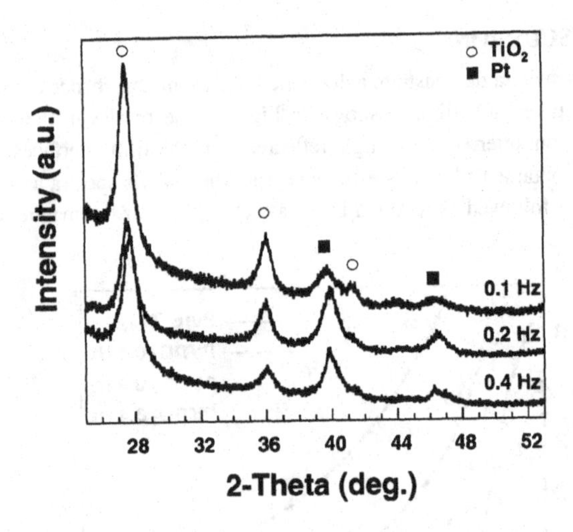

Fig. 4. The XRD spectra of annealed Pt/TiO_2 nanocomposite thin films prepared at different laser repetition rates

TiO_2 thin film, we obtain a value of ΔE = 3.66 eV. This value is shifted from the ΔE = 3.05 eV gap reported for single crystal titanium dioxide [10]. This shift in the absorption edge could be due to defects in our samples which are grown at room temperature and not annealed to preserve the original nanoparticle size distribution. The change from the pure TiO_2 film band gap can be attributed to platinum metal additions which could produce the some energy state in the band gap of TiO_2.

Because the thin films were deposited onto room temperature substrates, all as-deposited samples were amorphous as indicated by x-ray diffraction. To crystallize the samples, post-annealing in air at 600 °C for one hour was carried out. UV-vis spectra of the heated films showed only a very small red-shift in the absorption edge. After the post-annealing, a drastic improvement in the film crystallinity was observed as shown in Fig. 4. A number of crystalline peaks are present in the XRD spectra corresponding to Pt metal and rutile phase TiO_2. No indication of the anatase phase was seen in any of the samples. Annealing causes Pt particle growth whose size depends on the initial concentration of platinum in the composite, as well as, the temperature and time of the annealing. The concentration of Pt in the nanocomposite films should monotonically increases with the laser repetition rate. Indeed Pt/Ti atomic ratios of the composite films prepared at 0.1, 0.2 and 0.4 Hz were 0.24, 0.49 and 0.80, respectively. An estimate of crystallite size within the nanocomposite can be calculated from the x-ray peak width using Scherrer's formula [11]. Using this formula for the broadening of the platinum (111) peak near the 2-theta angle of 40 degrees the sizes of the annealed Pt nanoparticles can be estimated as shown in Fig. 5. It can be seen that the crystallite size of Pt nanoparticles in the post-annealed nanocomposite films increases from 6 to 9 nm with the repetition rate of the laser. This increase follows the expected total mass increase with increasing number of laser pulses as shown by the solid line through the data in Fig. 5. These results indicate that an increase of Pt mass results in an

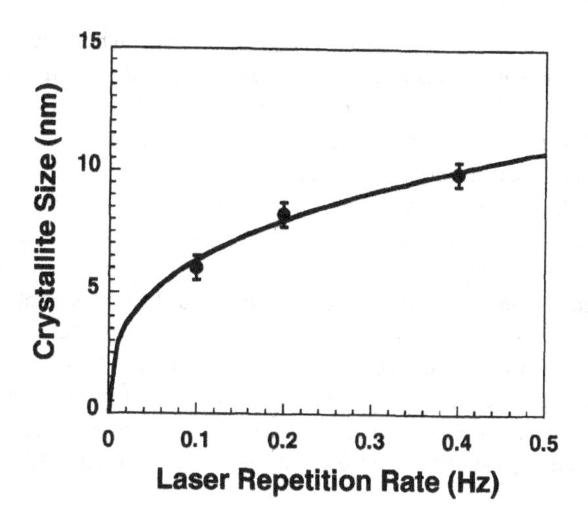

Fig. 5. Platinum crystallite size calculated from Scherrer's formula. The solid line shows the expected mass increase of platinum in the composite versus laser repetition rate.

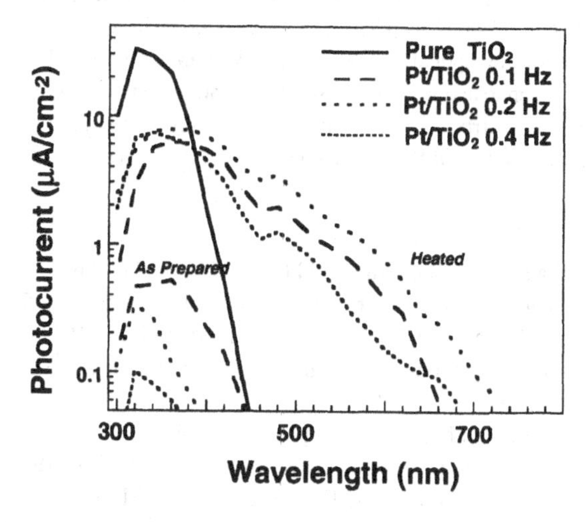

Fig. 6. Anodic photocurrent at 1.0 V vs. Ag/AgCl in 0.1 *M* Na_2SO_4 aqueous solution as a function of irradiating light wavelength for as prepared films and those annealed at 600 °C in air for 1 hour.

increase in the particle size rather than the number of particles in the nanocomposites after post-annealing.

Anodic photocurrents of Pt/TiO_2 nanocomposite and TiO_2 electrodes as a function of the irradiating light wavelength at 1.0 V vs. Ag/AgCl in 0.1 M Na_2SO_4 aqueous solution are shown in Fig. 6. Electrodes made from as prepared samples did not show visible light photoresponse. There is a dramatic enhancement in the response in both the ultraviolet and in particular the visible region upon annealing. XPS analysis indicates that the chemical state of Pt in the nanocomposite films was changed from oxide to metallic state after the annealing at 600 °C in air for 1 hour. These results suggest that the metallic Pt nanoparticles could play an important role in the photoresponse in the visible light range.

It is well known that a Schottky barrier is formed at the interface between a metal and semiconductor. The barrier height of the Pt-TiO_2 interface is estimated to be 1.64 eV from the work function of Pt (5.64 eV) [12] and the electron affinity of TiO_2 (4.0 eV) [1]. The photoexcitation of electrons from the Pt metal to the conduction band of TiO_2 could take place under the irradiation of light with energy exceeding 1.64 eV. The anodic photocurrent is clearly observed at wavelengths shorter than around 700 nm. These energy levels are a very reasonable energy range for the electron excitation from Pt to TiO_2. It is thought that holes could be generated on Pt nanoparticles via the photoexcitation of electrons from the Pt metal to the conduction band of TiO_2 in the visible light range and react with water.

CONCLUSIONS

Sputtering of TiO_2 and PLD of Pt were performed simultaneously to build the nanocomposite with homogenous dispersion of Pt in TiO_2 matrix. The films exhibited a pronounced decrease in the optical band gap energy with increasing Pt content and the electrode prepared from annealed nanocomposite films had the photoresponse in visible light. Pt nanoparticles in the matrix of TiO_2 can play an important role in the optical and photoelectrochemical properties of Pt/TiO_2 nanocomposite thin films.

REFERENCES

[1] G. Zhao, H. Kozuka and T. Yoko, Thin Solid Films, 277, 147 (1996).

[2] S. Tabata, H. Nishida, Y. Masaki and K. Tabata, Catalysis Letters 34, 245 (1995).

[3] L. Avalle, E. Santos, E. Leiva and V. Macagno, Thin Solid Films 219, 7 (1992).

[4] T. Sasaki, N. Koshizaki, S. Terauchi, H. Umehara, Y. Matsumoto and M. Koinuma, Nanostructured Materials, 8, 1077 (1997).

[5] T. Sasaki, N. Koshizaki and K.M. Beck, Appl. Phys. A, suppl. 69, 771 (1999).

[6] K.M. Beck, T. Sasaki and N. Koshizaki, Chem. Phys. Lett., 301, 336 (1999).

[7] J-W. Yoon, T. Sasaki, N. Koshizaki and E. Traversa, Scripta Materialia, 44, 1865 (2001).

[8] T.Sasaki, N. Koshizaki, Jong-Won Yoon, Kenneth M. Beck, J. Photochem. Photobio. A: Chem, 145, 11(2001).

[9] A.K. Abass, A.K. Hasen and R.H. Misho, J. Appl. Phys., 58, 1640 (1985).

[10] H. Tang, H. Berger, P.E. Schmid, F. Levy, Solid State Comm., 92, 267 (1994).

[11] B.D. Cullity, Elements of X-ray Diffraction, 2nd edition, Addison-Wesley, Reading, MA, 1978.

[12] H.B. Michaelson, J. Appl, Phys. 48 (1977) 4729.

Poster Session: Organometallic Optical Materials

Mater. Res. Soc. Symp. Proc. Vol. 846 © 2005 Materials Research Society DD9.1

Luminescence properties of $Eu^{3+}:Y_2O_3$ and $Eu^{3+}:Lu_2O_3$ nanoparticles, ceramics and thin films

Kai Zhang, D. Hunter, S. Mohanty, J.B. Dadson, Y. Barnakov and A.K. Pradhan
Center for Materials Research, Norfolk State University, 700 Park Avenue, Norfolk, VA 23504

ABSTRACT

Eu^{3+} doped Y_2O_3 and Lu_2O_3 nanocrystalline powders were synthesized via combustion technique using urea as a fuel and the metal nitrates as oxidants. The compacted nanopowders were vacuum sintered in order to form the translucent ceramics. A significant enhancement of emission characteristics was observed from the ceramics synthesized from the nanoparticles by controlling the vacuum-sintering conditions. Although the processed ceramics display superior emission characteristics, the nanocrystalline phosphor powders also display reasonably good emission characteristics. Highly epitaxial $Y_2O_3:Eu^{3+}$ and $Lu_2O_3:Eu^{3+}$ films were deposited on various substrates under different growth and optimization conditions using pulsed-laser deposition technique using high-density translucent ceramic target. Superior spectroscopic performance was obtained on films grown on sapphire substrates due to high-quality and epitaxial nature of the film.

INTRODUCTION

Oxide phosphors, such as Eu^{3+} doped Y_2O_3 have the potential of replacing the conventional displays [1-5]. On the other hand, Eu doped Lu_2O_3 is an important X-ray phosphor for digital radiographic imaging [6-8]. Although Lu_2O_3 is a structural analog of Y_2O_3, the major advantage of Lu_2O_3 over Y_2O_3 is its phenomenally high density, reaching 9.42 g/cm^2 that lead to a high stopping power for X-ray and other types of ionizing radiations. Due to $^5D_0 \rightarrow {}^7F_2$ transitions within europium, both $Y_2O_3:Eu^{3+}$ and $Lu_2O_3:Eu^{3+}$ shows luminescent properties and emits red light with a wavelength of 611 nm, and can be used as a red phosphor as well as scintillator for high-density material. Extensive research is currently pursued to develop rare earth-based oxide nanophosphors for superior performance such as large surface area, high sinterability at lower temperature and tuning of various properties related to a transition from solid-state matter to molecular structures as the particle size is scaled down to nanoscale. On the other hand, Eu^{3+} doped Y_2O_3 films have attracted considerable research interest due to its potential for application in displays. However, significantly improved performance of displays demands high-quality phosphors having sufficient brightness and long-term stability.

In this paper, we report on the fabrication and luminescent properties of $Y_2O_3:Eu^{3+}$ and $Lu_2O_3:Eu^{3+}$ nanopowders, ceramics and epitaxial films. Our results demonstrate superior microstructural and spectroscopic performance of nanopowders, ceramics and epitaxial films.

EXPERIMENTS

The nanocrystalline $Eu_2O_3:Eu^{3+}$ and $Lu_2O_3:Eu^{3+}$ powders were synthesized by combustion from their respective nitrate solutions. The Eu-doping was kept at as low as 5 at.% for all samples. The precursor mixture with urea (1:3 = Eu_2O_3:urea) was dehydrated completely before the combustion at a temperature of 600 °C, at which precursor powders comprising of nanoparticles were obtained. Then the powders were calcined 1100 °C. For ceramic and target processing, the precursor powders calcined at 1100 °C, were palletized into a circular disc, using a uniaxial pressure followed by calcination at 600 °C. The pellets were isostatically pressed using 400 MPa of pressure and the pellets were vacuum-sintered in the temperature range of

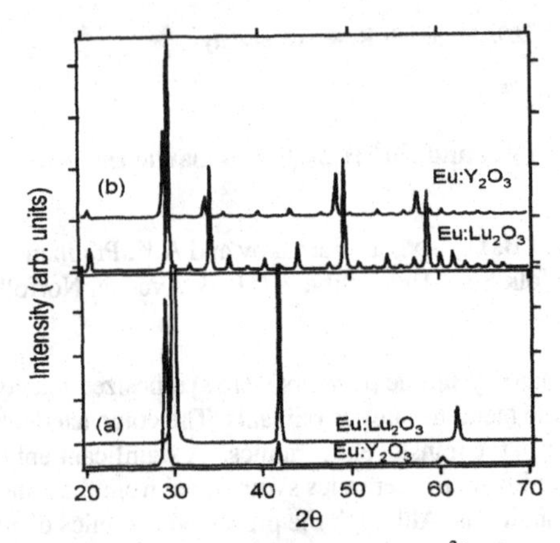

Figure 1. X-ray diffraction patterns of (a) $Y_2O_3:Eu^{3+}$ and $Lu_2O_3:Eu^{3+}$ nanopowders calcined at 1100 °C, (b) $Y_2O_3:Eu^{3+}$ and $Lu_2O_3:Eu^{3+}$ films grown on sapphire (0001) substrates at 200 mTorr of oxygen and substrate temperature of 800 °C.

1750 to 1800 °C in a W-mesh heater assembled vacuum chamber with a vacuum better than 5×10^{-7} T for 4 to 6 h depending on the size of the sample in order to form highly dense and very translucent targets.

Thin films were deposited using pulsed-laser deposition technique onto (0001) and (100) single crystalline sapphire and Si substrates, respectively. The targets were ablated using 248 nm KrF pulsed laser beam with an energy of 2 to 3 J/cm^2 at a repetition rate of 5Hz. with oxygen maintaining a desired pressure of 0.1 to 400 mTorr during depositions.

Phase identification of nanocrystalline $Eu_2O_3:Eu^{3+}$ and $Lu_2O_3:Eu^{3+}$ powders and thin film was performed by the X-ray diffraction (XRD) method on a Rigaku X-ray diffractometer using CuK_α radiation. Powder characterization was obtained by Transmission Electron Microscopy (TEM). The microstructure of films was analyzed using atomic force microscopy (AFM).

RESULTS AND DISCUSSION

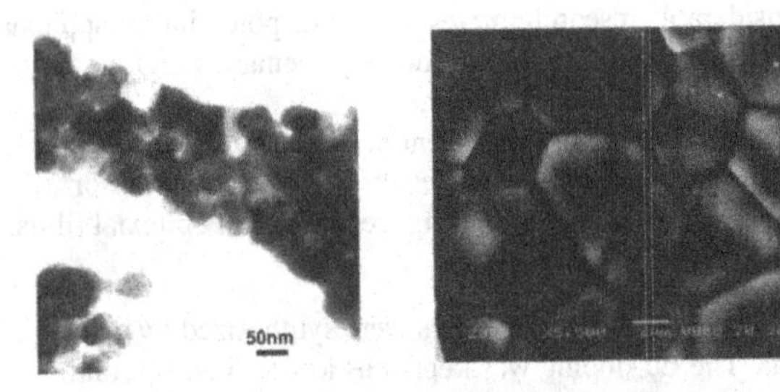

Figure 2. (Left) Transmission electron microscope image of $Y_2O_3:Eu^{3+}$ nanopowders and (right) SEM micrograph of $Y_2O_3:Eu^{3+}$ ceramic.

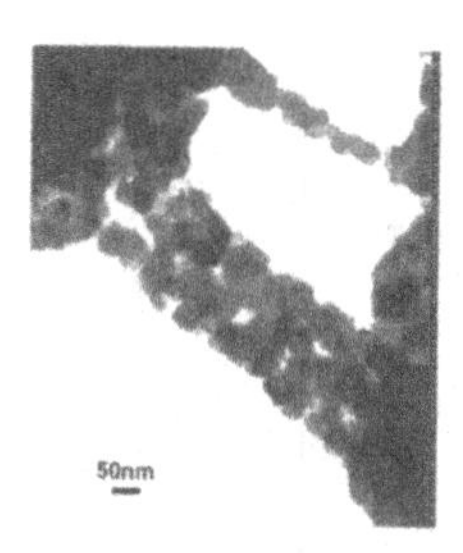

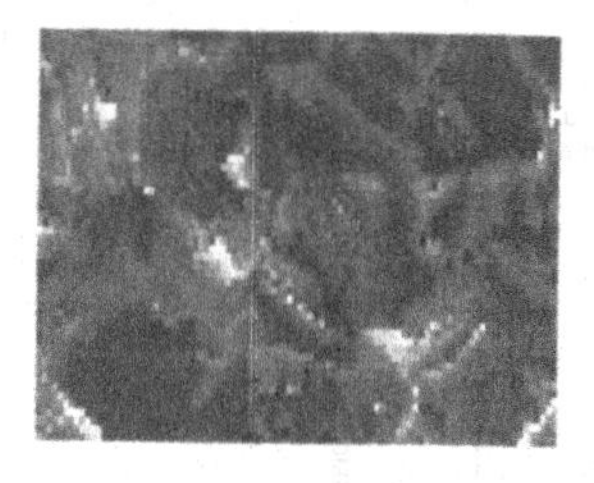

Figure 3. (Left) Transmission electron microscope image of $Lu_2O_3:Eu^{3+}$ nanopowders and (right) SEM micrograph of $Lu_2O_3:Eu^{3+}$ ceramic.

XRD patterns of $Y_2O_3:Eu^{3+}$ and $Eu_2O_3:Eu^{3+}$ nanocrystalline powders calcined at 700 to 1100 °C are single phase and are consistent with the cubic phase of Y_2O_3. Since the crystallite size increases rapidly with increasing calcinations temperature beyond 1200 °C, the powders calcined at 1100 °C were used for sintering the pellets in vacuum. The XRD for $Y_2O_3:Eu^{3+}$ and $Lu_2O_3:Eu^{3+}$ films reveal that the both types of films are highly epitaxial. The Bragg reflection corresponding to the positions (222) and (444) were only observed regardless of substrate, substrate temperature and oxygen partial pressure. It is very clear that the (222) reflection is very strong. The rocking curves at (222) for $Y_2O_3:Eu^{3+}$ and $Lu_2O_3:Eu^{3+}$ films for respective sapphire and Si substrates clearly illustrate the superior epitaxial film growth on sapphire substrates. The transmission electron microscopy image shown in Fig.2 reveals that the $Y_2O_3:Eu^{3+}$ nanopowders consist of particles less than 50 nm. Similar particle size was also noticed for $Lu_2O_3:Eu^{3+}$ nanopwders as shown in Fig. 3. However, the grain size and morphology for sintered ceramics of both samples are different and shown in Fig. 2 and 3. The grain size in $Y_2O_3:Eu^{3+}$ ceramic ranges from 2 to 5 μm in size, whereas the grain size in $Lu_2O_3:Eu^{3+}$ ceramic is irregular.

Figure 4 (a) and (b) show the AFM surface morphology of. $Y_2O_3:Eu^{3+}$ and $Lu_2O_3:Eu^{3+}$ films. The surface roughness of $Y_2O_3:Eu^{3+}$ and $Lu_2O_3:Eu^{3+}$ films increases with increasing oxygen partial pressure during film growth. However, films grown at the substrate temperature of 800 °C and 200 mTorr of oxygen exhibit remarkable surface morphology containing uniform grain sizes, illustrating the high-quality of the films.

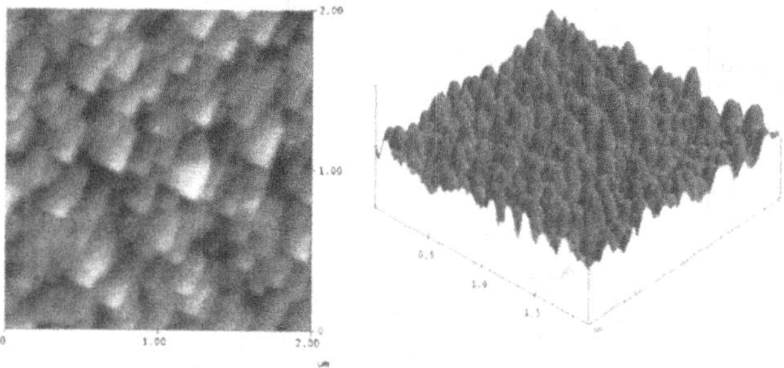

Figure 4. (left) Atomic force microscope image of $Y_2O_3:Eu^{3+}$ and (right) $Lu_2O_3:Eu^{3+}$ thin films grown at 800°C and 200 mTorr of oxygen.

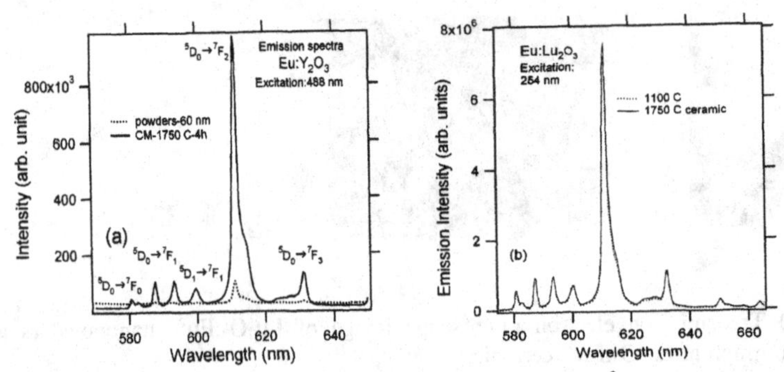

Figure 5. Emission spectra of (a) $Y_2O_3:Eu^{3+}$, and (b) $Lu_2O_3:Eu^{3+}$ nanopowders and ceramics.

Figures 5 (a) and (b) show the emission spectra of $Y_2O_3:Eu^{3+}$ and $Lu_2O_3:Eu^{3+}$ nanopowders and ceramics, illustrating their luminescence behavior. We have used excitation of 488 nm for measuring the emission spectra of $Y_2O_3:Eu^{3+}$ and 254 nm for $Lu_2O_3:Eu^{3+}$ nanopowders and ceramics because maximum intensity of photoluminescence was observed at this wavelength. However, excitation of 254 nm was used all thin films in order to obtain the maximum intensity. All the emission spectra illustrate that the Eu^{3+} ions are in the cubic symmetry exhibiting the characteristics of red-luminescent $Y_2O_3:Eu^{3+}$ and $Lu_2O_3:Eu^{3+}$ in which $^5D_0 \rightarrow {}^7F_2$ transition at ~610 nm is prominent, and the relatively weak emissions at the shorter wavelengths are due to the $^5D_0 \rightarrow {}^7F_1$ transitions. The nanopowders exhibit relatively high intensity in both samples. However, the remarkable enhancement in emission intensity was observed for their ceramic counterparts. The enhancement of luminescent property can be attributed to the efficient energy transfer rate from the luminescence centers due to reduced interfaces in the processed ceramics and correction of the oxygen defects. The most profound emission for the sintered ceramics compared to their nanocrystalline powder is attributed due to the increase of the crystallites, which make up the ceramics, and reduce the surface-to-volume ratio, which diminishes the negative processes related to the surface effects. In films, the increased volume for the photon-

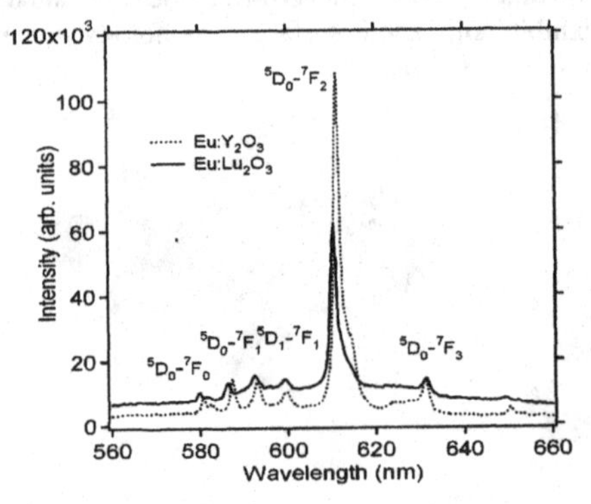

Figure 6. Emission spectra of (a) $Y_2O_3:Eu^{3+}$ films on sapphire substrate, and (b) Emission spectra of $Lu_2O_3:Eu^{3+}$ films on sapphire substrate.

solid interaction at the surface as well as improved crystallinity of the films is responsible for the enhanced emission intensity. However, there are other several potential factors affecting the emission intensity. These are (a) internal reflections or the light piping effect, and (b) surface morphology that restrict the internal reflections which reduce the emission intensity. The emissision characteristics for $Y_2O_3:Eu^{3+}$ and $Lu_2O_3:Eu^{3+}$ films grown on Si(100) substrates were poorer and they are not presented here.

Our results suggest that although the processed ceramics display superior emission characteristics, the nanocrystalline phosphor powders calcined at 1100 °C also display reasonably good emission characteristics, illustrating the possibility of their applications in display technology. On the other hand, the surface morphology of the epitaxial films controls the photoluminescence property of these phosphors film. Our results suggest that due to comparable photoluminescence behavior of $Lu_2O_3:Eu^{3+}$ films with $Y_2O_3:Eu^{3+}$ the former may be used in radiographic x-ray imaging on optimization.

CONCLUSION

Our results suggest that $Y_2O_3:Eu^{3+}$ and $Lu_2O_3:Eu^{3+}$ nanopowders, ceramics and films synthesized under optimization conditions exhibit enhanced photoluminescence behavior. The sintered ceramics and films show enhanced luminescence behavior due to better surface effects and improved crystallinity, respectively. The nanocrystalline powders, especially $Lu_2O_3:Eu^{3+}$, also exhibit moderate luminescence behavior and can be used for display technology. $Lu_2O_3:Eu^{3+}$ ceramics and films can be used in both display and radiographing imaging due to their very similar luminescence behavior found in $Y_2O_3:Eu^{3+}$ ceramics and films.

ACKNOWLEDGEMENTS

This work is supported by NASA and NSF for Center for Research Excellence in Science and Technology (CREST) grant HRD-9805059.

REFERENCES

1. K.C. Mishra, J.K. Berkowitz, K.H. Johnson, and P.C. Schmidt, *Phys. Rev.* B **45**, 10902 (1992).
2. R.C. Rop, *The chemistry of artificial lighting devices: lamps, phosphors, and cathode ray tubes,* (Elsevier, New-York, 1993).
3. S. Itoh, T. Kimizuka, and T. Tonegawa, *J. Electrochem. Soc.* **136**, 1819(1989).
4. T.H.C. Stewart, J.S. Sebastian, T.A. Trottier, S.L. Jones, and P.H. Horr, *J. Vac. Sci. Technol.* A **14**, 1697 (1996).
5. J. Lu, T. Mura, K. Takaichi, T. Uematsu, K. Ueda, H. Yagi, T. Yanagitani, and A.A. Kaminskii, *Jpn. J. Appl. Phys.*, **40**, L1277 (2001).
6. E. Zych, D. Hreniak, and W. Strek, *J. Alloys and Comp.*, **341**, 385(2002).
7. A. Lempicki, C. Brecher, P. Szupryczynski, H. Lingertat, V.V. Nagarkar, S.V. Tipnis, and S.R. Miller, *Nucl. Instrum. Meth.* A **488**, 579 (2002).
8. C. Brecher, R.H. Bartram, and A. Lempicki, *J. Lumines.* **106**, 159 (2004).
9. J.S. Bae, J.H. Jeong, S-S. Yi, and J-C. Park, Appl. Phys. Lett. **82**, 3629 (2003).
10. K.G. Cho, D. Kumar, D.G. Lee, S.L. Jones, P.H. Holloways, and R.K. Singh, Appl. Phys. Lett. **71**, 3335 (1997).

Mater. Res. Soc. Symp. Proc. Vol. 846 © 2005 Materials Research Society DD9.2

Layered Double Hydroxides as a Matrix for Luminescent Rare Earth Complexes

Natalia G. Zhuravleva[1], Andrei A. Eliseev[1], Alexey V. Lukashin[1], Ulrich Kynast[2], Yuri D. Tretyakov[1]
[1]Materials Science Department, Moscow State University, Moscow 119992, Russia
[2]Department of Chemical Engineering, Muenster University of Applied Science, 48565 Steinfurt, Germany

ABSTRACT

In the present work new luminescent materials with high quantum efficiencies based on layered double hydroxides (LDH) were obtained. Intercalation of complexes into the interlayer space of LDH doesn't affect their luminescent properties, forming non-volatile solid state material with good optical properties. The Coulomb interactions between LDH layers and complex can result in a change of complex structure in comparison with free complexes and in decreasing the number of the organic ligands per *Ln* atom. The energy transfer in the system was also studied.

INTRODUCTION

The synthesis and investigation of new luminescent materials combining excellent optical characteristics with physical and chemical stability is one of the important problems in modern material science [1-3]. The development of novel synthetic methods for the preparation of such materials usually involves combination of optical elements with stability of solid state matrices. In the present study we report synthesis of luminescent materials based on molecular complexes of Tb and Eu rare earths (RE) intercalated into the interlayer space of layered double hydroxides (LDH). Cheap and available magnesium-aluminum LDHs are known as a very convenient matrices for intercalation of different chemical compounds [4]. Optical transparency, chemical and photochemical stability of these matrices enables their possible application for preparation of various optical materials [5]. Besides the ability to control the quantity of complex intercalated and homogeneous distribution of anions in the interlayer space make LDH matrices very attractive for intercalation of different functional molecules in comparison with previously used zeolites and sol-gel matrixes. Moreover, intercalation of complexes into the interlayer space allows to investigate the structure and properties of complex in spatially constrained two-dimensional system of cavities, while the controlling the complex quantity of to adjust RE-RE distance. To determine the influence of the corresponding confinement on luminescent efficiency and energy transfer within the layers we performed synthesis of terbium and europium picolinate (pyridine-2-carboxilate, "pic") complexes embedded into layered double hydroxides with different Tb/Eu ratios.

EXPERIMENTAL DETAILS

The series of $Mg_nAl(OH)_{2n+2}[Ln(pic)_4{\cdot}mH_2O]$ samples with different terbium to europium ratio (Ln = Tb, Eu, $Tb_{0.98}Eu_{0.02}$, $Tb_{0.9}Eu_{0.1}$, $Tb_{0.5}Eu_{0.5}$) were synthesized. Mg-Al LDHs in carbonate form were obtained by co-precipitation of magnesium and aluminum nitrates

(Mg^{2+}:Al^{3+} = 2:1, 3:1, 4:1, 6:1, 9:1, 12:1) by the mixture of NaOH and Na_2CO_3 at pH = 10,0±0,1. [6]. To enhance the crystallinity of LDHs, they were aged at 80 °C for 4 days. The terbium and europium complexes were intercalated using the Chibwe procedure [7]. Carbonate LDHs were annealed at 550 °C to get rid of H_2O and CO_2. Obtained 'layered double oxides' were impregnated by the anionic complexes $Ln(pic)_4^-$ (Ln = Tb, Eu, $Tb_{0.98}Eu_{0.02}$, $Tb_{0.9}Eu_{0.1}$, $Tb_{0.5}Eu_{0.5}$) at 60°C for 12 hours. Synthesized LDHs were marked as (**Mg*n*Al_*Ln(pic)***) where n is the Mg:Al ratio and Ln = Tb, Eu, $Tb_{0.98}Eu_{0.02}$, $Tb_{0.9}Eu_{0.1}$, $Tb_{0.5}Eu_{0.5}$).

The emission, excitation and reflectance spectra were measured with ARC SpectraPro-300i spectrometer. The excitation measurements were performed with UG-5 filter, the system response was corrected with $BaMgAl_{10}O_{17}$:Eu (BAM), generally produced by Philips Research Laboratories. Reflectance spectra were obtained using commercial NaX as white and coal as black standards. Emission spectra were normalized to emission intensity of narrow-band luminescent standards for terbium ($LaPO_4$:Ce,Tb, NP220) and europium (Y_2O_3:Eu, U744). In case of **Mg*n*Al_*Ln(pic)*** samples, where Ln = $Tb_{0.98}Eu_{0.02}$, $Tb_{0.9}Eu_{0.1}$, $Tb_{0.5}Eu_{0.5}$ the normalization was performed to emission intensity of NP220:U744 mixture with the ratio corresponding to Tb/Eu ratio in the sample. The quantum efficiencies were calculated from normalized emission spectra by the method described in [8]. X-ray analyses was performed at Philips PW 1130/00 diffractometer (step 0.05°, 2-6 s) at CuK_α-irradiation (λ=1,54184 Å).

DISCUSSION

LDHs are hydrotalcite-like compounds with general formula $M^{2+}_{1-x}M^{3+}_x(OH)_2[(anion^{n-})_{x/n}\cdot mH_2O]$, where ($anion^{n-}$) is almost any anion, which does not form a stable complex with M^{2+} or M^{3+}. A structure of an LDH consists of positively charged hydroxide layers [4] linked by negatively charges anions, which occupy the interlayer space. Both europium and terbium are known to form anionic complexes with picolinate ligands, thus making possible incorporation them into the LDH matrix. One of the characteristic properties of LDHs is high mobility of anions between rigid hydroxide layers, which allows the exchange of anions without destruction of the layered structure. A major advantage of LDHs is the possibility to control the anion loading levels by variation of the M^{2+} to M^{3+} (in our case Mg to Al) ratio in the initial precursor [9]: each M^{3+} substituting an M^{2+} gives rise to a surplus positive charge on the lattice to be compensated by an interlayer anion. Thus, composites with different content of the RE could be obtained, allowing the investigation of the concentration dependence of the luminescent properties of the complex. Besides, 2D layered structure of matrix results in nearly square-law dependence between complex content and the distances between metal centers in the interlayer space, which enables one to study the energy transfer processes in the system as in a particular case of 2D system. Energy transfer channels $Tb^{3+}\rightarrow Eu^{3+}$ are well known in mixed complexes [8].

X-ray powder diffraction data of samples with different complex content indicates stepwise increase of the c parameter and the interlayer distance of the anion-substituted LDHs with decrease of complex content (table I). It should be noted that these parameters are similar for Eu- and Tb-containing samples with the same loading values (the same Mg:Al ratio in LDH). The formation of different types of complexes could be concluded: for higher complex content ***Ln:***pic ratio was found to be nearly 1:3, while for lower charge of the layers *Ln:pic* ratio equals 1:2. Apparently, insufficient negative charge is compensated by incorporation of extra OH^--groups into the interlayer space or binding of rare earth to the hydroxide layers. This was also confirmed by optical properties of the samples, which were discussed in detail in [10,11]. Here we will outline the major tendencies in optical behavior of prepared materials. The reflection spectra of samples exhibit two broad minima at 240-280 and 320-360 nm, which arise from excitation of the π-electrons of the aromatic ring. Besides the reflection minimum matches the main maximum in the excitation spectrum in all the samples, and corresponds the excitation transfer to the lanthanide atom. The structure of the reflection spectra for **Mg_nAl_*Ln*(*pic*)** samples doesn't depend on anion content.

Table I. X-ray diffraction and elemental analysis data for **Mg_nAl_Tb(*pic*)** composites.

Sample	Unit cell parameter *c*, Å	Interlayer space, Å (d_{003} - 4,8Å)	Content		Tb:*pic* ratio	Calculated distances *Ln-Ln*, Å
			Tb, weight %	C, weight %		
Mg_2Al_Tb(*pic*)	40,41±0,01	8,67±0,003	21,4	28,5	1:2,94	7,36±0,06
Mg_3Al_Tb(*pic*)	40,35±0,01	8,65±0,003	19,4	22,3	1:2,54	8,23±0,07
Mg_4Al_Tb(*pic*)	36,62±0,01	7,41±0,005	17,8	20,6	1:2,56	8,37±0,08
Mg_6Al_Tb(*pic*)	24,55±0,03	3,38±0,012	17,2	17,8	1:2,28	10,84±0,11
Mg_9Al_Tb(*pic*)	23,38±0,04	2,99±0,023	15	13,4	1:1,97	11,67±0,22
Mg_{12}Al_Tb(*pic*)	23,12±0,03	2,91±0,017	13,2	13,5	1:2,27	12,32±0,25

The values of quantum efficiencies for Tb-containing samples are rather high (30-50%) and comparable with the intensities of free molecular complexes, which give rise to possibility of their application in optics. Significantly, the Eu^{3+} environment shows very low efficiencies (quantum yield < 5%) with only minor changes on varying LDH composition in the analogous series of LDH-embedded complexes. The efficiency decrease indicates evidently the coordination of Eu central ion to OH-group or to the LDH lattice, where the excitation is quenched due to high frequency phonons (-OH-vibrations) or an Eu^{3+}→O charge transfer extending to low energies.

To provide further optimization of optical properties of the composites energy transfer studies were carried out in the system. It can be calculated that ***Ln-Ln*** distances increase with decreasing of complex concentration in the interlayer space. Two characteristic intervals could be pointed out: the first 3 samples with high charge of the layers have the distances of ~8 Å, while in case of lower complex content these values equals 11-12 Å correspondingly (table I). This difference obviously affects the disparity in energy transfer probability and, thus, quantum efficiencies. Figure 1 shows the dependence of quantum efficiencies of the samples as a function on Tb:Eu ratio and the overall charge of the layers calculated for the excitation at 254 nm (this wave length corresponds well to the maximum of Hg-lamp emission). The emission intensities drastically decrease already at very low Eu^{3+} levels. The minimal substitution value of 2% (**Mg_nAl_$Tb_{0.98}Eu_{0.02}$(*pic*)** samples) already leads to twice lower luminescence intensity, while at higher substitution levels the relative depreciation factor significantly increases. This proves a very effective energy transfer from Tb^{3+} to Eu^{3+}.

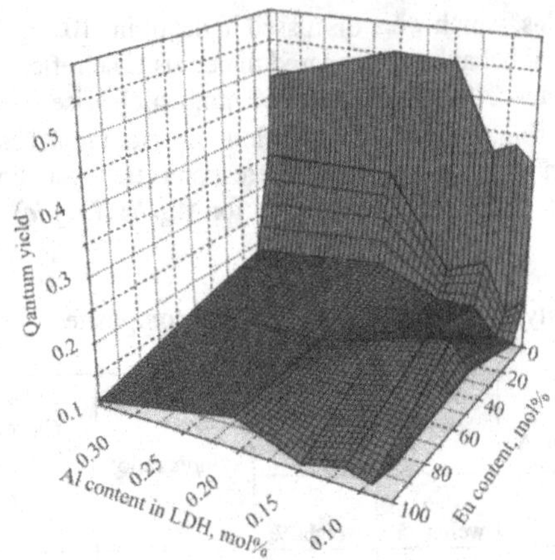

Figure 1. Quantum efficiencies of Tb- and Eu-containing layered double hydroxides as a function of Tb:Eu ratio and aluminum content in LDHs.

For the detailed characterization of energy transfer processes the luminescence decay measurements were performed (fig. 2, table II). It should be noted that life times of excited states in our system are much lower then in crystal picolinates for both Tb and Eu. According to the literature lifetime of the exited state of terbium ($^5D_4$) in picolinates equals 1,639 ms for Na[Tb(*pic*)$_4$] and 1,121 ms for [Tb_{16}(*pic*)$_{28}$] in zeolite X [3,8]. The lifetime of the excited state $^5D_0$ is about 1,603 ms in Na[Eu(*pic*)$_4$]. It can be explained by partially coordination of Rare Earth ion with OH-groups; quick enough quenching of excitation due to the -OH-vibratons decrease

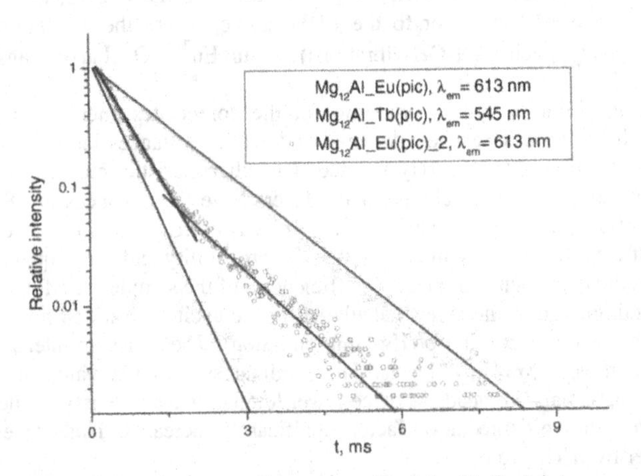

Figure 2. Luminescence decay curve for **Mg$_n$Al_*Ln(pic)*.**

the overall life times of excited state. Luminescence decay for LDH samples containing one type of Rare Earth is nearly exponential that proves formation of equal environment for all RE atoms in the system. The increase of lifetimes for **$Mg_nAl\_Eu_{0.02}Tb_{0.98}$*(pic)*** in comparison with Eu-containing samples confirms the energy transfer from Tb to Eu.

Table II. The life times of excited state in **Mg_nAl_Ln*(pic)*** composites.

Sample	τ_{exp}, ms	Sample	τ_{exp}, ms
Mg_2Al_Eu*(pic)*	0.417	**$Mg_{12}Al\_Eu$*(pic)***	0.408
$Mg_2Al\_Eu_{0.02}Tb_{0.98}$*(pic)*, λ_{em} = 615	0.559	**$Mg_{12}Al\_Eu_{0.02}Tb_{0.98}$*(pic)***, λ_{em} = 615	0.905
$Mg_2Al\_Eu_{0.02}Tb_{0.98}$*(pic)*, λ_{em} = 615	1.055	**$Mg_{12}Al\_Eu_{0.02}Tb_{0.98}$ *(pic)***, λ_{em} = 615	1.136
Mg_2Al_Tb*(pic)*	1.203	**$Mg_{12}Al\_Tb$*(pic)***	1.133

Considerable increase of lifetimes for **$Mg_{12}Al\_Eu_{0.02}Tb_{0.98}$*(pic)*** (0.905 ms) in comparison with **$Mg_2Al\_Eu_{0.02}Tb_{0.98}$*(pic)*** (0.559 ms) indicates the increase of distances between nearest Tb and Eu atoms. In other words, the energy transfer processes become slower with increasing distances between neighboring lanthanides.

Thus, the luminescence process in the system could be described as follows: UV-irradiation of samples leads to excitation of ligand electrons to π-orbital and energy transfer ligand→central atom take place; than, the energy transfer $Tb^{3+}\rightarrow Eu^{3+}$ appears due to rather high excited-state lifetime of Tb^{3+}.

CONCLUSIONS

In the present work new LDH-based luminescent materials with high quantum efficiencies were obtained. Intercalation of complexes into the interlayer space of LDH doesn't affect their luminescent properties, forming non-volatile solid state material with quantum yields up to 50 % and efficient excitation of picolinate ligand. It was shown that the Coulomb interactions between LDH layers and complex can result in a change of its structure and in decreasing the number of the organic ligands per *Ln* atom. The small enough distances between lanthanides lead to energy transfer from Tb^{3+} to Eu^{3+}. The presence of OH-groups and -O-layer bonds in the first coordination sphere of *Ln* results in additional participation of lattice → Eu CT states in energy transfer.

ACKNOWLEDGMENTS

The authors would like to thank T. Justel (Philips Research Laboratories) for luminescence decay measurements. This work was partially supported by "Leading Scientific Schools" program (2033.2003.3), Euler scholarship (DAAD), and grant from the President of the Russian Federation for young candidates of science (2274.2003.03). AAE would like to acknowledge the support from Haldor Topsoe and LG Chemical.

REFERENCES

1. D. Sendor, P. C. Junk and U. Kynast, *Solid State Phenom.* **90/91**, 521 (2003).

2. Q. Xu, L. Li, X. Liu and R. Xu, *Chem. Mater.* **14**, 549 (2002).
3. D. Sendor and U. Kynast, *Adv. Mater.* **14**, 1570 (2002).
4. V. Rives, M.A. Ulibarri, *Coord. Chem. Rev.* **181**, 61 (1999).
5. A.V. Lukahsin, A.A. Eliseev, N.G. Zhuravleva, A.A. Vertegel, Y.D. Tretyakov, O.I. Lebedev, G. Van Tendeloo, *Mend. Comm.* **4**, 174 (2004).
6. A.V. Lukashin, S.V. Kalinin, M.P. Nikiforov, V.I. Privalov, A.A. Eliseev, A.A. Vertegel and Yu.D. Tretyakov, *Dokl. Akad. Nauk*, 1999, **364**, 77 [*Doklady Chemistry*, 1999, **364**, 77].
7. K. Chibwe and W. Jones, *J. Mater. Chem.*, 1989, **1**, 489.
8. D. Sendor, M. Hilder, T. Juestel, P.C. Junk and U. Kynast, *New J. Chem.* **27**, 1070 (2003)
9. A.V. Lukashin, A.A. Vertegel, A.A. Eliseev, M.P. Nikiforov, P. Gornert, Yu.D. Tretyakov *J. of Nanoparticle Research* **5**, 455 (2003).
10. N.G. Zhuravleva, A.A. Eliseev, A.V. Lukashin, U. Kynast and Yu.D. Tretyakov, *Dokl. Akad. Nauk*, 2004, **396**, 126 (*Doklady Chemistry*, 2004, **396**, 126).
11. N.G. Zhuravleva, A.A. Eliseev, A.V. Lukashin, U. Kynast and Yu.D. Tretyakov, *Mend. Comm.* **4**, 176 (2004).

Novel and Efficient electroluminescent organo-iridium phosphorescent OLED materials

Heh-Lung Huang, Kou-Hui Shen, Miao-Cai Jhu, Mei-Rurng Tseng and Jia-Ming Liu*

Division of Organo-electronic Materials, Materials Research Laboratories, Industrial Technology Research Institute, Chutung, Hsinchu, Taiwan 310
E-mail: hehlung@itri.org.tw

ABSTRACT

We investigated and synthesized different orientation and substituted group of the thieno-pyridine framework organo-iridium complexes (**PO-01**, **PO-02**, **PO-03**). These materials exhibited yellow light. Fabrication of these emissive phosphorescent dopants with the device configuration of ITO/ NPB/ CBP: Ir-complex (4-6%) /BCP/Alq_3/LiF/Al showed the EL emitting peak from 532nm to 564nm. The electroluminescent efficiencies of the devices were from 9.62 lm/W to 30.41 lm/W and maximum brightness was all above 7000 cd/m^2. We demonstrated high-efficiency, high-brightness and saturated yellow Ir-complex phosphorescent materials for OLED.

INTRODUCTION

Since Tang and VanSlyke reported the first configuration of the double-layered organic light electroluminescent (EL) device[1], many research scientists have investigated and reported high performance organic EL devices. Organic EL materials have recently attracted much interest due to their different physical properties and potential applications in light emitting devices (LED) and flat panel displays. To realize full color display applications utilizing organic electroluminescent devices, the three elemental colors of red, green, and blue emitters with sufficiently high luminous efficiency and color purity are necessary. There is also a great demand for emitters that give a bright and high efficient color, such as yellow, orange or light blue color, for multicolor display usage.

Among the recent EL materials, transition organo-metallic complexes hold more concerns than the organic dyes due to their phosphorescent emission. OLEDs based on phosphorescent materials can significantly improve electroluminescent performance because strong spin orbital coupling induces an interchange between singlet and triplet states that strengthens phosphorescence. Theoretically, the upper limit of inner quantum efficiency for phosphors can reach 100%, which is four times higher than that of the devices using other fluorescent materials. Moreover, the power consumption of full color EL devices by using only red phosphorescent OLED materials can reduce up to 42% than using organic fluorescent dyes as an emitter.

The first reported phosphorescent red OLED material was $Eu(DBM)_3(Phen)$ [2]. In 1994, Kido, J. et al. reported triple-layer-type cells with the structure of a glass substrate/indium-tin oxide/triphenyl diamine derivative (TPD)/Eu complex: 1,3,4-oxadiazole derivative (PBD)/aluminum complex (Alq_3)/Mg: Ag exhibits bright red luminescence when a DC voltage is applied to it.

At a constant voltage of 16 V, the maximum brightness is 460 cd/m^2 at a 614 nm EL wavelength. The long lifetime of the electronically excited europium (III) (350 μs) leads to dominant triplet-triplet annihilation (T-T annihilation) at high currents, with a maximum efficiency of only 1.4%.

Later, Thompson, M. E. et al. reported the phosphorescent dye 2,3,7,8,12,13,17,18-octaethyl-21H,23H-porphine platinum(II) (PtOEP) [3] phosphorescent material in 1998. The electroluminescent efficiency of phosphorescent materials drops more quickly with current than does that of fluorescent materials, mainly because the longer phosphorescent lifetime results in saturation of the emissive sites as the current increases, triplet-triplet annihilation (T-T annihilation) at high currents.

In 2000, Thompson, M. E. and his co-workers reported another PtOEP [4] doped into 4,4'-*N,N'*-dicarbazole-biphenyl (CBP). The CBP is the host material of the main emitting layer. The phosphorescent lifetime of CBP is double that of Alq_3 and so the latter exhibits more efficient energy transfer. The use of BCP as a hole blocking layer constrains emission associated with the triplet state within the CBP layer, increasing the luminescent efficiency. These findings reveal an external quantum efficiency of $\eta_{ext} = 5.6\%$ for a red phosphorescent material, which is 1.6 % higher than that of the former one without BCP as a hole blocking layer. However, T-T annihilation is also inevitable at high current ($\eta_{ext} = 0.5\%$ at 100 mA/cm^2) because of the relatively long phosphorescent lifetime.

Recently, a more efficient red phosphorescent material has been reported. $Ir(btp)_2(acac)$ [5] (btp = (2-(2'-benzo[4.5-a]thienyl)-pyridinato; acac = acetylacetonate) achieved $\eta_{ext} = 7.0$ % at low current. The relatively short phosphorescent lifetime of $Ir(btp)_2(acac)$(4 μs) markedly improves the external efficiency at high current (η_{ext} = 2.5 % at 100 mA/cm^2).

Lecloux, D. D. et al. reported the orange to red color organo-iridium complexes.[6] The turn-on voltage is relatively high and the peak efficiency is also low.

RESULTS AND DISCUSSION

The UV absorption bands of **PO-01**, **PO-02**, and **PO-03** are shown in Table 1.

Table 1:

Complex	Absorption (nm)	Absorption (nm)
PO-01	282, 331	394, 432
PO-02	293, 330	398, 438
PO-03	302, 338	393, 427

The two UV absorption bands at 282, 331 nm of **PO-01** were assigned to a spin-allowed $^1\pi$–π^* transition of the cyclometalated ligand (Table 1), and the band at 394 nm was assigned to a spin-allowed metal-ligand charge transfer (^{1}MLCT). The most surprising feature of the complex is the strong intensity of the band near 432 nm, which is assigned to spin-forbidden ^{3}MLCT band. Likewise, the two UV absorption bands at 293, 330 nm of **PO-02** were assigned to a spin-allowed $^1\pi$–π^* transition and the band at 398 nm was assigned to a spin-allowed ^{1}MLCT. The strong intensity of the band near 438 nm was assigned to spin-forbidden ^{3}MLCT band. And the UV absorption bands at 302, 338 nm of **PO-03** were assigned to a spin-allowed $^1\pi$–π^* transition, and the band at 393 nm was assigned to a spin-allowed ^{1}MLCT. The strong intensity of the band near 427 nm was assigned to spin-forbidden ^{3}MLCT band. Therefore, the spin-forbidden ^{3}MLCT band indicates an efficient spin-orbit coupling that is prerequisite for phosphorescent emission.

The OLED material using **PO-01**, **PO-02**, or **PO-03** was grown on device glass substrate by vacuum (3×10^{-6} torr), thermal evaporation onto pre-cleaned glass substrates with ITO (Indium Tin Oxide) pattern on it. There are anode electrode (ITO), hole transport layer, dopant emitter (**PO-01**, **PO-02**, or **PO-03**), host emitter (CBP: 4,4'-*N,N'*-dicarbazole-biphenyl), hole blocking layer (BCP: 2,9-dimethyl-4,7-diphenyl-1,10-phenanthroline), electron transport layer (Alq_3: tris-(8-quinolino aluminum), LiF, and cathode electrode (Al) in this device configuration (Figure 1).

Figure 1: The Device Configuration for **PO-01**, **PO-02**, or **PO-03**.

Al
LiF
Alq_3
BCP
CBP: **PO-01**, **PO-02**, or **PO-03** (4-6%)
NPB
ITO
Glass Substrate

The luminance and electroluminescent efficiency data of **PO-01**, **PO-02**, and **PO-03** as a dopant material respectively are shown in Table 2, 3, and 4. The efficiency of these three yellow to orange dopant materials were all relatively high (between 8.09lm/W to 17.64 lm/W) at low current density. Although the turn-on voltage was a little bit high, the life time of the device was also very good under high voltage (above 15V).

Table 2: The luminance and power efficiency data of **PO-01** as a dopant material.

NPB (50nm)/4-6% PO-01 in CBP(30nm)/BCP(8nm)/Alq$_3$(20nm)/LiF(0.5nm)/Al(120nm)			
Max Peak	CIE, x	CIE, y	MAX Luminance
560	0.51	0.48	10999
Current Density 55 mA/cm^2			
V	cd/A	lm/W	Luminance cd/m^2
5.5	41.54	23.73	232
Best Performance			
V	cd/A	lm/W	Luminance cd/m^2
4.5	43.56	30.41	21

Table 3: The luminance and power efficiency data of **PO-02** as a dopant material.

NPB (50nm)/4-6% PO-02 in CBP(30nm)/BCP(8nm)/Alq$_3$(20nm)/LiF(0.5nm)/Al(120nm)			
Max Peak	CIE, x	CIE, y	MAX Luminance
532	0.40	0.58	7190
Current Density 93 mA/cm^2			
V	cd/A	lm/W	Luminance cd/m^2
11.0	26.76	7.64	250
Best Performance			
V	cd/A	lm/W	Luminance cd/m^2
9.5	29.10	9.62	46

Table 4: The luminance and power efficiency data of **PO-03** as a dopant material.

NPB (50nm)/4-6% PO-03 in CBP(30nm)/BCP(8nm)/Alq$_3$(20nm)/LiF(0.5nm)/Al(120nm)			
Max Peak	CIE, x	CIE, y	MAX Luminance
564	0.50	0.50	11005
Current Density 100 mA/cm^2			
V	cd/A	lm/W	Luminance cd/m^2
5.5	28.85	16.48	293
Best Performance			
V	cd/A	lm/W	Luminance cd/m^2
5.0	26.49	16.64	89

The brightness under different voltage is shown in Figure 2. We can easily see **PO-03** has highest brightness compared to **PO-01**.

Figure 2:

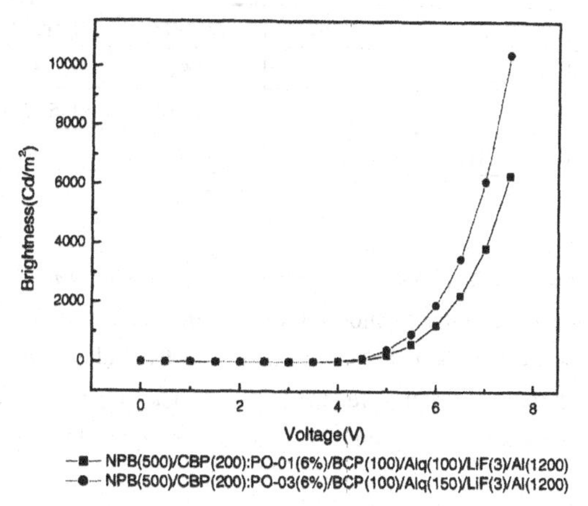

The power efficiency under different voltage is shown in Figure 3. We can easily see **PO-01** has highest power efficiency compared to **PO-03** owing to the stronger intensity of the UV absorption band near 432 nm, which is assigned to spin-forbidden ^{3}MLCT band. The trend of power efficiencies is downward with the increase of voltage and current. This is due to the triplet-triplet annihilation at high currents.

Figure 3:

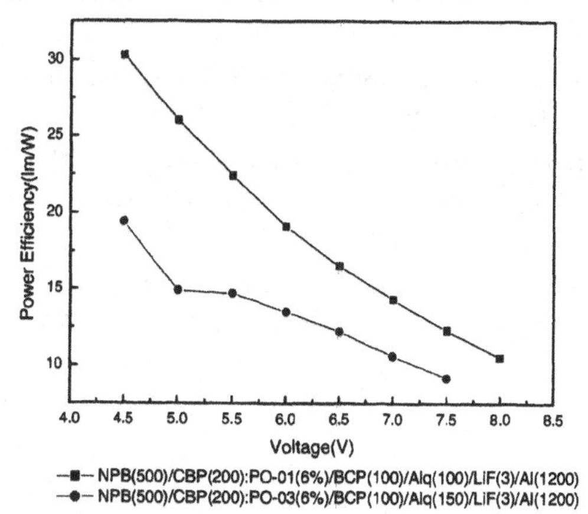

The CIE coordinates and power efficiencies of **PO-01**, **PO-02** and **PO-03** are shown in Figure 4. In 2002, Thompson, M. E. and his co-workers reported the yellow bt_2Ir(acac) [7] with wavelength of 565 nm. Its CIE coordinate is (0.51,0.49) and the power efficiency is 11.00 lm/W. **PO-01** holds higher power efficiency than bt_2Ir(acac) with the same CIE coordinate.

Figure 4:

	PO-01	**PO-02**	**PO-03**	***bt*₂Ir(acac)**
Peak wavelength (nm)	564	532	564	565
CIE (x, y)	(0.51,0.48)	(0.40,0.48)	(0.56,0.43)	(0.51,0.49)
P.E.(lm/W) at 1mA/cm^2	30.41	9.62	16.64	11.00

CONCLUSIONS

We established the easily synthetic route to prepare new yellow to orange iridium complex OLED materials based on the thieno-pyridine framework. According to the device data, **PO-03** holds the best luminance and **PO-01** holds the highest electroluminescent efficiency than the reported data.[7] The electroluminescent efficiencies of these emitters were good enough to apply on the multicolor display.

ACKNOWLEDGMENTS

We thank the Ministry of Economic Affairs of Taiwan for financial support on the project of the flat-panel display.

REFERENCES

1. C. W. Tang and S. A. VanSlkye, *Appl. Phys. Lett.* **27**, 713 (1987).
2. J. Kido, H. Hayase, K. Kongawa, K. Hagai and K. Okamoto, *Appl. Phys. Lett.* **65**, 2124 (1994).
3. M. A. Baldo, D. F. O'Brien, Y. You, A. Shoustikov, S. Sibley, M. E. Thompson and S. R. Forrest, *Nature* **395**, 151 (1998).
4. M. A. Baldo, M. E. Thompson and S. R. Forrest, *Nature* **403**, 750 (2000).
5. C. M. Adachi, A. Baldo, S. R. Forrest, S. Lamansky, M. E. Thompson and R. C. Kwong, *Appl. Phys. Lett.* **78**, 1622 (2001).
6. D. D. Lecloux, V. A. Petrov and Y. Wang, WO 03/040256 (2003).
7. P. E. Burrows, S. R. Forrest and M. E. Thompson, *J. Am. Chem. Soc.* **123**, 4311 (2001).

Poster Session: Linear and Nonlinear Optical Properties of Organic and Nanocomposite Materials

Mater. Res. Soc. Symp. Proc. Vol. 846 © 2005 Materials Research Society DD10.3

Fabrication of Micro-Optical Devices by Holographic Interference of High Photosensitive Inorganic-Organic Hybrid Materials (Photo-HYBIRMER)

Dong Jun Kang, Jin-Ki Kim and Byeong-Soo Bae
Laboratory of Optical Materials and Coating (LOMC), Department of Materials Science and Engineering, Korea Advanced Institute of Science and Technology (KAIST),
Daejeon 305-701, Korea

ABSTRACT

Sol-gel derived photosensitive inorganic-organic hybrid materials (Photo-HYBRIMER) containing a large quantity of photoactive molecules exhibit the large changes in both refractive index (over 10^{-2}) and volume (over 30%) on UV exposure. The materials could be used for direct fabrication of micro-optical devices using holographic interference. With the change of the beam number for holographic interference (1-beam, 2-beam, 3-beam and 4-beam interference), various typed micro-optical devices (Fresnel-type lens, 1D- and 2D-typed diffraction gratings) could be easily fabricated. Importantly, the fabricated micro-optical devices exhibited the very homogeneous surface structures and good optical performance.

INTRODUCTION

Micro-optical devices are the important component for beam dispersion and conversion, optical signal processing and modulation in compact and complicated optical data storage systems and optical integrated circuits [1-3]. Thus, the study of fabricating micro-optical devices has been intensively performed using a variety of methods, including a photo-mask, beam lithography, etching techniques, holographic interference. Contact imprinting using a photo-mask is not suitable for the fabrication of devices with a submicrometer period due to the diffraction limit between the mask and the samples. The beam lithography, which includes the etching process, is rather complex and needs several steps to reveal the precise surface structure. These disadvantages of a photo-mask and lithographic techniques cause limitations to easy fabrication and obtaining higher performance of micro-optical devices. However, the holography can produce fine and precise patterns easily. Thus, the holographic interference has great potential for the direct fabrication of micro-optical devices with good performance.

The materials as well as the fabrication processes are crucial factors in fabricating the micro-optical devices with good performance. In order to achieve higher performance in optical devices, materials must have high photosensitivity, meaning a large refractive index, and undergo a volume change upon light irradiation, because the direct photo-fabrication and the performance of an element depend on the photosensitivity of the materials used. In recent years, sol-gel hybrid materials (HYBRIMERs) [4-6] and photosensitive polymers [7,8] have been used as potential materials for fabricating highly efficient micro-optical elements. In particular, the photosensitive hybrid materials (Photo-HYBRIMERs) doped with large amounts of photoactive molecules were found to exhibit larger refractive index and volume change on UV exposure [9-11]. The photosensitivity of these materials could be increased by the simple addition of photoactive molecules. Also, these materials have the potential to be used in simple single-step photo-patterning due to the simultaneous changes in both refractive index and volume. Thus, these Photo HYBRIMERs are good candidates for highly efficient micro-optical devices.

In the present study, we report on the effect of UV dose and the functionality of the photoactive molecules on photosensitivity in HYBRIMERs, with a focus on increasing the photosensitivity. We make the various interference fringes by changing the beam number of holography for direct photo-fabrication of micro-optical devices. Finally, with both surface relief structures and refractive index modulation of HYBRIMERs, we manufactured directly well shaped micro-optical devices with different types and good optical performance by using various holographic interference fringes without any etching treatment. In addition, we investigated the effects of UV dose on the optical performance of fabricated micro-optical devices.

EXPERIMENTAL DETAILS

Transparent photo-HYBRIMERs were prepared using methacryloxypropyl-trimethoxysilane (MPTS, Aldrich), perfluoroalkylsilane (PFAS, Toshiba) and zirconium n-propoxide (ZPO, Aldrich) chelated with methacrylic acid (MAA, Aldrich) as precursors. All precursors were hydrolyzed with 0.01-N HCl. After 20 hour stirring for full sol-gel reaction, any residual products such as alcohols were removed at 50°C with an evaporator. Benzyldimethylketal (BDK, Aldrich) as a photoinitiator, and mono-, di-, and three-acrylate monomer (Aldrich) as a photoactive monomer, were added into the HYBRIMER solution prior to the coating. After stirring the solution for 1 hour at room temperature, a homogeneous photosensitive HYBRIMER solution was obtained. This solution was spin-coated on a cleaned glass substrate with 3000-rpm spinning speed. The coated films were exposed by a Hg UV lamp (wavelength 350-390nm) and a He-Cd laser (wavelength 325 nm) for photo-induced reactions and holographic interference. The exposed time was different for measuring the effects of the UV dose on the photosensitivity of the materials and then consolidated by baking at 150°C for 5 hours.

The changes in the refractive index and thickness of the films before and after illumination with the Hg lamp were measured using a prism coupler (Metricon 2010) at a wavelength of 632.8 nm. Holographic interference using a He-Cd laser was used for the fabrication of micro-optical devices. The images of the fabricated micro-optical devices were investigated using optical microscopy and atomic force microscope (AFM, Park Scientific Instruments, Autoprobe 5 M). The diffraction effects of micro-optical devices were monitored using a CCD camera.

RESULTS AND DISCUSSION

Figure 1 shows the refractive index and thickness changes of photo-HYBRIMERs as a function of UV dose and the acrylate functionality in the photoactive molecules. Most changes in refractive index and film thickness occurred during short UV doses and the photosensitivity was enhanced with the increase of acrylate functionality in photoactive monomers. Also, photo-HYBRIMERs with large amounts of photoactive molecules exhibit excellent photosensitivity, that is, the large changes in both refractive index (over 10^{-2}) and volume (over 30%) on UV exposure as shown in Fig. 1. These high photosensitivity are closely related to the various photo-induced mechanisms, which are photo-polymerization between sol-gel matrix and decomposed radicals of photoinitiator and photo-locking of photochemical species with UV exposure. Upon UV illumination, photochemical species, such as photo-initiators and photoactive monomers, are decomposed and are locked inside the sol-gel hybrid matrix or are dimerized themselves. Moreover, during space controlled UV illumination, these photo-HYBRIMERs undergo the

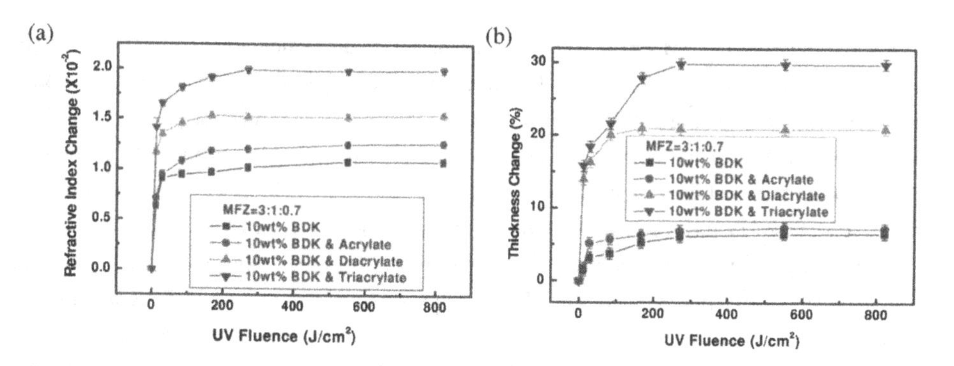

FIGURE 1. Refractive index (a) and thickness changes in Hg UV lamp (wavelength 350-390nm) exposed and baked photo-HYBRIMERs depending on UV dose and functionality of photoactive molecules.

photo-migration due to the concentration gradient of photoactive molecules between unexposed and exposed areas, which derives large changes in refractive index and volume between unexposed and exposed areas. Therefore, we can apply these photo-HYBRIMERs with high photosensitivity to direct photo-patterning for high-efficiency micro-optical devices. In particular, we used holographic interference for the fabrication of various type micro-optical devices.

Figure 2(a) shows an optical micrograph of a Fresnel-type lens patterned by 1-beam pinhole interference on photo-HYBRIMERs. The Fresnel type lens exhibited the higher odd zones and lower even zones by the negative-type sensitivity depending on the spatially controlled exposure of pinhole interference. Figure 2(b) and 2(c) show the focusing property and diffraction effects of a Fresnel type lens from a CCD camera. The diffracted laser beams exhibited the intensified focusing effect and the expanded diffraction effects were also strong, which lead to a very high efficiency approaching 85% of Fresnel type lens. This high-efficiency Fresnel type lens would be potentially useful for application in compact optical systems.

FIGURE 2. (a) Optical microscope, (b) CCD images of focusing effect and (c) diffraction beam of a patterned Fresnel-type lens by 1-beam interference.

Figure 3 shows (a) AFM image and (b) CCD images of diffraction effects of 1D diffraction gratings with periods of 1000 nm fabricated with the 2-beam interference on photo-HYBRIMERs. The grating profile was a perfect sinusoid, which is related the precise control of UV doses including the intensity of the beam and the exposure time. Importantly, the fabricated diffraction gratings exhibited a good diffraction performance. In addition, in order to investigate the effects of the UV doses on the diffraction efficiency of gratings, we measured the diffraction efficiency of gratings with increasing the exposure time of the photo-fabrication of the gratings. The maximum diffraction efficiency of 1D diffraction gratings was around 11.57 % and highly dependent on UV doses related to the intensity of the beam and exposure time. The exposure time for obtaining maximum diffraction efficiency agreed well with the optimum exposure time for obtaining the good-patterned gratings and the strongest diffraction effects. This UV exposure time dependence on diffraction efficiencies could be due to the line width broadening of the gratings, which results in the decrease of the grating depth and the refractive index changes between the high lines and the low lines in the sinusoidal gratings. A long irradiation time could also cause the distortion of the grating shape or surface, which could be another reason for the decrease in diffraction efficiency. Consequently, the diffraction effects and efficiencies of gratings were heavily dependent on the UV exposure time and would be decreased by much UV doses. Thus, the optimum UV doses in photo-fabrication of the diffraction gratings using photo-HYBRIMERs are the most important factors for obtaining the stronger diffraction effects of gratings.

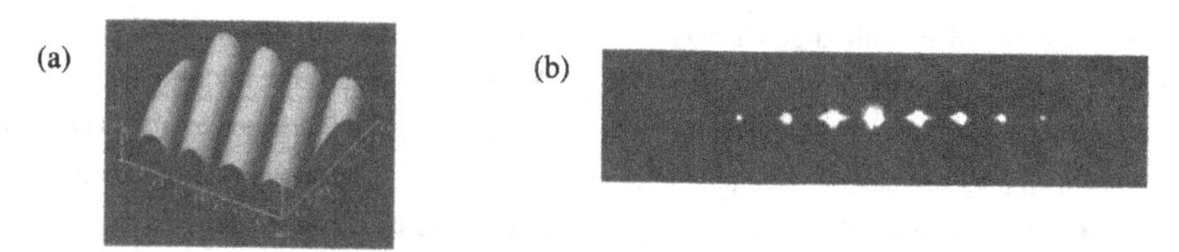

FIGURE 3. (a) AFM image and (b) CCD image of diffraction effects of 1D diffraction gratings fabricated by 2-beam interference.

Figure 4 shows optical microscope of (a) 2D hexagonal type diffraction gratings fabricated by 3-beam interference and (b) 2D rectangular type diffraction gratings fabricated by 4-beam interference. For forming the 3-beam interference, diffraction optical element with a triangle type lines was used and the laser beam through this element divided into three diffraction beams. These diffraction beams focused on one plane, where 3-beam interference is formed. The fringe shape of 3-beam interference is hexagonal type. Thus, 2D hexagonal type diffraction gratings were patterned on focal plane of three diffraction beams. In case of 4-beam interference, we used diffraction optical element with rectangular type lines. The principle of 4-beam interference formation is same as 3-beam interference. With the triangle and rectangular type diffraction optical elements, 3- and 4-beam interference could be easily formed and 2D type diffraction gratings with homogeneous surface structures could be fabricated precisely. Moreover, these well

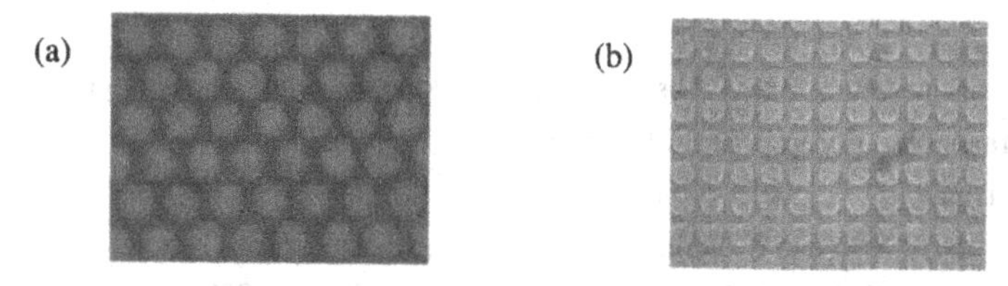

FIGURE 4. (a) Optical microscope of 2D hexagonal type diffraction gratings fabricated by 3-beam interference and (b) optical microscope of 2D rectangular type diffraction gratings fabricated by 4-beam interference

shaped 2D type diffraction gratings also showed 2D strong diffraction effects. 2D hexagonal and rectangular type diffraction gratings exhibited hexagonal and rectangular typed diffraction patterns, respectively.

CONCLUSIONS

The photo-HYBRIMERs containing large amounts of photoactive molecule exhibited very high photosensitivity with large changes in both refractive index and volume on UV exposure. With the high photosensitivity of photo-HYBRIMERs, various typed micro-optical devices (Fresnel-type lens, 1D- and 2D-typed diffraction gratings) could be fabricated easily by changing the beam number of holographic interference (1-beam, 2-beam, 3-beam and 4-beam interference). The diffraction effects and efficiencies of micro-optical devices depended heavily on UV doses related to the exposure time and the beam intensity. Importantly, the fabricated micro-optical devices exhibited the very homogeneous surface structure and good optical performance.

ACKNOWLEDGEMENTS

The authors gratefully acknowledge the financial support of the Sol-Gel Innovation Project (SOLIP) funded by Ministry of Commerce, Industry & Energy (MOCIE) in Korea.

REFERENCE

1. T. Fujita, H. Nishihara and J. Koyama, *Opt. Lett.*, **7**(12), 578 (1982)
2. V. Moreno, M. V. Pěrez and J. Liňares, *J. Mod. Opt.*, **39**(10), 2039 (1992)
3. D. Mendlovic, *Opt. Comm.*, **95**, 26 (1993)
4. B. S. Bae, O. H. Park, R. Charters, B. Luther-Davies and G. R. Atkins, *J. Mater. Res.*, **16**[11], 3184 (2001)
5. D. J. Kang, J. U. Park and B. S. Bae, J. Nishii and K. Kintaka, *Opt. Express*, **11**, 1144 (2003)
6. W. Yu and X. -. Yuan, *Opt. Express*, **11**, 1925 (2003)
7. D. Y. Kim, S. K. Tripathy, L. Li and J. Kumar, *Appl. Phys. Lett.*, **66**, 1166 (1995)
8. N. Zettsu, T. Ubukata, T. Seki and K. Ichimura, *Adv. Mater.*, **13**, 1693 (2001)
9. O. H. Park, J. I. Jung and B. S. Bae, *J. Mater. Res.*, **16**[7], pp2143-2148 (2001)
10. O. H. Park, S. J. Kim and B. S. Bae, *J. Mat. Chem.*, **14**, 1749 (2004)
11. D. J. Kang, P. V. Phong and B. S. Bae, **85**[19], *Appl. Phys. Lett.*, 4289 (2004)

Mater. Res. Soc. Symp. Proc. Vol. 846 © 2005 Materials Research Society

Fabrication of Polydiacetylene Nanocrystals Deposited with Silver Nanoparticles for a Nonlinear Optical Material

Tsunenobu Onodera[1,3], Hidetoshi Oikawa[1,3], Hitoshi Kasai[2,3], Hachiro Nakanishi[2,3] and Takashi Sekiguchi[1]
[1]Nanomaterials Laboratory, National Institute for Materials Science,
Namiki 1-1, Tsukuba 305-0044, JAPAN
[2]Institute of Multidisciplinary Research for Advanced Materials, Tohoku University,
Katahira 2-1-1, Aoba-ku, Sendai 980-8577, JAPAN
[3]Japan Science and Technology Agency/ Core Research for Evolutional Science and Technology (JST/ CREST),
Honcho 4-1-8, Kawaguchi 332-0012, JAPAN

ABSTRACT

Noble metal-coated polydiacetylene (PDA) nanocrystals are expected to enlarge effective third-order nonlinear optical susceptibility, owing to the enhancement of optical electric field induced by localized surface plasmon. The surface of PDA nanocrystals was decorated with silver nanoparticles, and the linear optical properties were investigated. The well-defined PDA nanocrystals were first prepared by the conventional reprecipitation method as an aqueous dispersion. The following two-steps of reduction methods, namely, seed deposition and seed growth, enabled higher silver coverage on the surface of PDA nanocrystals. Silver fine seeds were successfully deposited on the PDA surface by particularly employing an anionic surfactant as a binder. The extinction spectra of the PDA nanocrystals covered with silver exhibited the red-shift and broadening of surface plasmonic peak for silver nanoparticles, which would indicate the interaction among silver nanoparticles on the same PDA nanocrystal. In addition, the red-shift of excitonic peak for PDA nanocrystals was observed, resulting from the changes of dielectric properties around PDA nanocrystal. Both spectral changes would have an influence on nonlinear optical properties of PDA nanocrystal.

INTRODUCTION

Organic third-order nonlinear optical (NLO) materials are powerful candidates for highly efficient optoelectronic devices, because of their fast responsibility and easy processability in comparison with inorganics [1]. For example, π-conjugated polymers such as polydiacetylene (PDA) possess fast and large third-order NLO property attributed to delocalized π-conjugated electrons along the main chain backbone. PDA crystals can be obtained through topochemical solid-state polymerization of corresponding diacetylene monomer crystals by UV- or γ-irradiation or heat treatment. So far, the syntheses and optoelectronic properties of PDA derivatives have been extensively studied. However, the magnitude of effective third-order nonlinear optical susceptibility $\chi^{(3)}(\omega)$ is not still sufficient for the device application.

Neeves *et al.* [2] theoretically predicted that composite nanostructure consisted of PDA core and metal shell exhibited large effective $\chi^{(3)}(\omega)$, which is enhanced by electric field of localized surface plasmon. We have already established the reprecipitation method [3], which is a convenient technique to fabricate nanocrystals of thermally unstable organic compounds such as diacetylene monomer. On the other hand, metal deposition methods have been recently developed using a polymer binder for chemically surface-modified polystyrene and silica spheres [4, 5]. However, PDA nanocrystals have chemically inert surface. Namely, it is very important to mildly modify the surface of nanocrystals without damaging π-conjugated main chains of PDA as a core.

In the present study, we attempted to fabricate PDA nanocrystals coated with silver nanoshell. At first, PDA nanocrystals as a core were prepared by using the reprecipitation method. Silver seeds, then, could be deposited on the surface of PDA nanocrystal, on which was previously modified with anionic surfactant, followed by further reduction with hydroxylamine. In addition, the UV-Vis extinction spectra of the resulting nanocrystals will be discussed in relation to their morphologies.

EXPERIMENTAL

1,6-Di-(*N*-carbazolyl)-2,4-hexadiyne (abbreviated to "DA monomer") as a precursor of PDA was synthesized as described elsewhere [6]. Acetone (99.5%), sodium dodecylsulfite (SDS), silver nitrate, sodium borohydride, and hydroxylamine (99.999%, 50 wt% in water) were commercially available, and used without further purification. Water used was purified up to 18.2 MΩcm by a Millipore Super-Q system.

PDA nanocrystals as a core were fabricated by the reprecipitation method [7, 8] as follows: at first, DA monomer acetone solution (1 ml, 5 mM) containing saturated SDS (< 1 mM) was injected into vigorously stirred water (50 ml) at 318 K. After retaining for 30 minutes at a given temperature, the DA monomer nanocrystals dispersed in water were polymerized in solid state by irradiating UV ray with handy lamp (254 nm, 4 W) for 1 hour, providing PDA nanocrystals dispersion with dark blue and low scattering loss. Next, aqueous solution of sodium borohydride (0.3 ml, 1.11 mM) was dropwise added to the mixture of the PDA nanocrystals suspension (5 ml) and silver nitrate aqueous solution (0.1 ml, 22.2 mM). Silver cations were reduced, resulting in deposition of silver fine seeds on the surface of PDA nanocrystals. The subsequent reduction of silver cation (0.4 ml, 22.2 mM) with hydroxylamine (4 ml, 2.22 mM) was carried out in order to obtain higher silver coverage on the PDA surface.

The resulting nanocrystals were casted onto a silicon wafer, and then characterized to evaluate the crystal size and morphology by using scanning electron microscope (SEM; JEOL, JSM-6500F). The extinction spectra of the PDA nanocrystals dispersed in water were measured with UV-Vis spectrometer (Shimadzu, UV-160A).

RESULTS AND DISCUSSION

Formation of SDS-adsorbed PDA nanocrystal

The SEM image of SDS-modified PDA nanocrystals as a core is shown in figure 1 (a). It is found that the nanocrystals formed were cubic and *ca.* 70 nm in size, which was almost similar to the result measured with dynamic light scattering. This fact means that PDA nanocrystals were isolated in the water, not aggregated. Previously, we proposed the formation mechanism of PDA nanocrystals in the presence of SDS [8]. Just after injection of DA monomer acetone solution,

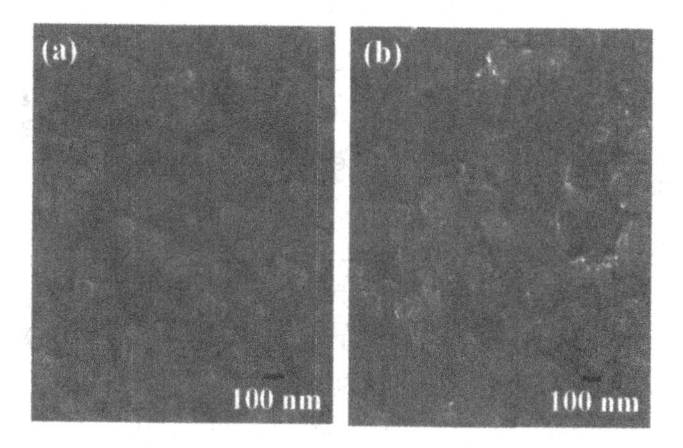

Figure 1. SEM images of (a) SDS-adsorbed PDA nanocrystals as a core and (b) silver seed-deposited PDA nanocrystals casted on the silicon wafer.

acetone droplets containing both DA monomer and SDS are first formed, acetone would immediately diffuse into water phase, and then DA monomer amorphous nanoparticles formed would be stabilized by SDS. When an anionic surfactant such as SDS co-exists, hydrophobic hydrocarbon chains of surfactant could be adsorbed onto or into the DA monomer amorphous nanoparticles, and anionic moiety of surfactant would be exposed to water phase. The following nucleation and crystallization of DA monomer in amorphous nanoparticles may occur, interestingly competing with collision between nanoparticles, which is sensitively influenced by water temperature. For example, SDS may protect and stabilize amorphous DA nanoparticles, and neither crystallization nor collision occurs below room temperature. On the other hand, amorphous DA nanoparticles and DA nanocrystals may stably co-exist at 333 K, owing to SDS, providing fibrous PDA nanocrystals, which would epitaxially grow through collision of amorphous nanoparticles to certain crystal face of already-formed nanocrystals. However, by controlling water temperature, only the cubic and SDS-adsorbed PDA nanocrystals were obtained successfully at 318 K as shown in figure 1 (a). The non-epitaxial crystal growth of DA monomer seems to be dominant at 318 K. The cubic PDA nanocrystals modified with SDS were available for the fabrication of core/shell nanocrystals.

Silver seed deposition and growth

Silver cations could be electrostatically trapped with sulfonate anion moiety of SDS adsorbed on the surface of PDA nanocrystals. Fine silver seeds, as expected, were deposited on the PDA surface only in the presence of SDS as shown in figure 1 (b). PDA nanocrystal dispersion liquid turned into bright-greenish blue during reduction.

Figure 2 shows the schematic illustration of the proposed mechanism of silver seeds deposition in the presence of both silver cation and surfactant. In the case (1), silver seeds are deposited onto the PDA surface, because SDS molecules are almost adsorbed on the surface of PDA nanocrystals and play a role of binders, capturing silver cations with electrostatic attraction. In the case (2), however, isolated silver seeds would be also formed, due to the excess amount of

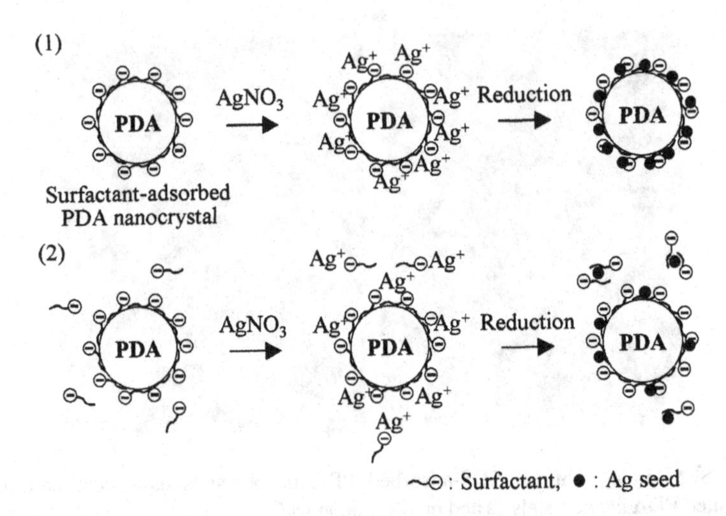

Figure 2. Schematic illustration of silver-seed deposition onto PDA nanocrystals: (1) surfactants are almost adsorbed on the surface of PDA nanocrystals; (2) in the presence of excess amount of surfactants.

SDS. Thus, it is important to exactly adjust the ratio of concentration of silver cation and surfactant to the number and the average size of PDA nanocrystals. As shown in figure 3 (a), a number of isolated silver nanoparticles were formed by adding excess SDS. Furthermore, we confirmed that metal deposition did not occur in the absence of SDS.

Finally, the subsequent reduction of silver cation with hydroxylamine was performed to obtain higher silver coverage on the surface of PDA nanocrystals. The color of the dispersion liquid

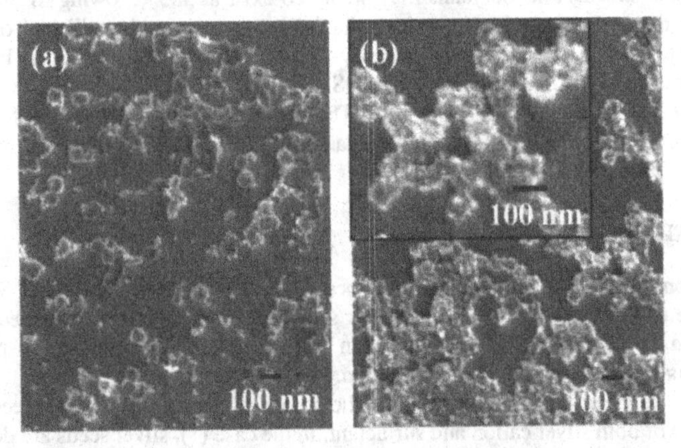

Figure 3. SEM images of seed-growth PDA nanocrystals: (a) in the presence of excess amount of SDS; (b) in the presence of proper amount of SDS.

changed into green. In figure 3 (b), silver seeds evidently grew to be 10-20 nm in size on the surface of PDA nanocrystals. It is worth noting that no new nucleations generated during reduction with hydroxylamine. In a word, hydroxylamine serves as a surface-catalyzed reducing agent for silver nanoparticle [9]. The average of coverage with silver nanoparticles was roughly 50% to 70% from SEM image in figure 3 (b) in the present case.

Extinction spectra

Figure 4 shows the UV-Vis extinction spectra of the PDA nanocrystals as a core, the seed-deposited PDA nanocrystals, and the seed-growth PDA nanocrystals. The extinction peak at 642 nm is assigned to the excitonic absorption (EA) of the π-conjugated backbones of PDA nanocrystals. After deposition of silver seeds, the extinction peak around 414 nm appeared, attributed to the localized surface plasmonic (LSP) peak of silver nanoparticles, and EA peak position was slightly red-shifted. Moreover, after treatment with hydroxylamine, LSP peak was broadened and red-shifted to 442 nm, and also EA peak position was red-shifted up to 649 nm. The red-shift of the LSP peak would be based upon increase in LSP-LSP interaction among silver nanoparticles adsorbed on the same PDA nanocrystal, because of shortened distance between silver nanoparticles with increasing volume of silver nanoparticles. This interaction also contributes to the enhancement of local electric filed between silver nanoparticles [10]. On the other hand, EA peak position is likely to be also red-shifted, owing to changes of dielectric properties around the PDA nanocrystal, involving a certain interaction between the exciton and the localized surface plasmon [11]. The silver deposition on the PDA nanocrystal would enlarge or at least influence internal electric field in PDA nanocrystal core, resulting in changes of NLO properties.

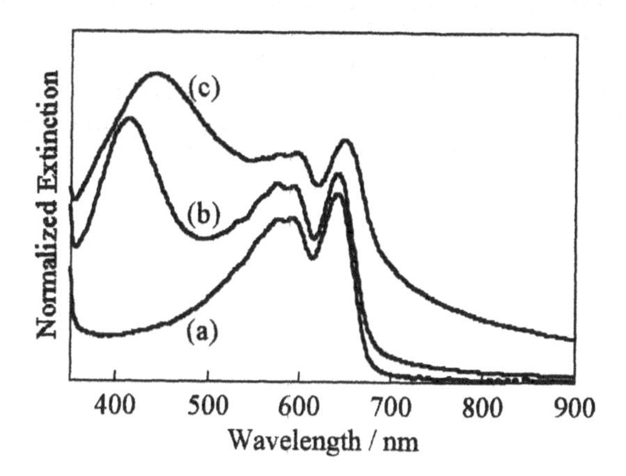

Figure 4. UV-Vis extinction spectra of (a) PDA nanocrystals as a core, (b) seed-deposited PDA nanocrystals, and (c) seed-growth PDA nanocrystals dispersed in water. The extinction was normalized by the number of PDA nanocrystals.

CONCLUSIONS

We succeeded in fabricating the PDA nanocrystals deposited with silver nanoparticles by using anionic surfactant as a binder. It is very important to optimize the amount of silver salt and surfactant to PDA nanocrystals. A surfactant with high affinity for PDA surface should be utilized in order to further raise coverage with silver. The extinction spectra of the PDA nanocrystals deposited with silver nanoparticles suggested a certain interaction between silver nanoparticles as well as changes of dielectric constant around the PDA nanocrystal, being clearly different from superimposed spectrum of PDA nanocrystals and silver nanoparticles. We would expect the enhancement of effective $\chi^{(3)}(\omega)$ for the PDA nanocrystals deposited with silver nanoparticles.

REFERENCES

1. *Organic Nonlinear Optical Materials*, Advances in Nonlinear Optics, vol. 1. ed. Ch. Bosshard, K. Sutter, Ph. Prêtre, J. Hulliger, M. Flörsheimer, P. Kaaatz and P. Günter (Gordon and Breach Science Pub., New York, 1995).
2. A. E. Neeves and M. H. Birnboim, *J. Opt. Soc. Am. B* **6**, 787 (1989).
3. H. Kasai, H. S. Nalwa, H. Oikawa, S. Okada, H. Matsuda, N. Minami, A. Kakuta, K. Ono, A. Mukoh and H. Nakanishi, *Jpn. J. Appl. Phys.* **31**, L1132 (1992).
4. F. Caruso, *Adv. Mater.* **13**, 11 (2001).
5. S. J. Oldenburg, R. D. Averitt, S. L. Westcott and N. J. Halas, *Chem. Phys. Lett.* **288**, 243 (1998).
6. K. C. Yee and R. R. Chance, *J. Polym. Sci. Polym. Phys. Ed.* **16**, 431 (1978).
7. H. Katagi, H. Kasai, S. Okada, H. Oikawa, K. Komatsu, H. Matsuda, Z. Liu and H. Nakanishi, *Jpn. J. Appl. Phys.* **35**, L1364 (1996).
8. T. Onodera, T. Oshikiri, H. Katagi, H. Kasai, S. Okada, H. Oikawa, M. Terauchi, M. Tanaka and H. Nakanishi, *J. Cryst. Growth* **229**, 586 (2001).
9. K. R. Brown and M. J. Natan, *Langmuir* **14**, 726 (1998).
10. H. Xu, E. J. Bjerneld, M. Käll and L. Börjesson, *Phys. Rev. Lett.* **83**, 4357 (1999).
11. I. Pockrand, J. D. Swalen, J. G. Gordon II and M. R. Philpott, *J. Chem. Phys.* **70**, 3401 (1979); I. Pockrand, A. Brillante and D. Möbius, *ibid.*, **77**, 6289 (1982).

Mater. Res. Soc. Symp. Proc. Vol. 846 © 2005 Materials Research Society DD10.8

Size-effect on Fluorescence Spectrum of Perylene Nanocrystal Studied by Single-particle Microspectroscopy Coupled with Atomic Force Microscope Observation

Hideki Matsune[1], Tsuyoshi Asahi[1], Hiroshi Masuhara[1], Hitoshi Kasai[2], and Hachiro Nakanishi[2]
[1]Department of Applied Physics, Osaka University, Japan
[2]Institute of Multidisciplinary Research for Advanced Materials, Tohoku University, Japan

ABSTRACT

This paper presents single particle fluorescence spectroscopy of perylene nanocrystals coupled with AFM observation of their topographic shapes. The fluorescence spectra of individual nanocrystals confirmed clearly and precisely a blue-shift in the excimer emission maximum on the reduction of their volume. The result can be explained in terms of "lattice softening", which makes intermolecular interaction weaker and modified the energy level of the excimer state in nanocrystal.

INTRODUCTION

Organic nanocrystals have attracted attentions in recent years, as they are just between single molecules and bulk materials. Their unique properties arise from their small size and large surface-to-volume ratio, and in general their distinct characteristics have been investigated as functions of particle size and shape. Size-dependent absorption and emission spectra are one of particularly important topics, and interesting size-dependent optical properties have been reported for several organic dye nanoparticles in their size range from several tens nanometers to a few hundred nanometers [1-12]. For example, fluorescence spectrum of perylene nanocrystal below the size of 200 nm show a blue-shift of excimer (E-) emission peak with decreasing of the crystal size and the enhanced monomer (M-) emission compared to the bulk fluorescence [3-5]. Previous experimental data, however, were obtained by ensemble measurement of the colloidal dispersion of the nanocrystals in water. Ambiguity in the size-dependent spectra was left due to distributions in the size and shape of nanocrystal in the colloidal solution, coexistence of amorphous particles, solvated molecules, and so on. In order to make these problems clear, single particle spectroscopy is important and indispensable [11-13]. In this paper, we have measured both fluorescence spectra and topographic shapes of individual perylene nanocrystals using a far-field optical microscope coupled with an AFM system, and characterized precisely the size dependence of their fluorescence spectra.

EXPERIMENTAL DETAIL

Perylene (Aldrich) was purified by vacuum sublimation. Colloidal perylene nanocrystals were prepared by the conventional reprecipitation method [3-5]; using a microsyringe, a perylene acetone solution (1 mM) was injected into pure water with stirring vigorously, leading to a dispersion of nanocrystals after a given aging period. We prepared samples for single particle measurement by taking one drop of the diluted dispersion onto the quartz substrate and then by drying it in air. Bulk crystals (several 100 μm to a few mm sizes) were grown from an ethanol solution.

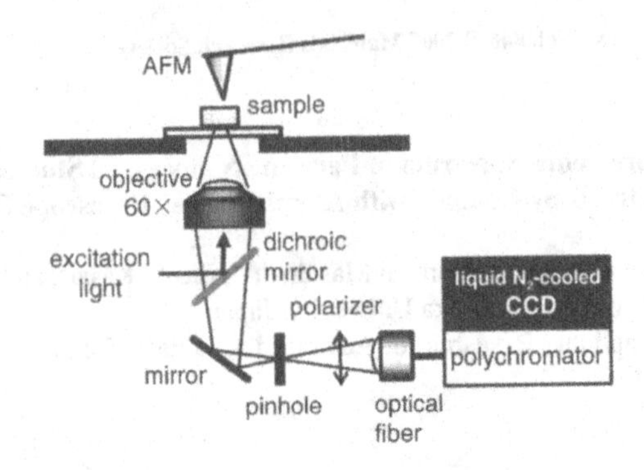

Figure 1. Schematic representation of a far-field optical microspectroscopy coupled with an AFM system.

Figure 1 shows the experimental setup for our far-field fluorescence microspectroscopy coupled with AFM measurement [13]. A nanocrystal sample was set on the stage of an inverted fluorescence microscope (IX-70, Olympus) mounted with an AFM scanner (Nanoscope IIIA Digital Instruments). A fluorescent light spot was first identified and located in the center of the scanning range of the AFM by moving the optical microscope stage, and then an AFM image of the nanocrystal was acquired. To measure the spectra, the AFM tip was moved up far from the sample, then UV light (405 nm) from a high pressure Hg-lamp was incident through a microscope objective. The fluorescence light was collected by the same objective, and the light from an area (about a 1.5 mm diameter) was selected by using an imaging pinhole, and introduced to a polychromator (Acton) with a liquid nitrogen cooled CCD camera (Roper Scientific). All fluorescence spectra were subtracted from the substrate background spectrum, and corrected with respect to the wavelength-dependent sensitivity of all the instrumental components, using fluoresce spectrum of N, N-dimethyl-4-nitroanilline/ethanol solution as a standard [14].

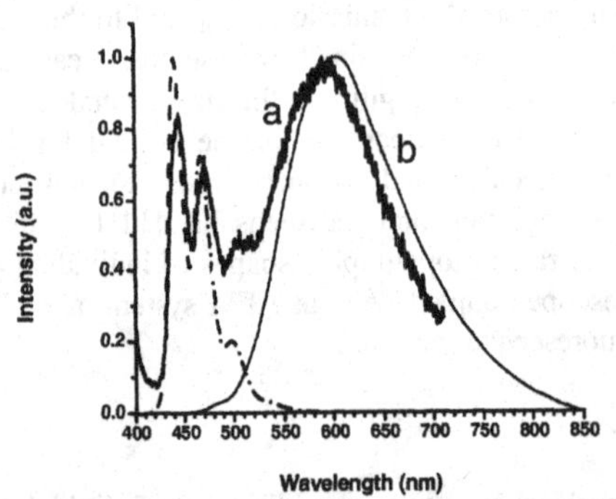

Figure 2. The fluorescence spectra of (a) perylene nanoparticles with the mean size of 120 nm dispersed in water and (b) a bulk crystal (solid lines), and molecular fluorescence in ethanol solution (a dashed line).

RESULTS AND DISCUSSION

Fluorescence of colloidal dispersion

Figure 2 shows the fluorescence spectrum of perylene nanoparticles (the mean size of 120 nm) dispersed in water and those of a bulk crystal and molecules in ethanol. The nanoparticle fluorescence consisted of a broad E-emission band around 600 nm and a structured M-emission band around 450 nm the E-emission is due to fluorescence of self-trapped exciton (excimer), which is characteristic to bulk crystal with the α-form composed by a sandwich-like pair of two molecules [15]. The emission peak position of nanocrystal was blue-shifted from bulk one. On the other hand, the M-emission intensity was comparable to the E-emission one in case of nanocrystals, and the spectral shape was very similar to the molecular fluorescence in ethanol solution. It is considered that perylene molecules and amorphous nanoparticles may disperse in the colloidal dispersion.

Single particle fluorescence

Figure 3 shows the fluorescence spectra and AFM images of two individual perylene nanocrystals on a quartz substrate, as a representative example. The fluorescence spectrum of a bulk crystal is also shown in Figure 3C for a comparison. All the spectra are corrected and normalized to unity at the maximum intensity. We examined about 20 nanocrystals with the different sizes and shapes. Most of nanocrystals had topological shapes of rectangular such as in A or rod-like such as B in figure 3. Their fluorescence exhibited a strong E-emission band characteristic to α-form bulk crystal, although their spectral shape differs from each other and also from bulk crystals. In addition, we found few bright, blue-emitted spots on a sample substrate. The fluorescence spectrum of the blue spot is shown in Figure 4. The spectrum was very similar to the fluorescence of the ethanol solution, and the AFM topography was a spherical-like shape. The blue-emission will be due to amorphous particles of perylene.

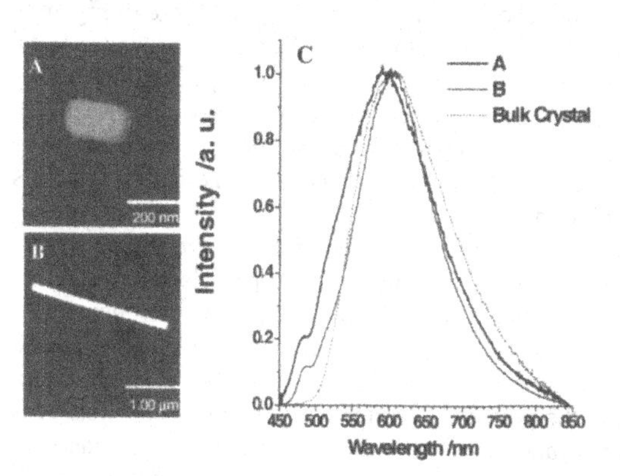

Figure 3. AFM images (A, B) and fluorescence spectra (C) of two individual nanocrystals and a bulk crystal.

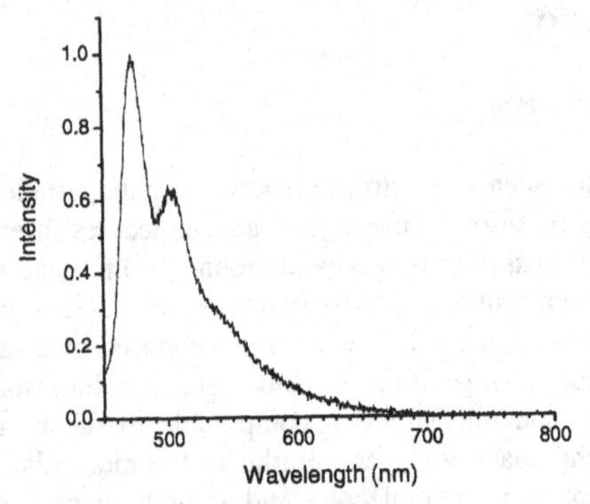

Figure 4. The fluorescence spectrum of a blue-emitted spot on a sample substrate.

It is obvious that the M-emission of single nanocrystals around 450 nm is very weak or negligible in the fluorescence spectra (Figure 3C) when compared to the colloidal solution (Figure 2) prepared by the reprecipitation method. This result indicates that perylene molecules and amorphous nanoparticles are dispersed in water, although their relative concentration to nanocrystal may be very small. Because the radiative transition rate constant of M-emission is about 100 times larger than that of E-emission, small amounts of molecular dispersion and amorphous particles will give apparently intense M-emission of the colloidal solution.

Figure 5 shows the relationship between the size and E-emission peak wavelength, λ^E_{max}, of 20 nanocrystals. Since each nanocrystal had different shapes, we used the size parameter defined by the following equation,

$$Size = \sqrt[3]{\text{length} \times \text{width} \times \text{height}} ,$$

where the length, width and height of nanocrystals were estimated from the cross section of the AFM topography images. The relation between λ^E_{max} and *Size* in Figure 5 clearly reveals that the peak wavelength shifts from 605 to 586 nm for decreasing of *Size* to 109 nm (the smallest crystal examined here), and that the nanocrystals with *Size* larger 300 nm exhibit the same fluorescence peak to the bulk one. It is now clearly and precisely confirmed from our single particle experiments that nanometer-sized crystalline state of perylene exhibits the blue-shift of λ^E_{max} in their size range from several tens to a few hundreds nanometers, which is qualitatively the same to that of colloidal solution.

It is known that the absorption and fluorescence spectra of semiconductor nanoparticles, such as CdS, changes drastically depending on the particle size when the size is less than 10 nm [16, 17]. The size effect is well explained by "a quantum confinement" of excitons. In contrast, the size dependent fluorescence of perylene was observed for larger nanoparticle sizes ranging from a few hundred nanometers, although the excited state is self-trapped in a dimmer unit of molecules. Therefore, the present size effect on perylene nanocrystal arises from quite different origin. It is considered that the large surface-to-volume ratio of nanocrystals induces instability of crystalline lattice structures in nanocrystal, resulting in the thermally "softened" crystal lattice [1, 6]. A certain change of intermolecular interaction in the surface area of nanocrystals may lead to the reduction of the self-trapping energy in nanocrystals.

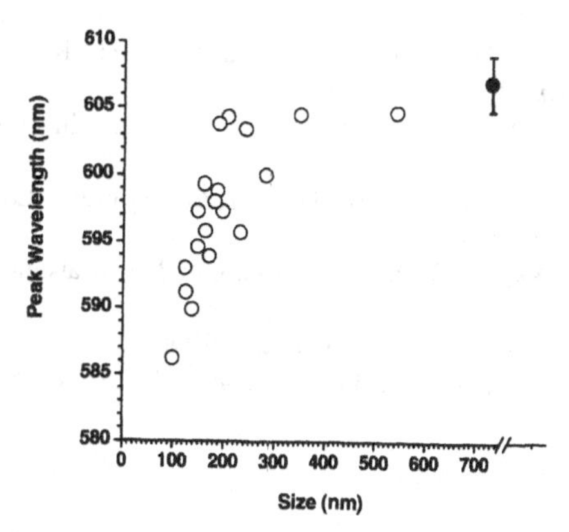

Figure 5. The relationship between the size and E-emission peak wavelength of 20 single nanocrystals of perylene.

ACKNOWLEDGMENT

The present work was supported by a grant from the Core Research for Evolution Science and Technology (CREST) of Japan Science and Technology Agency (JST).

REFERENCES

1. H. Oikawa, H. Nakanishi, "Optical Properties of Polymer Nanocrystals" ," *Single Organic Nanocpraticles*", ed. H. Masuhara, H. Nakanishi, and K. Sasaki (Springer-Verlag, 2003) pp. 169-183.
2. H. Katagi, H. Kasai, S. Okada, H. Oikawa, K. Komatsu, H. Matsuda, Z. Liu, and H. Nakanishi, *Jpn. J. Appl. Phys.* **35**, L1364 (1996).
3. H. Kasai, H. Oikawa, S. Okada, and H. Nakanishi, *Bull. Chem. Soc. Jpn.* **71**, 2597 (1998).
4. H. Kasai, H. Kamatani, S. Okada, H. Oikawa, H. Matsuda, and H. Nakanishi, *Chem. Lett.* 1181, (1997).
5. T. Seko, K. Ogura, Y. Kawakami, H. Sugino, H. Toyotama, and J. Tanaka, *Chem. Phys. Lett.* 291, 438-444 (1998).
6. H. Katagi, H. Oikawa, S. Okada, H. Kasai, A. Watanabe, O. Ito, Y. Nozue, and H. Nakanishi, *Mol. Cryst. Liq. Cryst.* **314**, 285 (1998).
7. B.-K. Lee, W. –K. Koh, W.-S. Chae, and Y.-R. Kim, *Chem. Comm.*, 138 (2002).
8. L. A. Kolodny, D. M. Willard, L. L. Carillo, M. W. Nelson, and A. V. Orden, *Anal. Chem.* **73**, 1959 (2001).
9. D. Xiao, L. Xi, W. Yang, H. Fu, Z. Shuai, Y. Fang, and J. Yao, *J. Am. Chem. Soc.* **125**, 6740 (2003).
10. B. K. An, S. K. Kwon, S. D. Jung, and S. Y. Park, *J. Am. Chem. Soc.* **124**, 14410 (2002).
11. H. Oikawa, T. Mitsui T. Onodera, H. Kasai, H. Nakanishi, and T. Sekiguchi, *Jpn. J. Appl. Phys.* **42**, L111-L113 (2003).

12. V. V. Volkov, T. Asahi, H. Masuhara, A. Masuhara, H. Kasai, H. Oikawa, and H. Nakanishi, *J. Phys. Chem. B*, **108**, 7674 (2004).
13. H. Matsune, T. Asahi, H. Masuhara, H. Kasai, and H. Nakanishi, submitted to *Chem. Phys. Chem.*
14. J. R. Lakowicz, "*Principles of Fluorescence Spectroscopy* 2nd Ed," (Kluwer Academic/ Premium Publishers, New York 1999).
15. H. Nishimura, T.Yamaoka, K. Mizuno, M. Iemura, and A. Matsuni, *J. Phys. Soc. Jpn.* **53**, 3999-4008 (1984).
16. A. P. Alivistors, *J. Phys. Chem.* **100**, 13226-13239 (1996).
17. B. C. Murray, C. R. Kagan, and M. G. Bawendi, *Annu. Rev. Mater. Sci.* **30**, 545-610 (2000).

Mater. Res. Soc. Symp. Proc. Vol. 846 © 2005 Materials Research Society DD10.10

Synthesis and Spectroellipsometric Characterization of Y_2O_3-stabilized ZrO_2-Au Nanocomposite Films for Smart Sensor Applications

George Sirinakis, Richard Sun[1], Rezina Siddique, Harry Efstathiadis, Michael A. Carpenter, and Alain E. Kaloyeros
College of Nanoscale Science and Engineering, The University at Albany-State University of New York, Albany, NY 12203, USA
[1]Sun International Inc, Acton, MA 01720, USA

ABSTRACT

Noble metal nanoparticles exhibit significant potential in all-optical, smart-sensing applications due to their unique optical properties. In particular, gold (Au) nanoparticles exhibit a strong surface plasmon resonance (SPR) band, the spectral position and shape of which depends on the size, shape, and density of the nanoparticles and the physical and chemical properties of surrounding environment. Embedding the nanoparticles in an yttria-stabilized zirconia (YSZ) matrix is believed to expand their range of operation to temperatures above 500 °C. YSZ is a material that has been proven suitable for optical applications due to its high refractive index, low absorption coefficient and high transparency in the visible and infrared regions. Thus, its use as a base platform for nanocomposite thin films is expected to provide significant benefits in the development of harsh environment multifunctional sensors.

In this work YSZ-Au nanocomposite films were synthesized from a YSZ and a Au target by the radio frequency magnetron co-sputtering technique in combination with a post-deposition annealing treatment in an argon atmosphere, with the annealing temperature being varied from 500-1000 °C in steps of 100 °C. The microstructure and the optical properties of the resulting films were characterized by x-ray diffraction spectroscopy, scanning electron microscopy and spectroscopic ellipsometry. Results on the effect of the Au particle size on the real and the imaginary part of the refractive index of the nanostructured composites are presented. Future smart sensor systems utilizing these multifunctional material sets for harsh environment sensing applications will likewise be outlined.

INTRODUCTION

Growing environmental concerns associated with fossil-fuel related emissions of green house gases have necessitated a tighter control over the various combustion processes for energy generation. Likewise, these concerns have led to the development of a new approach to energy generation which is based on the synergistic operation of existing and developing technologies such as turbines and solid oxide fuel cells (SOFC). In this respect there is an urgent need for smart sensor systems for the detection of H_2 and other exhaust gases such as CO that can reliably operate under aggressive environments. Au nanoparticles embedded in various oxide matrices have attracted significant interest as all-optical chemical gas sensors due to their unique optical properties [1]. Most of the experimental work on investigating the optical properties of Au nanocomposites has been focused on the behavior of the extinction coefficient of the nanocomposites which is associated with the surface plasmon resonance (SPR) band of the Au nanoparticles [2]. However the knowledge of the behavior of the real part of the refractive index of the nanocomposites is also necessary for the design of optical sensor devices.

In this work the optical properties of Au-YSZ nanocomposites synthesized by the rf magnetron co-sputtering technique in combination with an annealing treatment are investigated by spectroscopic ellipsometry (SE) in order to determine both the real and imaginary part of the refractive index of the nanocomposites as a function of Au nanoparticle size.

EXPERIMENTAL DETAILS

Depositions were carried out in a custom designed sputtering system, consisting of a 25.4 cm diameter five–way cross and two sputtering sources (US "MAK" sputter gun) in a co-focal configuration. Each sputtering source was driven by a 13.56 MHz r.f. power supply (AE RFX600). The targets (Williams Inc.) were a 5.08 cm diameter disk of ZrO_2 stabilized with 5 w.t.% Y_2O_3 (99.9% purity), and a 5.08 cm diameter disk of Au (99.99% purity). Argon was used as a sputtering gas at a process pressure of $5x10^{-3}$ Torr. Before deposition the targets were cleaned by sputtering with the shutters closed for 60 min at 200 W of rf power for the YSZ target and for 15 min at 20 W of rf power for the Au target. Silicon substrates, which were ultrasonically cleaned prior to deposition for 5 min first in ethanol and then in acetone, were used. The as-deposited films were ex-situ annealed at 600, 700, 800, 900, and 1000 °C for 2h in an Ar atmosphere.

The microstructure of the resulting films was determined by X-ray diffraction spectroscopy (XRDS) using a Scintag XDS 2000 x-ray diffractometer equipped with a Cu K_α X-ray source. Spectroscopic ellipsometry (SE) measurements were performed using a SOPRA GES5 ellipsometer in the wavelength region from 250 to 2000nm. Film thickness measurements employed a LEO 1550 microscope using a 5keV primary electron beam. The Au content was found to be approximately 10 at.% measured using a Perkin-Elmer Phi 600 Auger Electron Spectroscopy (AES) system. A Maxwell-Garnett (M-G) effective medium theory was employed to model the optical response of the nanocomposite films. The dielectric function of the YSZ matrix was described by a standard Cauchy dispersion, while the dielectric function of Au was modeled by a Drude model for the free electron and two lorentz peaks to describe the interband transition of Au with energies around 3 and 4 eV [3]:

$$\varepsilon_{Au} = \varepsilon_{\mathrm{interband}} - \frac{\omega_p^2}{\omega^2 + i\gamma\omega} \quad (1)$$

where ω_p is the plasma frequency and γ is a damping constant related to the mean free path of the free electron. For particles with sizes smaller than the bulk mean free path γ is modified according to the following expression in order to account for the additional surface scattering [4]:

$$\gamma(R) = \gamma_b + A\frac{\upsilon_F}{R} \quad (2)$$

where γ_b is the damping constant of the bulk conduction electrons (γ_b=$3.3*10^{13}sec^{-1}$ for bulk Au), υ_F is the Fermi velocity (υ_F=$1.4*10^{15}$nm/sec for Au), R is the particle radius, and A is a coefficient determined by the details of the scattering processes at the surface of the particle[5].

RESULTS AND DISCUSSION

Figure 1 shows typical fitted SE spectra for the sample that was annealed at 800 °C measured at incident angles of 65°, 70°, and 75°. A good agreement between the measured data

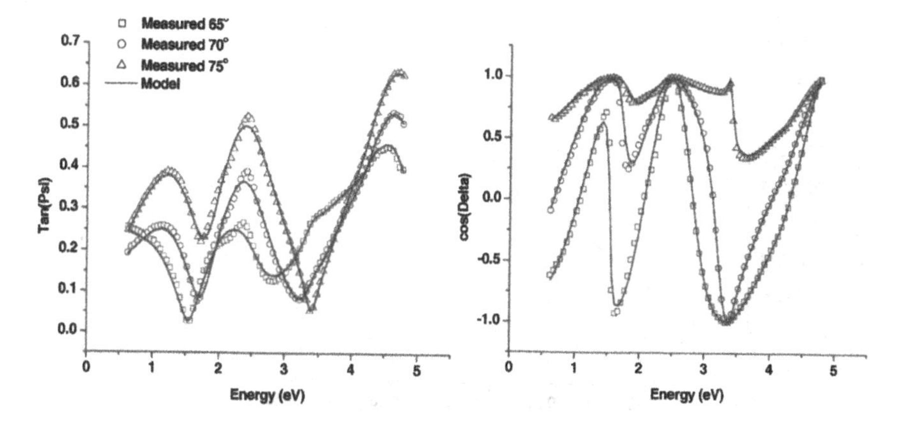

Figure 1. Ellipsometry data and model fit for the sample that was annealed at 800 °C at 65°, 70°, and 75°.

for all the samples and the prediction of the model was achieved and the results from this analysis are summarized in Table 1 along with the XRD data on the average Au particle size obtained using the Scherrer formula [6].

The thickness of the films (sum of surface roughness and layer thickness) seems to be fairly constant for all the samples and in good agreement with the value of ~100nm experimentally obtained from SEM cross-sectional measurements. Similarly the Au volume fraction seems to be fairly constant for all the samples and in reasonable agreement with the expected value of 14 vol.%, which was calculated from the AES data assuming fully dense films comprised of metallic Au and YSZ.

Figure 2 plots the damping constant predicted by the model versus the inverse particle radius as determined by the XRD analysis. It is observed that the damping constant increases linearly with the inverse particle radius. However, by performing linear regression analysis on the data according to equation 2, a negative value of $-3.88*10^{14}$ sec^{-1} for the bulk scattering frequency and a value of 3.2 for the A parameter is obtained. It should be noted that the physical

Annealing T (°C)	Roughness (nm)	Thickness (nm)	Au vol.%	$\gamma(sec^{-1})$	XRD Au average grain size, D (nm)
600	4.0	93.0	10.1	$1.39*10^{15}$	4.7
700	4.4	88.3	11.1	$1.14*10^{15}$	7.7
800	3.7	80.3	10.9	$5.5*10^{14}$	8.9
900	4.6	86.8	10.2	$3.12*10^{14}$	11.4
1000	12.0	80.9	9.0	$1.13*10^{14}$	14.8

Table 1: Summary of SE and XRD results for Au-YSZ films

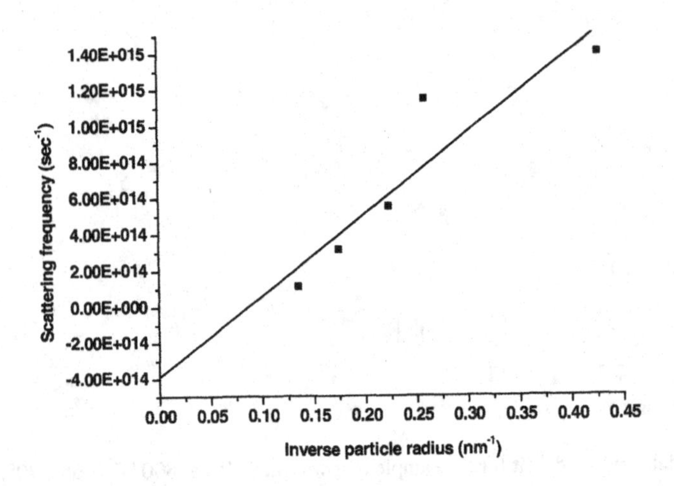

Figure 2. Damping constant versus inverse particle radius

meaning of the A parameter is not very well understood and can not be expressed in terms of physical constants or materials related parameters, such as the dielectric functions of the Au and/or the host material. However, experimental data suggest that for Au nanoparticles embedded in rather inert matrices, such as SiO_2 or Al_2O_3, A has a value smaller or close to unity [7,8]. Recent XPS work indicates that the Au-YSZ system can also be considered inert [9]. In this context, the value of 3.2 obtained for the A parameter by the model, which would be indicative of a strongly coupled Au-YSZ system, is higher than would be expected.

Accordingly, the above analysis suggests that although qualitatively the evolution of particle size as function of annealing temperature is successfully predicted, the particle size is significantly underestimated by the model. Therefore, additional refinements in the model are required in order to more accurately describe the interband transitions of Au, especially the 5d-6sp transitions which have an onset energy of ~2.4eV (520nm). This is expected to result in a more quantitative agreement of the particle size with the experimental data, which would lead to γ_b and A parameter values in agreement with the modern literature.

Figure 3 shows the extinction coefficient of the films extracted from the SE analysis. The extinction coefficient exhibits a peak associated with the SPR band of the Au nanoparticles around 600nm. Films without Au nanoparticles are not shown in this figure as YSZ has a negligible contribution to the extinction coefficient within this wavelength region. As the annealing temperature increases, the band shifts to longer wavelengths and becomes more intense and sharp. These findings when coupled to the increasing particle size as the annealing temperature increases, indicate that the predictions of the model can be qualitatively understood within the context of the Mie theory and its size-dependent dielectric function [4]. Furthermore, by using bulk dielectric functions for Au [10] Mie theory predicts that the peak position of the SPR band of Au nanoparticles embedded in a YSZ matrix occurs at 594nm as indicated by the dashed line in figure 3 and is in reasonable agreement with the experimental results.

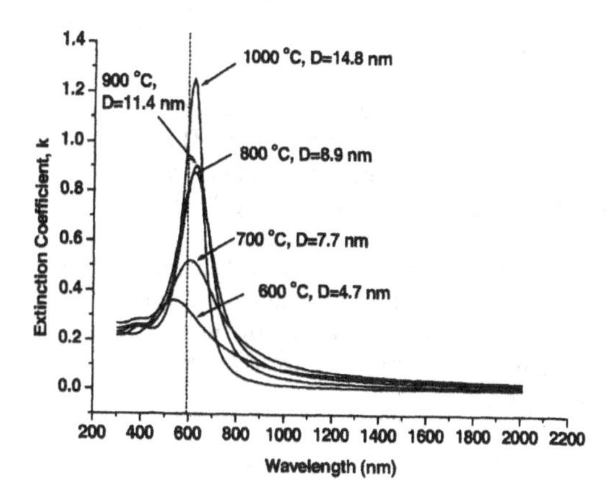

Figure 3. Extinction coefficient of Au-YSZ nanocomposite films as a function of annealing temperature. The dashed line indicates the position of the SPR band as predicted from the Mie theory using bulk dielectric functions for Au.

Figure 4 shows the index of refraction for all the films obtained from the SE analysis. The index of refraction of YSZ is also shown for comparison. At the immediate wavelength region where the extinction coefficient takes its maximum, the index of refraction was found to increase with an increase in wavelength. This experimentally observed anomalous dispersion is consistent with the Kramers-Kronig relations [11]. Furthermore at longer wavelengths the index of refraction was found to decrease with wavelength and leveling off to ~2.45 for all the samples.

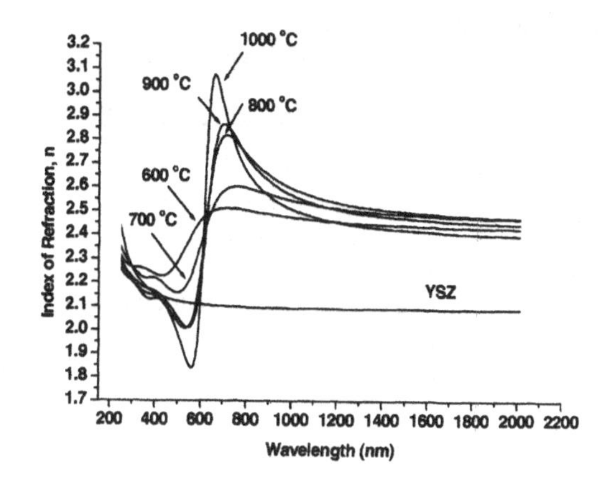

Figure 4. Index of refraction of Au-YSZ films as a function of annealing temperature.

CONCLUSIONS

Au-YSZ nanocomposite films were synthesized by the rf magnetron co-sputtering technique in combination with a post-deposition annealing treatment. The resulting films were optically characterized with the spectroscopic ellipsometry technique. It was found that a M-G effective medium theory in combination with a simple model using 2 lorentz peaks for the interband transitions of Au and the Drude law for the free electrons was successful in qualitatively describing the optical response of the Au-YSZ films. The index of refraction of the nanocomposites was found to undergo an anomalous dispersion in the immediate wavelength vicinity where the extinction coefficient takes its maximum. Furthermore at longer wavelengths it was found that the addition of Au in the YSZ matrix leads to an increase of the index of refraction with respect to that of the YSZ matrix.

ACKNOWLEDGEMENTS

This work was supported by the United States Department of Energy National Energy Technology Laboratory under contract number DE-FG26-02NT41542, and the New York State Office of Science, Technology and Academic Research (NYSTAR contract # 3538479). This support is gratefully acknowledged.

REFERENCES

1. M. Ando, T. Kobayashi, S. Iijima, M. Haruta, Sens.Actuators B 96, 589 (2003).
2. D. Dalacu, L. Martinu, J. Appl. Phys. 87, 228 (2000).
3. U. Kreibig, M. Vollmer, "Optical Properties of Metal Clusters" (Springer, New York, 1995) pp.15.
4. H. Hövel, S. Fritz, A. Hilger, U. Kreibig, and M. Vollmer, Phys. Rev. B 48, 18178 (1993).
5. N. W. Ashcroft and N. D. Mermin, Solid state physics (Saunders College Publishing, 1976) pp.10.
6. B. D. Cullity and S.R. Stock, Elements of X-ray diffraction 3rd ed, (Prentice-Hall, 2001) pp.388.
7. D. Dalacu, L. Martinu, J. Appl. Phys. 87, 228 (2000).
8. Y. Hosoya, T. Suga, T. Yanagawa, Y. Kurokawa J. Appl. Phys. 81, 1475 (1997)
9. S. Zafeiratos, S. Neophytides, S. Kennou Thin Solid Films 386, 53 (2001)
10. P.B. Johnson, R.W. Christy, Phys. Rev. B 6, 4370 (1972).
11. C.F. Bohren, D.R. Huffman, Absorption and Scattering of light by small particles, (Wiley, New York, 1983) pp.232.

Mater. Res. Soc. Symp. Proc. Vol. 846 © 2005 Materials Research Society DD10.11

Synthesis of Hydrophilic Two-Photon Absorptive Fullerene-diphenylaminoflourene Dyads for Molecular Self-assembly in Water

Sarika Verma[1]; Tanya Hauck[2]; Prashant A.Padmawar[1]; Taizoon Canteenwala[1]; Long Y.Chiang[1] and Kenneth P.H.Pritzker[2]
[1]Department of Chemistry, Institute of Nanoscience and Engineering, University of Massachusetts, Lowell, MA 01854, U.S.A.
[2]Pathology and Laboratory Medicine, Mount Sinai Hospital and University of Toronto, Toronto, ON M5G1X5, Canada.

ABSTRACT

Synthetic conditions for the preparation of amphiphilic two-photon absorptive poly(ethylene glycolated) diphenylaminofluorene[60]fullerene conjugates C_{60}(>DPAF-PEG_6) were investigated. UV-vis spectra of C_{60}(>DPAF-PEG_6) collected at the concentration of 1.0×10^{-5} M in water showed absorption bands with a slight red shift, broadened, and in a less intensity as compared with those of the bands obtained in $CHCl_3$. These features are important characteristics of C_{60}(>DPAF-PEG_6) molecular aggregation in water. Transmission electron micrographs taken at the concentration of 1.0×10^{-5} M in H_2O clearly displayed images of spherical aggregates in a diameter of 200–400 nm. That substantiated high hydrophobic intermolecular interactions between these amphiphilic molecules for assemblies in water.

INTRODUCTION

A wide range of electronically conducting, magnetic, photochemical, and electrical features of C_{60} derivatives make them suitable optoelectronic materials for practical applications [1, 2]. Considering the unique high electron accepting ability of [60]fullerene, a large variety of donor-linked fullerene conjugates have been designed, synthesized, and studied for their photoinduced electron- and energy-transfer processes [3–5]. These processes make possible myriad of potential uses in photophysical and biomedical areas. Over the past decade, two-photon absorption (TPA) phenomena of organic chromophore molecules have attracted a considerable attention due to related promising applications, such as fluorescence imaging, data storage, microfabrication and photodynamic therapy. A number of chromophores including diphenylaminofluorene (DPAF) derivatives were investigated for the structural relationship to their multi-photon absorptivity. Relatively high TPA cross-sections were demonstrated on some branched DPAF-derived chromophores by covalent linking with different π-acceptors in a conjugate structure [6, 7]. Utility of diphenylaminofluorene subunit in the design of new chromophores is reasoned for its inherently high thermal and photochemical stability and easy functionalization at 2-, 7-, and/or 9-carbon positions. Taking on similar approaches, our group reported the first highly TPA-active diethylated C_{60}-DPAF derivative possessing a one-dimensional A-sp^3-D molecular linkage with a close coupling of the electron accepting (A) fullerene cage and the electron donating (D) diphenylaminofluorene moiety. Accordingly, C_{60}-DPAF dyads 7-(1,2-dihydro-1,2-methano[60]fullerene-61-carbonyl)-9,9-diethyl-2-diphenylaminofluorene [C_{60}(>DPAF-C_2)] exhibits large TPA cross-sections $\sigma_2' = 196 \times 10^{-48}$ cm^4-sec in nanosecond regime, which is among the highest value known for many DPAF-derived TPA molecules [8].

A major drawback on the use of fullerene derivatives in biomedical applications arises from the extremely hydrophobic nature of the C_{60} cage. Several groups including our own have described the synthesis of amphiphilic fullerene derivatives for the enhancement of water-solubility and the study of their self-assemblies into the form of micelles, vesicles, nanospheres, and nanorods in aqueous solution [9–13]. Considering remarkable biocompatibility properties of poly(ethylene glycol) (PEG), several amphiphilic C_{60} derivatives consisting of poly(ethylene glycol) chains as hydrophilic end groups were synthesized and analyzed for the aggregation behavior in aqueous medium [14, 15]. Furthermore, two-photon excitation process of responsive chromophores may serve as a potential alternative treatment technique for photodynamic cancer therapy (PDT) because of its ability to focus on a confined small treatment area of diseased tissue in a greater depth using a spectral window of 800–1100 nm in mammalian tissue. Therefore, synthesis of amphiphilic fullerene-DPAF conjugates containing polar hydrophilic functional groups, such as carboxylic acid and poly(ethylene glycol), leading to water-soluble fullerene derivatives is of our interest to improve their potency in biomedical treatments.

Here, we report the synthetic procedure of amphiphilic TPA-active poly(ethylene glycolated) diphenylaminofluorene[60]fullerene conjugates, namely, C_{60}(>DPAF-PEG_6). Structural characterization of these new fullerenic chromophores was made using fourier transform infrared spectroscopy (FTIR) and thermogravimetric analysis (TGA). Subsequently, we investigated molecular self-assembly behavior of C_{60}(>DPAF-PEG_6) in dilute aqueous concentrations resulting in the formation of nano- to submicron sized spherical vesicles. These spherical vesicles were characterized by ultraviolet-visible spectroscopy (UV-vis) and transmission electron microscopy (TEM).

EXPERIMENTAL DETAILS

Materials and instrumentation methods

1,3-Dicyclohexylcarbodiimide (DCC), 4-dimethylaminopyridine (DMAP), poly(ethylene glycol) bis(carboxymethyl) ether (molecular weight 600), and anhydrous dichloroethane were purchased from Aldrich. C_{60}-methanocarbonyl-9,9-diethanol-2-diphenylaminofluorene [C_{60}(>DPAF-OH)] 1 (figure1) was prepared according to the reported procedure [8, 16] with chemical modification at C-9 group to incorporate alcohol functionality.

Infrared spectra were recorded as KBR pellets on a Thermo Nicolet 370 FT-IR spectrometer [except for poly(ethylene glycol) bis(carboxymethyl) ether]. For poly(ethylene glycol) bis(carboxymethyl) ether, a small amount of the sample was placed between a pair of KBr pellets and scanned for infrared absorption. ^{1}H and ^{13}C NMR spectra were recorded on a Bruker 200 spectrometer. Thermogravimetric analysis (TGA) measurements were made with a HI-Res TGA 2950 Thermogravimetric Analyser. In TGA measurements, the sample (about 7.0 mg) was heated at a rate of 10 °C per minute from room temperature to 800 °C under a dynamic nitrogen atmosphere. UV-vis absorption spectra were collected on a computer-controlled Perkin Elmer UV/VIS/NIR Lambda 9 series spectrophotometer. Transmission electron microscopic (TEM) images were taken on a Philips Tecnai 20 TEM at 200 kV.

Synthesis of amphiphilic poly(ethylene glycolated) diphenylaminofluorene-[60]fullerene [2, C_{60}(>DPAF-PEG_6)]

Poly(ethylene glycol) bis(carboxymethyl) ether (molecular weight 600, 463 mg, 0.08 mmol) was taken in a round bottom flask (100 ml) and stirred under reduced pressure in oil bath maintained at 80 °C for 5.0 h. The reaction flask was cooled to room temperature and added 1,3-dicyclohexylcarbodiimide (DCC, 159 mg, 0.08 mmol), 4-dimethylaminopyridine (DMAP, 94 mg, 0.08 mmol) and C_{60}-methanocarbonyl-9,9-diethanol-2-diphenylaminofluorene (**1**, 150 mg, 0.013 mmol) in sequence. To the reaction mixture was then added anhydrous dichloroethane (50 ml) and stirred in oil bath maintained at 65 °C for an additional 2.0 h until complete disappearance of the starting material on thin-layer chromatographic plate (TLC). The resulting reaction mixture was filtered and the filtrate dried on a rotary evaporator. The crude semi-solid product was then dissolved in a solvent mixture of THF-H_2O (1:1, 10 ml) and dialyzed against distilled water using a dialysis membrane MWCO 1000 for effective removal of residual impurities and unreacted poly(ethylene glycol). After completion of dialyses and solvent evaporation by the freeze-dry technique, the product of poly(ethylene glycolated) diphenylaminofluorene-[60]fullerene **2** (figure2) as C_{60}(>DPAF-PEG_6) was obtained as semi-solids in 40% yield.

The spectroscopic data of **2**: ^{1}H NMR (200 MHz, $CDCl_3$, ppm) δ8.58 (dd, J = 8 Hz, J = 1.6 Hz, 1 H), 8.43 (d, J = 1.6 Hz, 1 H), 7.88 (d, J = 8 Hz, 1 H), 7.69 (d, J = 8 Hz, 1 H), 7.40–7.10 (m, 12 H), 5.77 (s, 1 H), 4.33 (t, J = 4 Hz, 4 H), 4.18 (s, 4 H), 3.88 (s, 4 H), 3.68 (broad, 80 H), and 2.26–2.61 (m, 4 H). FT-IR (KBr) ν_{max} 3423, 2872, 1746, 1671, 1592,1489, 1466, 1425, 1349, 1277, 1247, 1200, 1109, 950, 844, 756, 695, and 526 cm^{-1}.

Preparation of spherical vesicles of C_{60}(>DPAF-PEG_6)

Preparation of C_{60}(>DPAF-PEG_6) derived vesicles was made by dissolving **2** (1.4 mg) in THF–DMSO (1:1, 0.1 ml) with ultrasonication for approximately 5.0 min, followed by the addition of H_2O (1.9 ml) with vigorous stirring to form a solution of 3.0 x10^{-4} M in concentration. A portion of this solution was diluted to solutions with a concentration of 1.0 x10^{-5} M and 5.0 x10^{-6} M for subsequent transmission electron microscopic measurements.

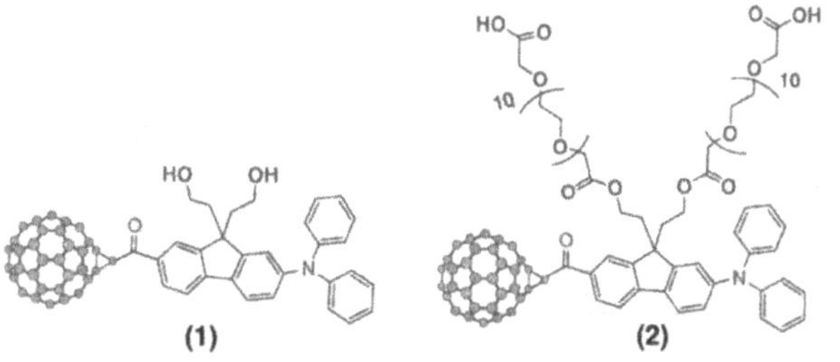

Figure 1. Molecular structure of precursor compound [C_{60}(>DPAF-OH)] **1** and amphiphilic poly(ethylene glycolated) diphenylaminofluorene-[60]fullerene conjugate compound **2**.

DISCUSSION

Synthesis of C_{60}(>DPAF-PEG$_6$) **2** was carried out by the reaction of C_{60}-methanocarbonyl-9,9-diethanol-2-diphenylaminofluorene **1** with poly(ethylene glycol) bis(carboxymethyl) ether (M.W. 600, 6.0 equiv.) in the presence of 1,3-dicyclohexylcarbodiimide and 4-dimethylamino-pyridine at 80 °C for a period of 5.0 h to afford the crude semi-solid products. Owing to a relatively high molecular weight of **2**, chromatographic separation method is not effective for the purification. Therefore, the crude products were dissolved in a solvent mixture of THF-H_2O (1:1) and subjected to dialysis treatment against distilled water using a dialytic membrane MWCO 1000 for effective removal of residual unreacted poly(ethylene glycol). The technique allows isolation of **2** as semi-solids in 40% yield. The compound **2** is soluble in a variety of solvents including THF, DMSO, H_2O, and $CHCl_3$.

Structural characterization

Structure of the product C_{60}(>DPAF-PEG$_6$) was fully characterized by various spectroscopic methods including ^{1}H NMR, ^{13}C NMR, FT-IR, UV-visible spectroscopy, and thermogravimetric analysis. Confirmation of poly(ethylene glycol) moiety in the structure of **2** was made by its IR spectrum (figure 2c) showing a optical absorption profile superimpose with a combination of the IR profiles of poly(ethylene glycol) bis(carboxymethyl) ether (figure 2a) and C_{60}(>DPAF-OH) (figure 2b). Poly(ethlene glycol) gave the strongest symmetrical ether stretching band (-C-O-C-) at 1109 cm^{-1}. In figure 2c, a strong band centered at 1746 cm^{-1} corresponds to the absorption of C=O functional groups indicating the presence of ester and acid moieties that is consistent with the chemical conversion of hydroxyl groups of C_{60}(>DPAF-OH) to the corresponding ester in **2**. Characteristic optical absorption of carbonyl stretching that bridges fluorene and fullerene moieties of **2** is also detected at 1670 cm^{-1}. Other significant IR bands include fullerenic signals appearing at 750, 695 and 534 cm^{-1}.

Thermogravimetric profile of C_{60}(>DPAF-PEG$_6$) is shown in figure 3a. A sharp weight loss was observed above 450 °C with a relatively constant weight remaining after 500 °C. Since pristine C_{60} is thermally stable below 600 °C, we assume the main weight loss being attributed

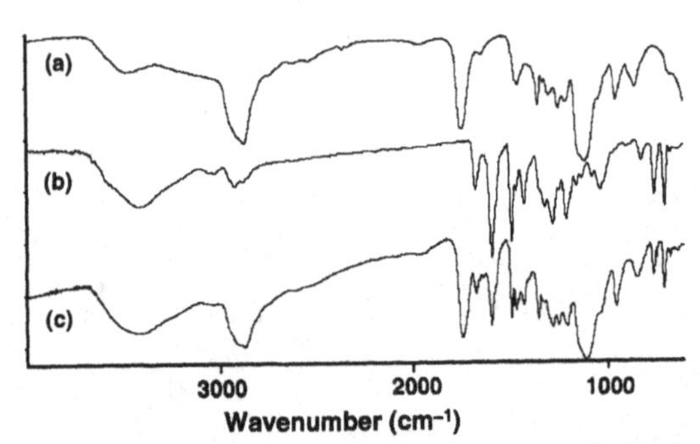

Figure 2. FT-IR spectra of (a) poly(ethylene glycol) bis(carboxymethyl) ether, (b) C_{60}-methanocarbonyl-9,9-diethanol-2-diphenylaminofluorene as C_{60}(>DPAF-OH), and (c) poly(ethylene glycolated) diphenylaminofluorene-[60]fullerene as C_{60}(>DPAF-PEG$_6$).

from the bond cleavage of poly(ethylene glycol) moieties. This weight loss corresponds to a 54% mass of **2** that is consistent with the theoretical value of 53% in weight ratio of two poly(ethylene glycol) diacid subunits in **2.**

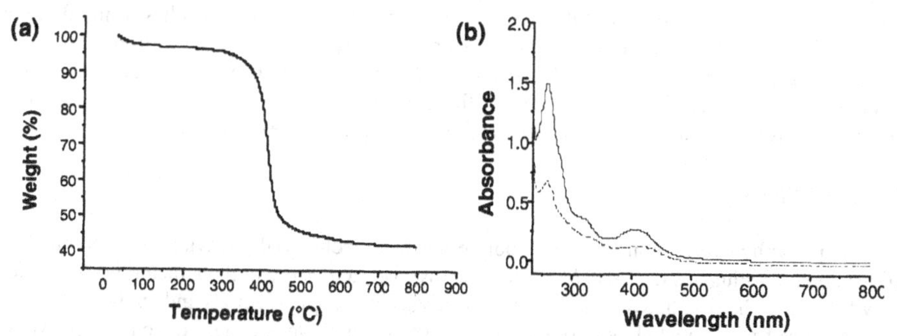

Figure 3. (a) Thermogravimetric profile of C_{60}(>DPAF-PEG$_6$) and (b) UV-vis absorption spectra of C_{60}(>DPAF-PEG$_6$) in $CHCl_3$ (solid line, 1.0 x10^{-5} M) and C_{60}(>DPAF-PEG$_6$) in H_2O (dotted line, 1.0 x10^{-5} M).

Molecular self assembly of C_{60}(>DPAF-PEG$_6$)

Molecular self assembly of amphiphilic C_{60}(>DPAF-PEG$_6$) was investigated in aqueous solution. During the vesicle preparation, a small quantity of THF–DMSO (1:1, 2.4 % by volume) was applied. UV-visible absorption spectra of C_{60}(>DPAF-PEG$_6$) at the concentration of 1.0 x 10^{-5} M in chloroform and water are shown in figure 3b. High solubility of **2** in $CHCl_3$ gave a clear presentation of molecular absorption maxima at 258, 322, and 409 nm (solid line of figure 3b). These data agree well with those reported [16] for spectroscopic characteristics of monofunctionalized fullerenes with the peak assignment of the bands centered at 258 and 322 nm for the optical absorption of the fullerene cage and the band centered at 409 nm for the DPAF moiety. Interestingly, in water medium, these absorption peaks showed a slight red shift to 260, 327, and 417 nm, respectively, with broadened bands in a less intensity. These features are

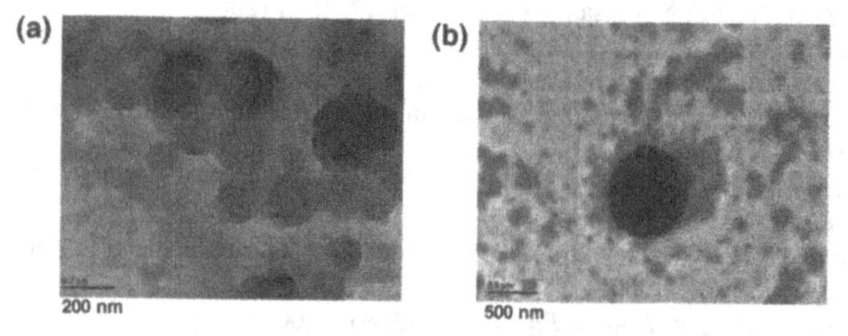

Figure 4. Transmission electron micrographs of C_{60}(>DPAF-PEG$_6$) taken at the concentration of 1.0 x10^{-5} M in H_2O showing spherical aggregates with an average diameter of 200–400 nm.

important characteristics indicating molecular aggregation of C_{60}(>DPAF-PEG_6) in water.

Molecular self assembly of C_{60}(>DPAF-PEG_6) was further supported by transmission electron micrographs taken at the concentration of 1.0 x10^{-5} M in H_2O (figure 4). TEM images displayed spherical aggregates with an average diameter of 200–400 nm. The size of these spheres falls within the range of fullerene vesicles reported that typically reaches from 50 nm to 5.0 μm [13]. On close examination of some imperfect spheres, there is an indication of bilayer membrane formation as the shell of the vesicles with a hollow core. Considering the amphiphilic structure of C_{60}(>DPAF-PEG_6), formation of the nanosphere is most likely due to the balance between hydrophobic and hydrophilic interactions among molecules in water.

CONCLUSIONS

Amphiphilic two-photon absorptive poly(ethylene glycolated) diphenylaminofluorene-[60]fullerene conjugates C_{60}(>DPAF-PEG_6) was synthesized. Structural characterization of these new fullerenic chromophores was made by various spectroscopic methods and analyses. Molecular self-assembly behavior of C_{60}(>DPAF-PEG_6) in dilute aqueous concentrations was investigated. Transmission electron micrographs taken at the concentration of 1.0 x10^{-5} M in H_2O clearly displayed images of spherical aggregates in a diameter of 200–400 nm. That substantiated high hydrophobic intermolecular interactions between these amphiphilic molecules for assemblies in water.

REFERENCES

1. N. Martin, L. Sanchez, B. Illescas and I. Perez, *Chem. Rev.* **98**, 2527 (1998).
2. M. T. Rispens and J. C. Hummelen, *Developments in Fullerene Science* **4**, 387 (2002).
3. F. Effenberger and G. Gunther, *Synthesis* **9**, 1372 (1998).
4. H. Imahori and Y. Sakata, *Adv. Mater.* **9**, 537 (1997).
5. D. M. Guldi, C. Luo, A. Swartz, R. Gómez, J. L. Segura, N. Martin, C. Brabec and N. S. Sariciftci, *J. Org. Chem.* **67**, 1141 (2002).
6. E. Zojer, D. Beljonne, P. Pacher and J. L. Brÿdas, *Chem. Eur. J.* **10**, 2668 (2004).
7. R. Kannan, G. S. He, L. Yuan, F. Xu, P. N. Prasad, A. G. Dombroskie, B. A. Reinhardt, J. W. Baur, R. A. Vaia and L. S. Tan, *Chem. Mater.* **13**, 1896 (2001).
8. L. Y. Chiang, P. A. Padmawar, T. Canteenwala, L. S. Tan, G. S. He, R. Kannan, R. Vaia, T. C. Lin, Q. Zheng and P. N. Prasad, *Chem. Comm.* **17**, 1854 (2002).
9. C. H. Tan, P. Ravi, S. Dai, K. C. Tam and L. H. Gan, *Langmuir* **20**, 9882 (2004).
10. J. Hao, H. Li, W. Liu, A. Hirsch, *Chem. Comm.* **5**, 602 (2002).
11. C. Burger, J. Hao, Q. Ying, H. Isobe, M. Sawamura, E. Nakamura and B. Chu, *J. Colloid and Interface Sci.* **275**, 632 (2004).
12. U. S. Jeng, T. L. Lin, C.S. Taso, C. H. Lee, T. Canteenwala, L.Y. Wang, L. Y. Chiang and C. C. Han, *J. Phys. Chem. B* **103**, 1059 (1999).
13. V. Georgakilas, F. Pellarini, M. Prato, D. M. Guldi, M. Melle-Franco and F. Zerbetto, *Proc. Natl. Acad. Sci.* **99**, 5075 (2002).
14. Z. Li, P. Shao and J. Qin, *J. Appl. Polym. Sci.* **92**, 867 (2004).
15. X. D. Huang, S. H. Goh and S. Y. Lee, *Macromol. Chem. Phys.* **201**, 2660 (2000).
16. P. A. Padmawar, T. Canteenwala, S. Verma, L.-S. Tan, and L. Y. Chiang, *J. Macromol. Sci. A, Pure Appl. Chem.* **41**, 1387 (2004).

Mater. Res. Soc. Symp. Proc. Vol. 846 © 2005 Materials Research Society DD10.17

Effects of Deposition Angle on the Optical Properties of Helically Structured Films

Jason B. Sorge, Andy C. van Popta, Jeremy C. Sit, and Michael J. Brett
Department of Electrical and Computer Engineering, University of Alberta
Edmonton, Alberta, T6G 2V4, Canada

ABSTRACT

Glancing-angle deposition (GLAD) is a fabrication method capable of producing thin films with variable porosity. The GLAD process exploits substrate shadowing and limited adatom diffusion to create isolated columns of material that collectively comprise a highly porous thin film. GLAD can be used to create chiral or helical structures with a wide range of porosity through variation of the substrate tilt angle and controlled substrate rotation. We present the effect of the deposition angle on the selective transmittance of circularly polarized light in helical thin films fabricated with the GLAD process. Transmission measurements of titanium dioxide helical films reveal two regimes of enhanced selective transmittance: one corresponding to a substrate tilt angle that produces a maximum circular birefringence and another corresponding to strong anisotropic scattering.

INTRODUCTION

Thin films grown by physical vapor deposition (PVD) at oblique incident vapor flux angles exhibit a columnar microstructure that is oriented towards the vapor source. At highly oblique angles of incidence, self-shadowing becomes the dominant growth mechanism, resulting in highly porous thin films composed of isolated columns. The GLAD technique combines self-shadowing and minimal adatom diffusion with computer-controlled substrate motion to manipulate thin film morphology on the nanometer scale. Using GLAD, columnar thin films may be shaped into helices, square spirals, vertical posts and zig-zags [1-4]. The GLAD process is compatible with evaporation, pulsed laser deposition [5] and long-throw, low pressure sputtering [6]. Applications incorporating GLAD thin films include ultra-fast humidity sensors [7,8], thermal barrier coatings [9], three-dimensional photonic crystals [4,10], gradient-index optical filters [11,12] and antireflection coatings [13]. Helical films have potential applications as circular polarization filters, but before any devices can be realized, the optical response must be well understood and optimized by studying the impact of film composition and structure. For this study, a set of helical films was fabricated to determine the effect of film porosity on the selective response to circularly polarized light and to identify the dominant mechanisms involved.

EXPERIMENTAL DETAILS

Glancing angle deposition of helical thin films

GLAD films were grown in an electron beam evaporator fitted with two computer-controlled stepper motors to allow substrate motion with two degrees of freedom. The apparatus and growth mechanisms are shown in figures 1a and 1b, respectively. One stepper motor controls the incident vapor flux angle or substrate tilt α, while the other controls substrate rotation φ.

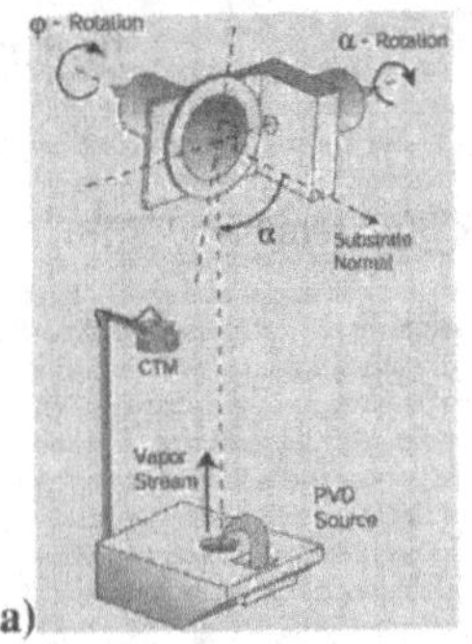

a)

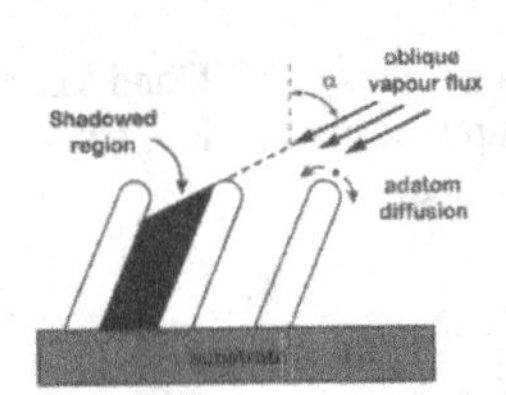

b)

Figure 1. a) Schematic of setup used in glancing angle deposition, indicating the incident vapor flux angle α and substrate rotation angle φ. b) Thin film growth mechanisms governing GLAD: self-shadowing and limited adatom diffusion.

By dynamically manipulating these two parameters, α(t) and φ(t), using in-house control software, a large number of thin film structures can be produced. To fabricate helical films for this study, α was kept constant and the substrate was rotated at a constant rate with respect to the growth rate of the film. This was accomplished by providing feedback from the crystal thickness monitor (CTM) to the control computer.

In this study all of the films were fabricated in titanium dioxide because of its high refractive index and transparency in the visible regime, and were deposited onto glass substrates (Corning 7059). Samples were grown at 14 substrate tilt angles ranging from 30° to 87°. Select SEM micrographs of these films are shown in figure 2. During each deposition, the substrates were rotated counter clockwise to produce right-handed helical films consisting of 3 full turns, a helical pitch of 330 nm and a total film thickness of approximately 1 µm. The deposition pressure was kept constant at approximately 7×10^{-3} Pa by adding O_2(g) to the vacuum chamber. The deposition rate ranged from 10-15 Å/s as measured by the CTM.

Optical characterization

Much like cholesteric liquid crystals, helical films exhibit circular Bragg effects. A helical film will preferentially reflect circularly polarized light of equal handedness, while transmitting circularly polarized light of opposite handedness. This means a left-handed film will preferentially transmit right circularly polarized (RCP) light while reflecting left circularly polarized (LCP) light, and a right-handed film will tend to transmit LCP light while reflecting RCP light. Circular Bragg effects occur at wavelengths of light that are proportional to the helical pitch of the film ($\lambda \sim np$, where n is the average refactive index and p is the pitch of the film). To measure the selective spectral transmittance of circularly polarized light, a variable angle spectroscopic ellipsometer (VASE) (model V-VASE from J. A. Woolam Co. Inc.), operating in transmission mode, was used. First, ellipsometric scans were performed to determine the m_{14} (1st row, 4th column) Mueller matrix coefficient. Then, the selective transmittance of circularly polarized light was calculated using the following equation.

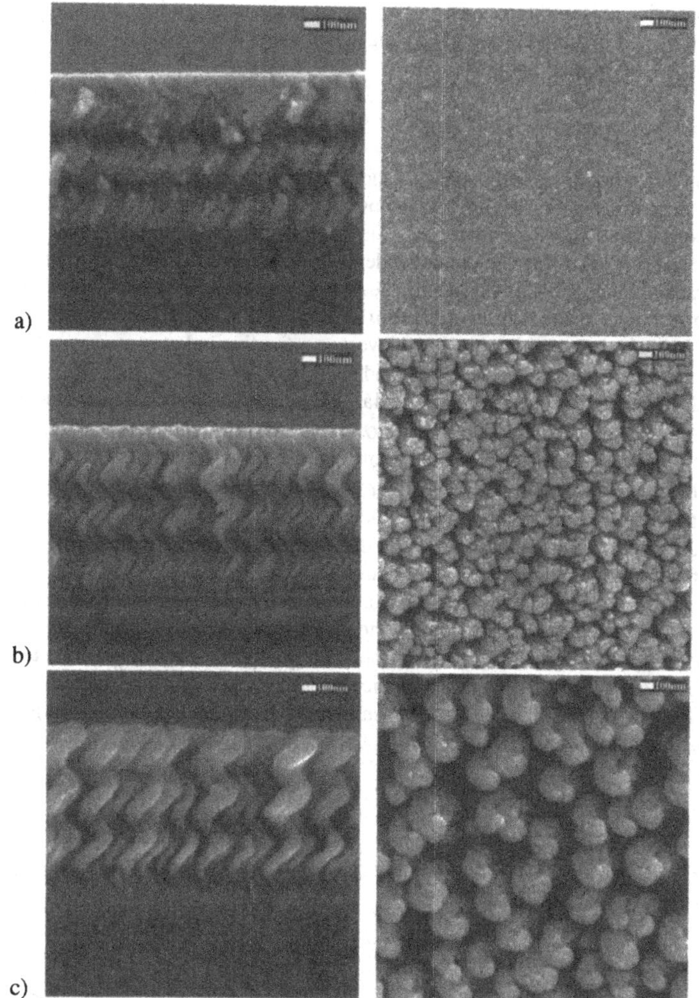

Figure 2. Helical GLAD films deposited at a) 65°, b) 80° and c) 87° illustrating the variation in porosity resulting from adjusting the incoming vapor flux angle. The left hand column are side views and the right hand column are top views of each film.

$$\%T(LCP) - \%T(RCP) = -2m_{14}(\lambda) \quad (1)$$

At deposition angles of $\alpha \geq 80°$, GLAD thin films noticeably scatter light, as they no longer appear transparent upon visible inspection. To quantify the amount of diffusely scattered light and its contribution to the selective transmittance over a broad range of wavelengths, a dual-beam Perkin Elmer Lambda 900 UV/VIS/NIR Spectrometer was used. To create LCP and RCP light, a linear polarizer and an achromatic quarter-wave plate, with its fast axis oriented at ±45°

with respect to the transmission axis of the polarizer, were placed in the beam path. The sample was placed against the front port of a broadband integrating sphere, which was used to measure both the spectral and diffuse transmitted light.

DISCUSSION

The wavelength at which maximum selective transmittance occurs for a particular film strongly depends on the deposition angle the film was deposited at. Figure 3a shows a typical selective transmittance spectrum with a maximum occurring at a specific wavelength. Figure 3b describes the dependence of this peak wavelength on the deposition angle of the film. In figure 3b, the peak selective transmittance wavelength steadily decreases with increasing deposition angle. This decrease corresponds to a reduction of the film density with increased deposition angle (as verified by SEM analysis), resulting in a lower average refractive index.

The maximum difference in the transmittance of LCP and RCP light for a series of right-handed TiO_2 films of varying porosity is shown in figure 4a. The selective spectral transmittance data collected by the VASE has two peaks at $\alpha = 65°$ and $\alpha = 87°$. At 65°, both the data from the VASE and the spectrometer give the same magnitude of selective transmittance. This is because there is negligible diffuse scattering from a film grown with an incoming vapor flux of $\alpha \leq 65°$ and the selective transmittance through such a film is completely due to form birefringence. This result agrees with previously reported data stating that a maximum form birefringence occurs for substrate tilt angles between 55° and 65° [14,15].

Once the substrate tilt angle is increased to 80° and above, the selective transmittance collected with the integrating sphere begins to decrease relative to the measurements made with the VASE. This is due to the films undergoing a transition from a dense, tightly-packed structure to a highly porous film of isolated helices. Separation of helical columns causes diffuse scattering that must be accounted for when measuring selective transmittance. Even though circular Bragg effects diminish at deposition angles > 65°, a strong chiral optical response is still observed in

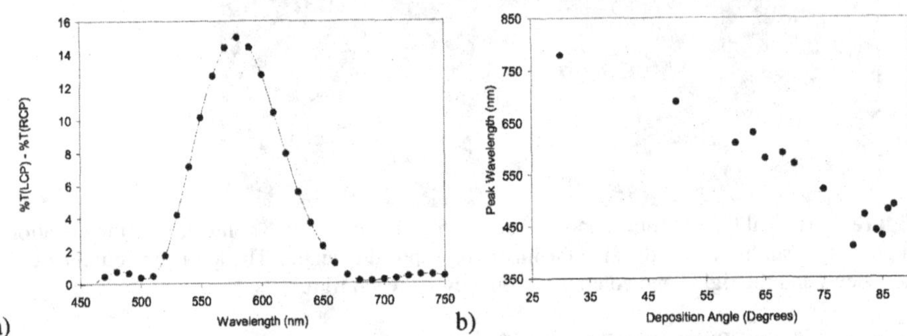

Figure 3. a) Selective transmittance spectrum for a right handed helical TiO_2 film deposited at $\alpha = 65°$; b) Peak selective transmittance wavelength versus deposition angle for helical TiO_2 thin films.

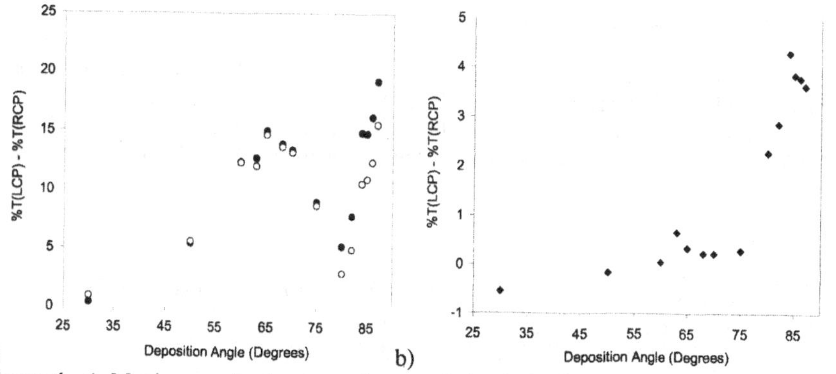

Figure 4. a) Maximum selective transmittance of circularly polarized light in helical TiO_2 films grown at various deposition angles ; (•) – spectral transmission collected; (○) – spectral and diffuse transmission collected; b) Maximum selective diffuse transmittance of circularly polarized light in helical TiO_2 films grown at various deposition angles.

films with deposition angles > 80°, but instead of circular birefringence, anisotropic scattering is the dominant mechanism. Due to this anisotropy in scattering, when only the spectral transmittance is recorded, the diffusely scattered light contributes to the selective transmittance because it is not collected by the detector. By using an integrating sphere to capture both the spectral and diffusely scattered light, the diffusely transmitted RCP light partially negates the spectrally transmitted LCP light and the resulting selective transmittance is reduced. This is supported in figure 4a by the VASE data indicating a larger selective transmittance than the spectrometer data for incident vapor flux angles greater than about 80°. This effect is also illustrated in figure 4b where the two sets of data are subtracted from each other, leaving the diffuse scattering contribution. Also, at the 87° peak, when the selective transmittance due to anisotropic scattering is neglected, the selective transmittance drops to a value that is comparable to that at 65°. It is expected that this value would drop even further if the diffusely reflected light were accounted for as well. Measurements of diffusely reflected light are the subject of future experiments.

CONCLUSION

In summary, helical GLAD films selectively transmit circularly polarized light of opposite handedness to the helical film structure. The magnitude of selective transmittance of circularly polarized light peaks at two deposition angles. In the tightly-packed film regime, typically at deposition angles < 80°, selective transmittance is dominated by form birefringence, while in the highly porous film regime, typically at deposition angles > 80°, selective transmittance is dominated by anisotropic scattering. For deposition angles > 80°, the onset of scattering directly corresponds to the transition from tightly packed helices into individual isolated structures as witnessed by SEM analysis.

ACKNOWLEDGEMENTS

The authors would like to acknowledge the financial support provided by the Natural Sciences and Engineering Research Council of Canada (NSERC), the Alberta Informatics Circle of Research Excellence (iCORE), and Micralyne Inc. The authors would also like to thank George Braybrook for the exceptional SEM imaging and the University of Alberta Micromachining and Nanofabrication Facility.

REFERENCES

1. K. Robbie, M.J. Brett, and A Lakhtakia, *J. Vac. Sci. Technol.*, **A 13**, 2991 (1995).
2. K. Robbie, M.J. Brett, and A Lakhtakia, *Nature*, **384**, 616 (1996).
3. K. Robbie, and M.J. Brett, *J. Vac. Sci. Technol.*, **A 15**, 1460 (1997).
4. S.R. Kennedy, M.J. Brett, O. Toader, and S. John, *Nano Lett.*, **2**, 59 (2002).
5. D. Vick, Y.Y. Tsui, M.J. Brett, and R. Fedosejevs, *Thin Solid Films*, **350**, 49 (1999).
6. J.C. Sit, D. Vick, K. Robbie, and M.J. Brett, *J. Mater. Res.*, **14**, 1197 (1999).
7. K.D. Harris, A. Huizinga, and M.J. Brett, *Electrochem. Solid-State Lett.*, **5**, H27 (2002).
8. J.J. Steele, K.D. Harris, and M.J. Brett, *Mat. Res. Soc. Symp. Proc.*, **788**, L11.4.1 (2004).
9. K.D. Harris, D. Vick, E.J. Gonzalez, T. Smy, K. Robbie, and M.J. Brett, *Surface & Coatings Technol.*, **138**, 185 (2001).
10. S.R. Kennedy, M.J. Brett, H. Miguez, O. Toader, and S. John, *Photonics and Nanostructures.*, **1**, 37 (2003).
11. A.C. van Popta, M.H. Hawkeye, J.C. Sit, M.J. Brett, *Opt. Lett.*, **29**, 2545 (2004).
12. K. Kaminska, T. Brown, G. Beydaghyan, and K. Robbie, *Appl. Opt.*, **42**, 4212 (2003).
13. S.R. Kennedy, and M.J. Brett, *Appl. Opt.*, **42**, 4573 (2003).
14. I. Hodgkinson, and Q. Wu, *Advanced Materials*, **13**, 889 (2001).
15. I. Hodgkinson, Q. Wu, and S. Collett, *Appl. Opt.*, **40**, 452 (2001).

Poster Session:
Organic and Hybrid
Light Emitting Devices:
Electrical and Optical Properties

Mater. Res. Soc. Symp. Proc. Vol. 846 © 2005 Materials Research Society DD11.2

Fast-Switching, High-Contrast Electrochromic Thin Films Prepared Using Layer-by-Layer Assembly of Charged Species

Jaime C. Grunlan
Department of Mechanical Engineering and Polymer Technology Center,
Texas A&M University, College Station, TX 77843-3123, U.S.A.

ABSTRACT

Thin films were prepared by depositing alternating layers of tungstate anion (WO_4^{2-}) and poly(4-vinylpyridine-co-styrene) (PVP-S) onto an electrode from aqueous solutions. These films have very high contrast (CR > 8) relative to equivalent films prepared using poly(ethylene dioxythiophene) (PEDOT), but suffer from slow color change due to poor electrical conductivity. The switching time of the tungstate-based films was decreased by an order of magnitude, from 30 seconds down to three, by adding layers of indium tin oxide (ITO) particles stabilized with poly(diallyldimethylammonium chloride) (PDDA). In this case, a four-layer repeating structure was created (i.e., PVP-S and PDDA-ITO were each deposited every fourth layer). Unlike tungstate, ITO has a high intrinsic conductivity (~ 10^4 S/cm) that accounts for the dramatic increase in the switching speed. It is only through the nanometer-scale control of film architecture, provided with the layer-by-layer (LbL) deposition process, that switching speed and contrast ratio can be optimized simultaneously.

INTRODUCTION

A variety of functional films can be produced using the layer-by-layer (LbL), or electrostatic self-assembly (ESA) technique [1-3]. LbL-based thin films are currently being evaluated for a variety of applications that include controlled drug delivery [4], molecular sensing [5], solid battery electrolytes [6], and photovoltaics [7]. Thin films, typically < 1μm thick, are created by alternately exposing a substrate to positively- and negatively-charged molecules or particles, as shown in Figure 1. In this case, steps 1 – 4 are continuously repeated until the desired number of "bilayers" (or cationic-anionic pairs) is achieved. Bilayers are typically 1 – 10nm thick, but they can become much thicker if particulate materials are deposited [8]. Some of the key factors affecting polyelectrolyte thickness include molecular weight [9], charge density [10], temperature [11], deposition time [12], counterion type [10], solution pH [13] concentration [12,14], and ionic strength [12,14]. The ability to control coating thickness down to the nm-level, easily insert variable thin layers without altering the process, economically use raw materials (due to thin nature), self-heal, and process under ambient conditions are some of the key advantages of this deposition technique [1]. A variety of electrochromic systems have already been prepared using the LbL process to make use of these benefits [3,15-18]. The systems with the highest contrast ratios contain inorganics and take several seconds to change color due to low conductivity in one or both states. Tungstate species (WO_4^{2-}) have been shown to exhibit excellent electrochromic behavior when complexed with polyvinylpyridine (PVP), but suffer from very slow switching (> 10s) [19]. In the present study, a high-contrast, fast-switching PVP-tungstate thin film is created via LbL assembly. Switching speed is improved by more than an order of magnitude by incorporating ITO nanoparticles to enhance conductivity.

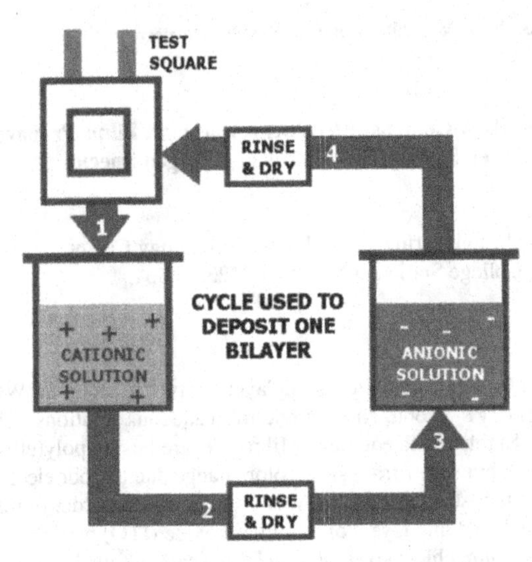

Figure 1. Schematic of layer-by-layer deposition process used to produce electrochromic thin films. Steps 1 – 4 are repeated until the desired number of cationic-anionic pairs are deposited.

EXPERIMENTAL DETAILS

Materials

Poly(4-vinylpyridine-co-styrene) (PVP), sodium tungstate dihydrate (Na2WO4_2H2O), poly(diallyldimethylammoniumchloride) (PDDA), and lithium chloride (LiCl) were purchased from Aldrich. Indium tin oxide (ITO), with a particle diameter of 50 nm, was purchased from Nanophase Technologies Corporation (Romeoville, IL). Poly(3,4-ethylene dioxythiophene)-poly(styrene sulfonate) (PEDOT-PSS), tradename Baytron P, was provided by H. C. Starck (Newton, MA). White, conductive 'test square' electrodes were printed by Avery Dennison's Industrial & Automotive Products Division (IAPD) (Wichita, KS). These squares have a 3:1 electrode area ratio and were prepared using a build-up of silver ink, followed by carbon ink, and finally ATO-TiO_2 ink, as described previously [20].

Thin Film Deposition

Test squares were first dipped into a 0.25wt% PVP-S solution that was made cationic by adding 1wt% 1M HCl. After dipping, test squares were rinsed with deionized water and dried with nitrogen. Next, the test square was dipped into an aqueous solution containing 1wt% of the tungstate species and 6wt% 1M HCl. After rinsing and drying again, this entire procedure was repeated to achieve the desired number of 'bilayers', as shown in Figure 2(a). Other films were prepared by using a solution of PDDA (0.02wt%) – stabilized ITO (1wt%) to substitute for the PVP-S solution (Fig. 2(c)). Some films were built up with 'quad-layers' that used the PDDA-stabilized ITO to alternate for the PVP-S solution every other bilayer (Fig. 2(d)). Each dip was sustained for 60 seconds regardless of chemistry used. Reference films were produced using

0.25wt% aqueous PEDOT-PSS alternated with 0.25wt% aqueous PDDA (low MW) (Fig. 2(b)). The performance of PEDOT-based electrochromes has already been described in literature [21].

Contrast Ratio Determination

Electrochromic films deposited onto test squares were biased ±3V in 1M LiCl solution (pH = 2.15 adjusted with HCl). Digital pictures were taken of coated test squares with a fixed illumination and fixed distance from the sample. The resulting color pictures were converted to black-and-white and a CCD was used to measure the intensity of the light and dark areas of a test square. Contrast ratio was then taken as the ratio of light intentsity to dark intensity (I_{light}/I_{dark}).

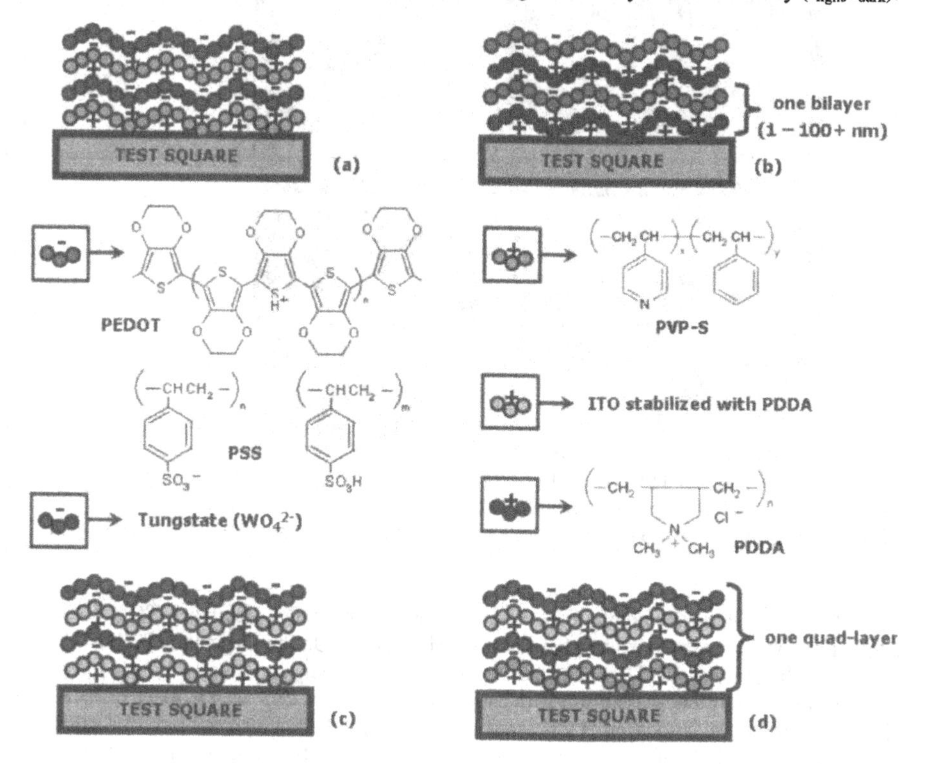

Figure 2. Schematics of four different electrochromic architectures deposited on test squares.

RESULTS AND DISCUSSION

Figure 3 shows one version of the tungstate-based eletrochromic film, produced via LbL deposition, using 60 bilayers of WO_4^{2-} alternated with PVP-S. Sodium tungstate readily dissolves in water and PVP-S is made soluble with the addition of dilute HCl. Prior to the application of an electric field, the test-square shown in Figure 3 was completely white (i.e., the deposited film was colorless). Applying a 1.5V electric field produces only a light blue color

(Fig. 3(a)), but with the application of 3V a deep, dark blue color is formed on one electrode (Fig. 3(c)), while pure white is maintained on the other. Additional bilayers do not enhance the contrast seen here, although with only 60 bilayers a CR > 8 was achieved. Furthermore, the intensity of the blue color can be tuned with applied voltage. This feature may be useful for a "smart window" application when used in conjunction with a transparent electrode (e.g., sputtered ITO). Unfortunately, this film has a very low intrinsic conductivity, resulting in very slow switching time (~ 25 s).

Figure 3. Test square coated with 60-bilayers of tungstate (WO_4^{2-}) alternated with poly(vinyl pyridine-co-styrene) [PVP-S] (schematic shown in Figure 2(a)). This test square was biased with 1.5 (a), 2.5 (b), and 3V (c). The reversibility of this film is also demonstrated by reversing the 3V bias (d).

In an effort to speed the switching of the tungstate-based film, PDDA-stabilized ITO nanoparticles were incorporated into the assembly (Fig. 2(c)). Unlike tungstate, ITO has a high intrinsic conductivity (~ 10^4 S/cm) that can be used to dramatically improve switching speed. Figure 4 shows a tungstate-based film with cationically-stabilized ITO used in place of PVP-S. This film achieves full coloration in less than one second, although it does not have the same intensity as the film in Figure 3. The CR of this ITO-substituted film varied from 3 – 5 and the full coloration appeared more gray than blue. In this film, the polycation (PDDA) is a poor complexing agent for tungstate, relative to PVP-S. This may be due in part to an excessive charge density along its chain. A 30 "quad-layer" film was prepared to combine the benefits of both PVP-S and PDDA-ITO. Figure 5 shows the result of depositing thirty repeat units of WO_4^{2-}:PVP-S:WO_4^{2-}:PDDA-ITO. This film is able to achieve full coloration in about three seconds and shows comparable intensity (CR > 8) to the film in Figure 3. Direct comparisons can be made between Figures 3, 4, and 5 due to all having 60-layers of the active electrochromic species (WO_4^{2-}).

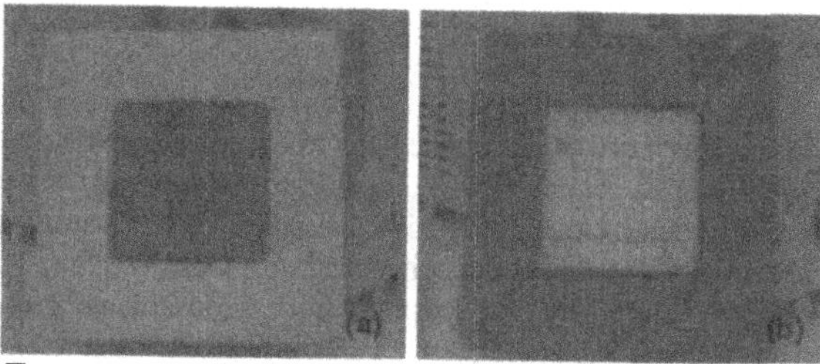

Figure 4. Test square coated with 60-bilayers of tungstate alternately deposited with PDDA-stabilized ITO particles (schematic shown in Figure 2 (c)). The films reversibility is shown by biasing with 3V.

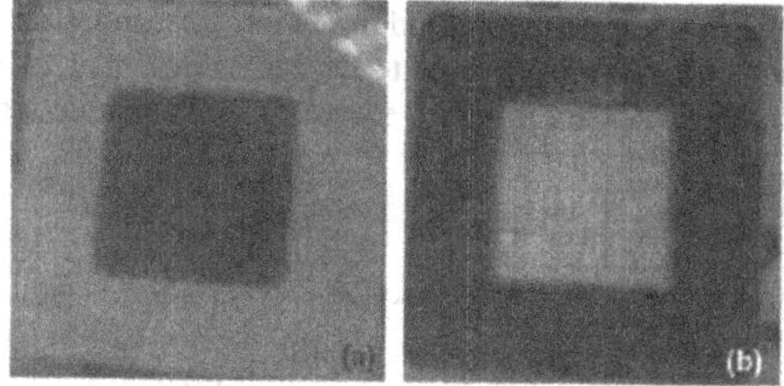

Figure 5. Test square coated with 30-quadlayers of tungstate alternately deposited with PVP-S and PDDA-stabilized ITO particles every other bilayer (schematic shown in Figure 2 (d)). The film's reversibility is shown by biasing with 3V.

PDDA-ITO qualitatively improves the switching speed and appearance of PEDOT-based films, as shown in Figure 6(a) and (b). These films, made with 30-bilayers of PDDA-ITO alternated with PEDOT-PSS, achieve full coloration in less than one second and have contrast comparable to 60-bilayer PEDOT-based films in the absence of ITO. Unlike tungstate-based films, PEDOT-PSS assemblies show increasing contrast ratio with number of bilayers [22]. Unfortunately, PEDOT never becomes completely colorless and residual blue worsens with number of bilayers in the "bleached" state. The use of ITO may serve to improve the usefulness of this system, providing greater contrast and switching speed with fewer bilayers.

Figure 6. Test squares coated with 60-bilayers of PEDOT-PSS alternated with PDDA (schematic shown in Figure 2(b)) (a) and 30-bilayers of PEDOT-PSS alternated with PDDA-stabilized ITO (b). The 30-bilayer film is reversible with a 1.5V bias (c).

CONCLUSIONS

Nanoparticulate indium tin oxide stabilized with PDDA is a convenient component to use when attempting to increase switching speed of LbL-assembled electrochromes. Stabilizing the ITO with PVP-S, which would eliminate the need for PDDA, could further improve the behavior of the tungstate-based system studied here. Other conductive nanoparticles (e.g., antimony tin oxide) could also be polyelectrolyte-stabilized and incorporated into these systems. As these possibilities are further explored it is likely that low-cost, possibly disposable uses (e.g., advertising media) will become more feasible.

ACKNOWLEDGEMENTS

The author would like to thank Avery Dennison and the Avery Research Center (ARC) for financial support of this research. Thanks are also paid to Dr. James Coleman, of Albion LLC, and Li Shu, of the Avery Research Center, for a variety of assistance in conducting this work.

REFERENCES

1. P. Bertrand, A. Jonas, A. Laschewsky, and R. Legras, *Macromol. Rapid Comm.*, **21**, 319 (2000).
2. *Multilayer Thin Films*, edited by G. Decher and J. B. Schlenoff (Wiley-VCH, 2002).
3. P. T. Hammond, *Advanced Materials*, **16**, 1271 (2004).
4. C. M. Nolan, M. J. Serpe, and L. A. Lyon, *Biomacromolecules*, **5**, 1940 (2004).
5. J. H. Kim, S. H. Kim, and S. Shiratori, *Sens. Actuat. B*, **102**, 241 (2004).
6. D. M. DeLongchamp and P. T. Hammond, *Langmuir*, **20**, 5403 (2004).
7. G. J. Yao, B. Q. Wang, Y. P. Dong, M. F. Zhang, Z. H. Yang, Q. L. Yao, L. J. W. Yip, and B. Z. Tang, *J. Polym. Sci. A Polym. Chem.*, **42**, 3224 (2004).
8. Y. Lvov, K. Ariga, M. Onda, I. Ichinose, and T. Kunitake, *Langmuir*, **13**, 6195 (1997).
9. Z. Sui, D. Salloum, and J. B. Schlenoff, *Langmuir*, **19**, 2491 (2003).
10. O. Mermut and C. J. Barrett, *J. Phys. Chem. B*, **107**, 2525 (2003).
11. K. Buscher, K. Graf, H. Ahrens, and C. A. Helm, *Langmuir*, **18**, 3585 (2002).
12. H. Zhang and J. Ruhe, *Macromolecules*, **36**, 6593 (2003).
13. S. S. Shiratori and M. F. Rubner, *Macromolecules*, **33**, 4213 (2000).
14. S. T. Dubas and J. B. Schlenoff, *Macromolecules*, **32**, 8153 (1999).
15. S. Liu, D. G. Kurth, H. Mohwald, and D. Volkmer, *Adv. Mater.*, **14**, 225 (2002).
16. D. M. DeLongchamp, M. Kastantin, and P. T. Hammond, *Chem. Mater.*, **15**, 1575 (2003).
17. F. Huguenin, M. Ferreira, V. Zucolotto, F. C. Nart, R. M. Torresi, and O. N. Oliveira, *Chem. Mater.*, **16**, 2293 (2004).
18. D. M. DeLongchamp and P. T. Hammond, *Adv. Func. Mater.*, **14**, 224 (2004).
19. S. J. Babinec in *Electrochromic Materials*, Proceedings of the Electrochemical Society, New Orleans, LA, October, 1993; edited by K. C. Ho and D. A. MacArthur (Electrochemical Society: Pennington, NJ, 1994) p. 30.
20. J. P. Coleman, A. T. Lynch, P. Madhukar, and J. H. Wagenknecht, *Sol. Energy Mater. Sol. Cells*, **56**, 375 (1999).
21. D. DeLongchamp and P. T. Hammond, *Adv. Mater.*, **13**, 1455 (2001).
22. J. C. Grunlan, J. P. Coleman, and L. Shu, *Polym. Mater. Sci. Eng.*, **90**, 589 (2004).

Mater. Res. Soc. Symp. Proc. Vol. 846 © 2005 Materials Research Society DD11.7

A Novel Light-Emitting Mixed-Ligand Iridium(III) Complex With a Polymeric Terpyridine-PEG Macroligand: Synthesis And Characterization

Elisabeth Holder, Veronica Marin, Michael A. R. Meier, Dmitry Kozodaev and Ulrich S. Schubert*
Laboratory of Macromolecular Chemistry and Nanoscience,
Eindhoven University of Technology and Dutch Polymer Institute (DPI),
P.O. Box 513, 5600 MB Eindhoven, The Netherlands

ABSTRACT

The synthesis of mixed ligand orthometalated iridium(III) complexes as well as the characterization of their electro-optical properties is briefly illustrated. The cyclic voltammetry results exemplify an enhanced stability of the complex with a poly(ethylene glycol (PEG) side chain compared to the corresponding organometallic model material. Less degradation of the electrically active centers was found to occur on the electrode surface. Furthermore the morphology of thin films of the described polymeric material was investigated with respect to potential applications in thin film devices.

INTRODUCTION

Recently iridium(III) complexes have become increasingly interesting due to their valuable electroluminescent properties as a result of their high-potential application in light-emitting devices. The importance of such light-emitting materials is based on their ability to change their emission color depending on the introduced ligand set and their relatively short phosphorescence lifetime. A further application possibility arises due to their phosphorescence-based emission. In light-emitting devices (LEDs), the compatibility of host materials with iridium(III) phosphors is of importance for the collection of all emissive (host and triplet emitter) excitones at the phosphors.[1] These properties make them ideal candidates for technological relevant developments where very high electron-to-light efficiencies are envisioned, e.g. in portable electronics.

The charged iridium(III) complexes with applications in light-emitting electrochemical cells (LECs) are those based on chelating ligands, such as 2,2'-bipyridine and terpyridine. These materials can reveal color changes based on the introduced cyclometalating ligand set, as their neutral counterparts used in LEDs. Due to the relatively small bias window where these complexes do not decompose, the charged complexes are placed in polymer matrices in order to avoid the irreversible oxidation on the electrode surface.[2] In the present study we briefly show the synthesis and the optical and electrical characterization of a yellow light-emitting iridium(III) complex with a polymeric terpyridine-PEG macroligand. The detailed synthesis and characterization of the structures of the presented material can be found elsewhere.[3]

EXPERIMENTAL DETAILS

UV and emission spectra were recorded on a Perkin Elmer Lamda-45 and a Perkin Elmer LS50B Luminescence spectrometer, respectively (1 cm cuvettes, CH_2Cl_2).

Electrochemical experiments were performed using an Autolab PGSTAT30 model potentiostat. A standard three-electrode configuration was used, with platinum-bead working and auxiliary electrodes and a Ag/AgCl reference electrode. Ferrocene was added at the end of each experiment as an internal standard. The potentials are quoted versus the ferrocene/ferrocenium couple (Fc/Fc^+). The solvent used was CH_2Cl_2 (freshly distilled from CaH_2), containing 0.1 M n-Bu_4PF_6. The scan rate was 100 mV/sec.

AFM measurements were performed on a Solver LS produced by NT-MDT, Russia. Cantilever: NSG01s, produced by NT-MDT, Russia.

DISCUSSION

The discussion is structured in a brief description of the general synthetic procedure to obtain polymeric and non-polymeric iridium(III) complexes, followed by a summary of characterization methods including optical spectroscopy, cyclic voltammetry as well as morphology studies of the respective films. The initial investigated parameters are found to be helpful to classify the general functionality of novel materials for potential application in thin film devices.

Synthesis

The preparation of the mixed ligand orthometalated iridium(III) complexes **1** and **2** (Figure 1) was carried out according to previously described procedures.[3-5] The reaction of the orthometalated iridium(III) dimer $[Ir(N^\wedge C\text{-}CHO^-)_2Cl]_2$ (where ($N^\wedge C$-CHO^-) is ppy-CHO^-) with the chelating polymeric terpyridine gave access to the monomeric mixed-macroligand iridium(III) complex **1** $[Ir\text{-}(N^\wedge C\text{-}CHO^-)_2(tpy\text{-}PEG)]^+$ *via* a bridge-splitting reaction.

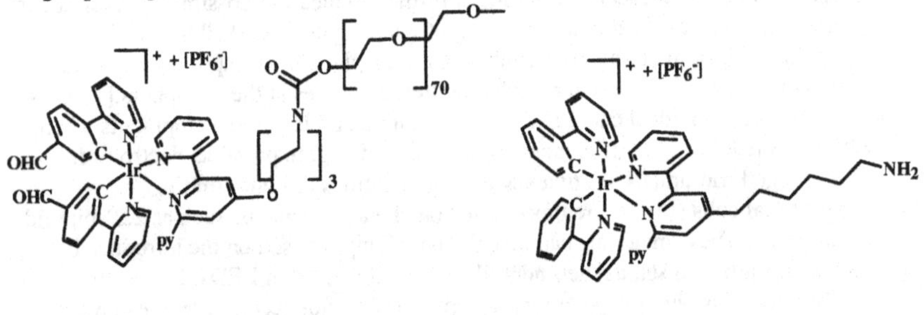

1 **2**

Figure 1. Schematic representation of the synthesized *mono*terpyridine iridium(III) compound **1** and its corresponding model complex **2**.

The same reaction mechanism was applied for the preparation of the model complex **2**. Methatesis with $[NH_4][PF_6]$ yielded the *mono*terpyridine iridium(III) compounds **1** and **2**. The yellow-orange *mono*cationic complexes **1** and **2** are soluble in organic solvents of medium to high polarity.

UV-vis and emission properties

The UV-vis spectra of the terpyridine (tpy)-PEG macroligand (dashed line), complex **1** (solid line) and complex **2** (dotted line) in dichloromethane solutions are shown in Figure 2. For the PEG-macroligand no MLCT was observed. The iridium(III) macroligand complex **1** and the model complex **2** displayed strong absorption bands at about 270 nm and 300 nm characterizing ligand centered $\pi \rightarrow {}^*\pi$ transitions on the diimine as well as the 4-(2-pyridyl)benzaldehyde (ppy-CHO$^-$) and the phenylpyridine (ppy$^-$) ligands, respectively. The absorption bands at lower energy (~ 370 nm) are due to spin-allowed metal-to-ligand charge-transfer transitions (^{1}MLCT, (dπ(Ir)$\rightarrow {}^*\pi$) diimine and ppy-CHO$^-$ transitions or ppy$^-$ transitions). The band tailing to 440 nm are spin-forbidden ^{3}MLCT (dπ(Ir)$\rightarrow {}^*\pi$) diimine and ppy-CHO$^-$ transitions or ppy$^-$ transitions.[6,7]

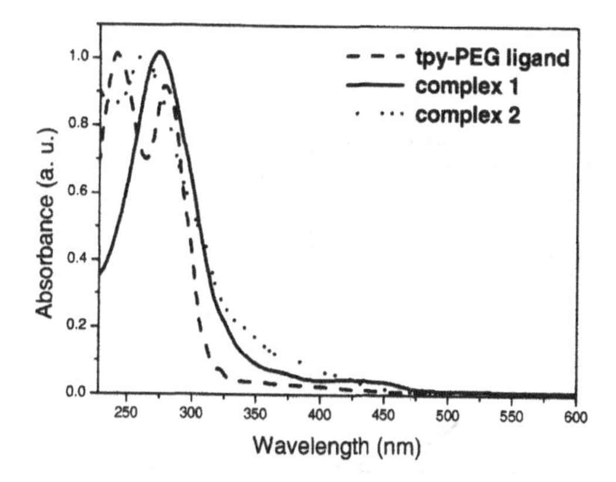

Figure 2. Absorption spectra of the PEG-macroligand and of the *mono*terpyridine iridium(III) compounds **1** and **2** (recorded in CH_2Cl_2).

Excitation (λ_{ex} = 370 nm) of compound **1** lead to a broad emission band with a maximum at 535 nm (shoulder with a maximum at 570 nm, Figure 3, top). A yellow emission color was observed upon excitation of a film of complex **1** with a standard UV-lamp (λ_{ex} = 365 nm). The emission spectra (λ_{ex} = 370 nm) of the polymer film of **1** is less pronounced and both maxima were about 10 nm red-shifted compared to the emission of a corresponding dichloromethane solution of **1** (λ_{em} @ 535 nm and @ 570 nm). The observed emission (λ_{ex} = 370 nm) of the model complex **2** shows a maximum @ 575 nm. The solution spectra of complex **2** (Figure 3, bottom) revealed a clear red shift of 40 nm as compared to material **1** (main emission band @ 535 nm) as a result of the different ligand set (ppy-CHO$^-$ *versus* (ppy$^-$) on the iridium(III) metal center. A variation of the introduced ligand set provokes changes of the emission color due to different triplet emission energies in solution.

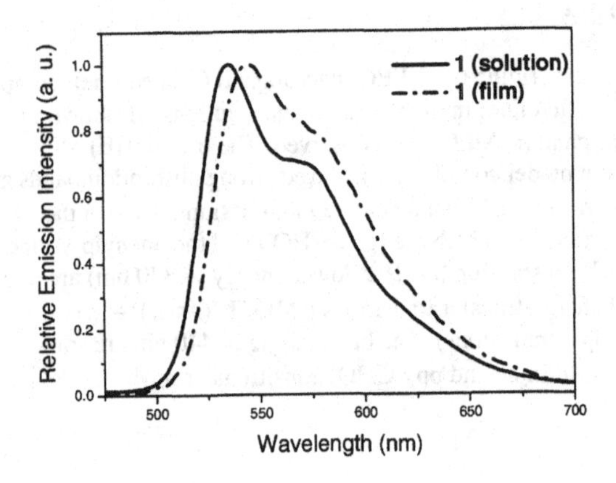

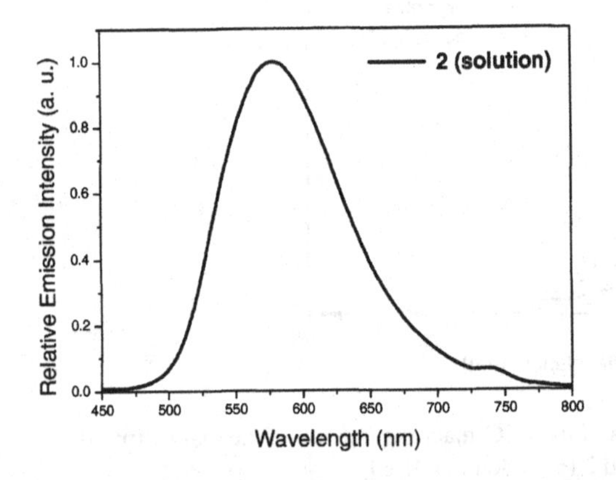

Figure 3. Top: Thin film (spin coated from a $CHCl_3$ polymer solution, fully covered substrate, rotation speed: 2000 rpm) and solution emission spectra of **1** (solution spectrum recorded in CH_2Cl_2). Bottom: Solution emission spectrum of **2** (recorded in CH_2Cl_2).

Cyclic voltammetry studies

Cyclic voltammetry investigations were carried out for stability investigations of materials **1** and **2**. The *mono*terpyridine iridium(III) compound **1** revealed two oxidation potentials at 1.45 and 1.64 V.[8,9] Throughout three oxidation cycles no significant change in the stability of the material was observed. This is in contrast to the model complex **2**, which revealed quick degradation and a non-reversible oxidation potential. In the case of the PEG-modified iridium(III) complex such a behavior was not observed and might be due to the shielding effect of the PEG ligand. This effect might prevent non-reversible degradation processes on the electrode surface and might be of importance to obtain films that reveal less quick degradation and will therefore lead to devices with an improved lifetime.

Morphology studies by AFM

The morphology of a thin film of **1** was investigated by AFM (Figure 4). The film spin coated from *o*-dichlorobenzene revealed a non-crystalline relatively smooth active layer, which were not obtained using low boiling point solvents such as e.g. chloroform or dichloromethane. Using such low boiling point solvents domains lead to crystalline domains, which are not favored in thin film devices since these effects will reduce performance and device lifetime.

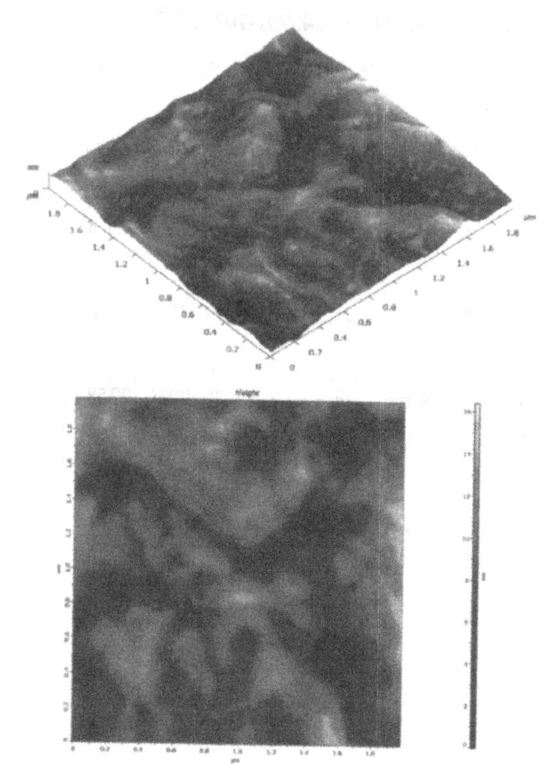

Figure 4. Top: 3D AFM image of material **1**. Bottom: Height image of the investigated polymer **1**.

CONCLUSIONS

From the initial study of the optical properties, valuable emission characteristics were observed. Furthermore, the study revealed some improvements on the stability of a polymer modified iridium(III) complex compared to the neat material as being proven by cyclic voltammetry. It was shown that rather smooth films could be obtained, which might be suitable to prepare devices with a sufficient film quality.

ACKNOWLEDGEMENTS

This work is part of the *Dutch Polymer Institute (DPI)* research program (project 360, 324 and 399). We thank the *Fonds der Chemischen Industrie* for financial support. We are grateful to Prof. R. Janssen (TU/e) for granting access to the optical set-ups.

REFERENCES

1. A. Köhler, J. S. Wilson, R. H. Friend, *Adv. Mater.* **14**, 701-707 (2002).
2. J. Slinker, D. Bernards, P. L. Houston, H. D. Abruna, S. Bernhard, G. G. Malliaras, *Chem. Commun.* 2392-2399 (2003).
3. E. Holder, V. Marin, M. A. R. Meier, U. S. Schubert, *Macromol. Rapid Commun.* **25**, 1491-1496 (2004).
4. V. Marin, E. Holder, R. Hoogenboom, U. S. Schubert, *J. Polym. Sci., Part A: Polym. Chem.* **42**, 4153-4160 (2004).
5. E. Holder, V. Marin, A. Alexeev, U. S. Schubert, *J. Polym. Sci., Part A: Polym. Chem.* **43**, (2005), in press.
6. F. Neve, A. Crispini, S. Campagna, S. Serroni, *Inorg. Chem.* **38**, 2250- 2258 (1999).
7. K. K.-W. Lo, C.-K. Chung, T. K.-M. Lee, L.-H. Lui, K. H.-K. Tsang, N. Zhu, *Inorg. Chem.* **42**, 6886-6897 (2003).
8. M. Polson, S. Fracasso, V. Bertolasi, M. Ravaglia, F. Scandola, *Inorg. Chem.* **43**, 1950-1956 (2004).
9. A. Mamo, I. Stefio, M. F. Parisi, A. Credi, M. Venturi, C. Di Pietro, S. Campagna, *Inorg. Chem.* **36**, 5947-5950 (1997).

Mater. Res. Soc. Symp. Proc. Vol. 846 © 2005 Materials Research Society DD11.11

Degradation of $Ru(bpy)_3^{2+}$-based OLEDs

Velda Goldberg[1], Michael Kaplan[1,2], Leonard Soltzberg[2], Joseph Genevich[1], *Rebecca Berry [1,2], *Alma Bukhari[1,2], *Sherina Chan[1,2], *Megan Damour[1,2], *Leigh Friguglietti[1,2], *Erica Gunn[1,2], *Karen Ho[1,2], *Ashley Johnson[1,2], *Yin Yin Lin[1,2], *Alisabet Lowenthal[1,2], *Seiyam Suth[1,2], *Regina To[1,2], *Regina Yopak[1,2], Jason D. Slinker[3], George G. Malliaras[3], Samuel Flores-Torres[4], and Hector D. Abruña[4]
[1]Physics, Simmons College, Boston, Massachusetts; [2]Chemistry, Simmons College, Boston, Massachusetts; [3]Materials Science and Engineering, Cornell University, Ithaca, New York; [4]Chemistry and Chemical Biology, Cornell University, Ithaca, NY.
*Undergraduate student authors.

ABSTRACT

Analysis of the possible mechanisms of degradation of $Ru(bpy)_3^{2+}$-based OLEDs has led to the idea of quencher formation in the metalloorganic area close to the cathode. It has been suggested that the quencher results from an electrochemical process where one of the bipyridine (bpy) groups is replaced with two water molecules [1] or from reduction of $Ru(bpy)_3^{2+}$ to $Ru(bpy)_3^0$ [2]. We have tested these and other degradation ideas for $Ru(bpy)_3^{2+}$-based OLEDs, both prepared and tested with considerable exposure to the ambient environment and using materials and procedures that emphasize cost of preparation rather than overall efficiency. In order to understand the mechanisms involved in these particular devices, we have correlated changes in the devices' electrical and optical properties with MALDI-TOF mass spectra and UV-vis absorption and fluorescence spectra.

INTRODUCTION

In their simplest form OLEDs consist of a single organic semiconductor sandwiched between electrodes. Under forward bias, the anode injects holes into the highest occupied molecular orbital (HOMO) of the organic layer and the cathode injects electrons into the lowest unoccupied molecular orbital (LUMO) of the organic layer. These charges migrate in the opposite directions and when they meet, they may form an exciton, and the radiative decay of a fraction of these excitons produces light. For such a simple device to produce a high yield, the single organic layer must easily transport both electrons and holes. Since for many materials this is not the case, high efficiency OLEDs often employ multiple organic layers, which increases the cost and the processing steps.

As an alternative to the use of organic semiconductors in OLEDs, ionic transition metal complexes have created much interest because of their charge transport properties. We have focused on a particular example of these materials—$Ru(bpy)_3(PF_6)_2$. In this material $[Ru(bpy)_3]^{2+}$ is surrounded by $(PF_6)^-$ ions. Under forward bias, holes are injected into the t_{2g} orbital of the predominantly-metal (HOMO) and electrons are injected into the π^* orbitals with mostly ligand-type character. In addition, the counter ions also contribute to ionic conductivity. These materials show promise because their performance is now close to that for very good OLED devices.[3] They show stability in multiple redox states so both electrons and holes are injected efficiently, and they show strong spin-orbit coupling leading to intersystem crossing and a higher probability of exciton decay through radiative processes.[4]

OLEDs in general, degrade over time, and $Ru(bpy)_3^{2+}$-based OLEDs are not immune to this problem. Some work has been done on the degradation of these materials,[3,5-7] and recent results suggest that quenching sites may develop during operation, due to the presence of moisture or due to the instability of the counter ion.[8] On the other hand, these materials show higher stability when exposed to the ambient environment, as compared to some of the leading polymers used in OLED fabrication. If the degradation could be reduced, these materials then might have a potential use as low cost devices intended for short-term use.

Hence, our aim in this work is to investigate the degradation of $Ru(bpy)_3(PF_6)_2$–based OLEDs, handled with extensive exposure to the ambient environment. Materials and procedures were chosen to emphasize cost of preparation over overall efficiency. In order to understand the mechanisms involved, we examined electrical and optical properties of these devices and correlated these observations with the devices' MALDI-TOF mass spectra and UV-vis absorption and fluorescence spectra. The electrical and optical properties of our devices were also compared to those prepared and tested with no exposure to the ambient environment.

EXPERIMENTAL DETAILS

Chemical synthesis and device preparation

The ruthenium complex, $Ru(bpy)_3(PF_6)_2$ (where bpy = bipyridyl) was synthesized from $Ru(bpy)_3Cl_2$ and NH_4PF_6.[9] All chemicals were purchased from Aldrich and used without further purification. Distilled water solutions of ammonium hexafluorophosphate (100mg/1.5 ml) and of $Ru(bpy)_3Cl_2$ (100 mg/1.5 ml) were mixed together. The resulting precipitate was filtered, and recrystallization was then done in acetone with a small amount of water added very gradually to induce precipitation. The recrystallized product was finally filtered and washed with distilled H_2O. The final product was tested using negative ion MALDI TOF mass spectrometry which confirmed the replacement of the Cl^- by PF_6^-.

OLED devices were made with this starting material sandwiched between a patterned ITO anode and an evaporated Al cathode. ITO coated glass substrates were purchased from Alpha Aesar and using photolithography, a bottom electrode approximately of 25 mm x 3 mm was patterned in the ITO. The Ru-complex was dissolved in acetonitrile (24 mg/ml) and was spun onto the ITO layer at 1000 rpm for 60 seconds. The substrates were then annealed overnight in a vacuum oven at 80°C after which a top 20 nm Al layer was deposited by vacuum evaporation at 5 Å/s and 2.0 x 10^{-6} torr. None of these procedures were carried out in a glove box and devices were exposed to the ambient environment during spin coating, storage, and measurement.

Electrical and Optical Measurements

Electrical measurements were done using a Keithley Sourcemeter 2400. Devices were exposed to a constant 4.0 Volts for 2 hours immediately after preparation and then at other times over the following day. Figure 1 shows the current vs voltage results for a representative sample exposed to a constant 4.0 volts for 1 hour immediately after preparation. The device then rested in the ambient environment for ten hours before being measured again. After resting for an additional 4 hours, the device was measured for a final time.

At the same time these electrical measurements were being recorded on this particular sample, changes in the relative intensity of the light output were monitored, as shown in Figure 2. In this case all intensity measurements were normalized to the maximum intensity for the fresh device.

As shown in Figure 1, the maximum current found in a representative fresh device was 8.06 mA. After running for 2 hours and then resting for 10 hours, the maximum current dropped to 3.96 mA (49% of the initial) and the light intensity dropped to 38% of the maximum reading on the fresh sample. After resting for an additional 2 hours, the maximum current was 3.35 mA (42% of the initial), and the light intensity was 27% of the original maximum reading for the fresh device. In addition, the current for the fresh device declined rapidly during the first 3-4 minutes and was then relatively constant for the next 60 minutes. Ten hours later, the current did not show this initial decline. Instead, it spiked to 1.4-1.5 times its initial reading during the first minute, increased gradually over the next 20 minutes, and was essentially constant for the final 40 minutes.

Figure 2 depicts the normalized light intensity over time and shows that the device took about 2 minutes to turn on. The light intensity then increased in all cases and showed a marked increase in the fresh sample and reached a maximum in about 8 minutes. For the remainder of the 60 minute measurement, the current changed little in the aged devices; however, in the fresh sample, it decreased by about 50%.

Similar devices (ITO/$Ru(bpy)_3(PF_6)_2$/Al or ITO/$Ru(bpy)_3(PF_6)_2$/Au prepared and tested with no exposure to the ambient environment showed similar turn on times.[9,10] However, in these devices the current and light intensity initially increased and both reached a steady state in about 8-10 minutes. In our fresh devices, the current initially decreased and then approached a steady

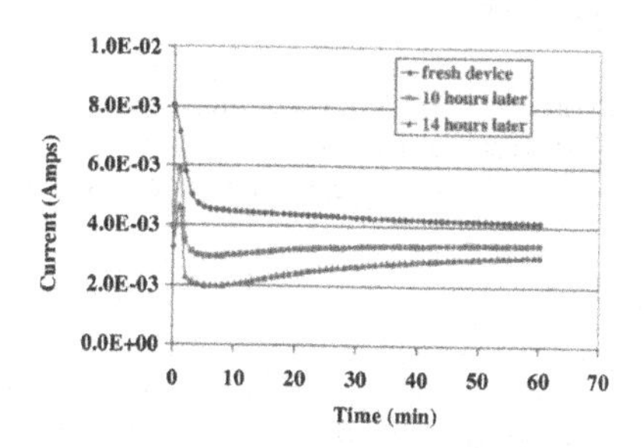

Figure 1. Current vs time measurements for a Al/ $Ru(bpy)_3(PF_6)_2$/ ITO device—freshly prepared and 10 hours and 14 hours later.

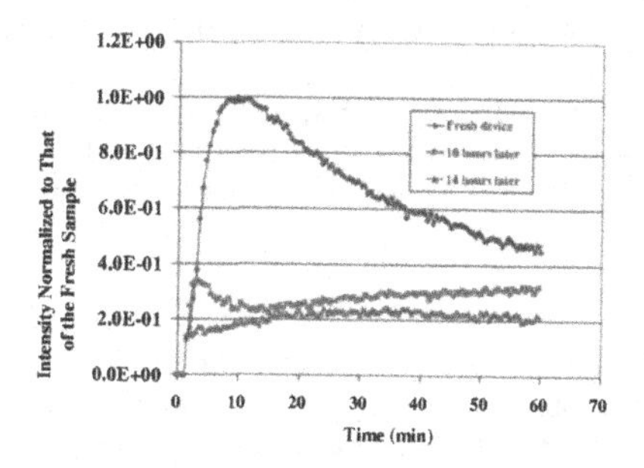

Figure 2. Normalized light intensity vs time for the same Al/ $Ru(bpy)_3(PF_6)_2$/ITO device shown in Figure 1.

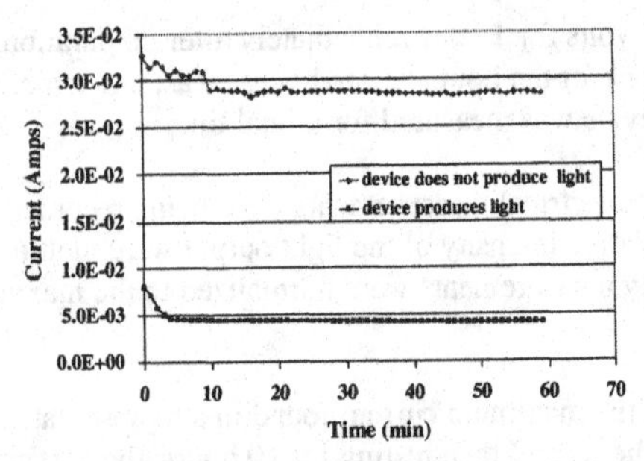

Figure 3. Current vs time measurements for a fresh Al/ $Ru(bpy)_3(PF_6)_2$/ITO device that lights and a fresh device that does not light. hours later.

state in about 3 minutes, while the light intensity reached a maximum in about 8 minutes and then gradually declined to approximately 50% of its maximum value in approximately 60 minutes. Approximately 25% of the devices prepared as described produced no light. In these devices, when fresh, the variation of the current as a function of time had the same character as that for a fresh sample that lights, although the current is higher in the device that does not light—initially 32.7 mA compared to 8.06 mA. This is shown in Figure 3.

MALDI-TOF Mass Spectra

MALDI-TOF mass spectrometry is notable for its ability to detect femtomole quantities of analyte as well as its gentle ionization mechanism. Taking advantage of these attributes, we have developed a method for recovering small amounts of material from individual OLED devices and running mass spectra to compare the composition of material that has carried current and emitted light versus undisturbed material from nearby locations. Mass spectra of recently operated OLED material show significant amounts of a $Ru(bpy)_2$ species, while those of undisturbed material show only the $Ru(bpy)_3$ species. (Figure 4).

Since MALDI-TOF spectra rarely show multiply charged ions, we infer that the various species observed are all +1 ions. This inference, combined with the characteristic isotope distribution of ruthenium, suggests that more than one oxidation state of ruthenium is represented in these spectra.

MALDI-TOF mass spectra of material recovered from ruthenium based OLEDs compliment other observations regarding the stability and performance of these devices. Taken together, these findings suggest a coherent picture for the degradation of such OLEDs.

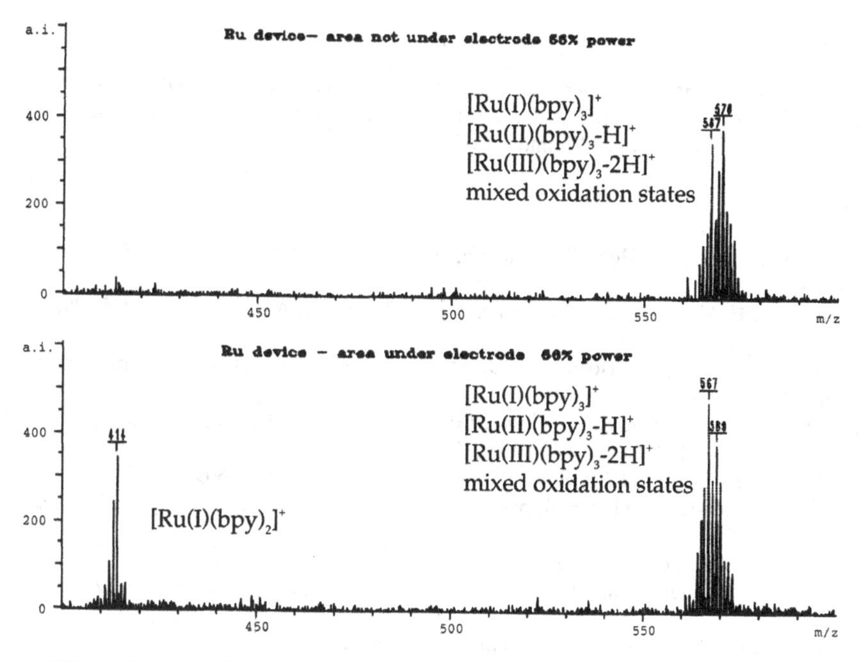

Figure 4. Comparison of MALDI-TOF mass spectra of material recovered from $Ru(bpy)_3(PF_6)_2$ OLEDs - unoperated versus operated.

Work from other groups provides a theoretical framework for these observations. Of special importance are the suggestion that light emission arises from the following process [1]:

$$Ru(bpy)_3^{+} + Ru(bpy)_3^{3+} \rightarrow 2\,Ru(bpy)_3^{2+} + h\nu \tag{1}$$

and the idea that a specific excited electronic state of such ruthenium complexes is unstable with respect to loss of a ligand molecule [11].

CONCLUSIONS

It seems plausible that electrical drive in an operating OLED produces a state of the ruthenium complex that is unstable with respect to loss of one ligand. This instability may result from electron-vibrational interaction leading to Jahn-Teller distortion of the Ru complex and subsequent loss of one bipyridyl group. Our work provides direct support for the suggestion [1] that a bisbipyridyl ruthenium species serves as a quencher in these devices. If the lost ligand does not disappear from the ruthenium environment, it may recombine after the device is turned off, accounting for the recovery observed in such devices.

ACKNOWLEDGEMENTS

This project is supported by support from the National Science Foundation (DMR-0108497 and CHE-0216268) and the George I. Alden Trust. In addition generous support has been provided by Cornell University and Simmons College.

REFERENCES

1. G. Kalyuzny, M. Buda, J. McNeil, P. Barbara, and A. Bard. J. Am.Chem.Soc. **125**, 6272-6283 (2003).
2. K.M. Maness, H. Masui, R.M. Wightman, and R.W. Murray, J. Am. Chem. Soc. **119**, 3987-3993 (1997).
3. J. Slinker, D. Bernards, P.L. Houston, H.D. Abruña, S. Bernhard and G.G. Malliaras, Chem. Comm. **19**, 2392 (2003).
4. J.C. Scott and G.G. Malliaras, in Conjugated Polymers, edited by G. Hadziioannou and P.F. van Hutten, (Wiley-VCH, New York, 1999), Chap. 13..
5. M. Buda, G. Kalyuzhny and A.J. Bard, J. Am. Chem. Soc., **124**, 6090-6098 (2002).
6. H.Rudmann, S. Shimada and M.F. Rubner, D.W. Oblas and J.E. Whitten, J. Appl. Phys., **92**, 1576-1581 (2002).
7. G. Gao and A.J. Bard, Chem. Mater. **14**, 3465-3470 (2002)
8. G. Kalyuzhny, M. Buda, J. McNeill, P. Barbara and A.J. Bard, J. Am. Chem. Soc. **125**, 6272 (2003).
9. H.Rudmann, and M.F. Rubner, J. Appl. Phys, **90**, 4338 (2001).
10. J.D. Slinker, G.G. Maliaras, S. Flores-Torres, H. D. Abruña, W. Chunwachirasiri and M.J. Winokur, J. Appl. Phys. **95**, 4381 (2004).
11. B. Carlson, G. D. Phelan, W. Kaminsky, L. R. Dalton, X. Jiang, S. Liu, and A. K. Y. Jen J. Am. Chem. Soc., **124**, 14162-72 (2002).

Mater. Res. Soc. Symp. Proc. Vol. 846 © 2005 Materials Research Society DD11.12

An efficient top-emitting electroluminescent device on metal-laminated plastic substrate

L.W.Tan, X.T.Hao, K.S.Ong, Y.Q Li, and F.R. Zhu*
Institute of Materials Research and Engineering, No.3 Research Link Singapore, 117602

ABSTRACT

An efficient flexible top-emitting organic light-emitting device (OLED) was fabricated on an aluminum-laminated polyethylene terephthalate substrate. A spin-coated light-emitting polymer layer was sandwiched between a silver anode and a multi-layered semitransparent cathode. The performance of polymer OLEDs was analyzed and compared with that of the devices having a conventional structure. An optical microcavity formed in the device enables to tune the emission color by varying the thickness of the active polymer layer. The OLEDs having a 110-nm-thick active polymer layer exhibited superior electroluminescence performance, with a turn-on voltage of 2.5V and a luminance efficiency of 4.56 cd/A at an operating voltage of 10V.

INTRODUCTION

Organic light emitting devices (OLEDs) have recently attracted attention as display devices that can replace liquid crystal displays because OLEDs can produce high visibility by self-luminescence and they can be fabricated into lightweight, thin and flexible displays[1-4]. The conventional structure of OLEDs consists of a metal or metal alloy cathode and a transparent anode on a transparent substrate, whereby light can be emitted from the transparent substrate. The OLEDs may also have a top-emitting structure that has a relatively transparent top electrode so that light can emit from the side of the top electrode, which can be formed on either an opaque or a transparent substrate[5-7]. The top-emitting OLED structures increase the flexibility of device integration and engineering and are desirable for high-resolution active matrix displays. OLEDs have usually been built on rigid glass substrates due to their low permeability to oxygen and moisture. Over the past few years, ultrathin glass sheets[8-10] and transparent plastic substrates[11, 12] have been considered as the possible substrate choices for flexible OLEDs. Ultrathin glass sheets, however, are very brittle and OLEDs formed on ultrathin glass sheets have limited potential as flexible OLED displays. To make OLEDs that are lighter, thinner, more rugged and highly flexible, plastic substrates, e.g. polyethylene terephthalate (PET) and polyethylene naphthalate (PEN), have been used for flexible OLEDs. It is apparent that PET, PEN and other commonly used plastic foils do not have sufficiently high impermeability for OLEDs. Accordingly, efforts have been made to develop highly effective barrier against oxygen and moisture permeation and hence to minimize degradation of the devices on plastic substrates[13]. Multilayer barrier approaches have been used to improve the barrier property of plastic substrates, such as using alternative multi-layers of organic-inorganic structures, incorporation of getter materials, thick capping metals, pinholes reduction, etc. Further optimization is required to avoid the possible exfoliation between the organic and inorganic SiON layers [14]. Alternatively, metal laminated plastic substrates are promising for flexible display application due to their high mechanical flexibility and the barrier property. Metal laminated plastic foils have the potential to meet permeability standards in excess of the most demanding display and organic electronics requirements.

In this work, we demonstrated the feasibility of fabricating a flexible OLED on aluminum-laminated PET (Al-PET, 0.1mm) substrate using top emission device structure. In this flexible OLED, a modified Ag anode is used, OLEDs with this architecture do not require the process of ITO deposition that is needed in the conventional OLED fabrication. The ITO-free OLED may be of practical importance towards the low-cost flexible OLED displays. The Fabry-Perot planar cavity structure thus formed between a reflective Ag anode and the semitransparent cathode is also discussed. The results also show that the emission properties of OLEDs can be modified by microcavity effect [15-17]. The color tuning and efficiency enhancement observed in the flexible OLEDs were examined by choosing different emitting layer thicknesses.

EXPERIMENTS

The surface of Al-PET film (400 Gauge Mylar 453) was cleansed sequentially with acetone, methanol, and de-ionized water. The plastic foils were then coated with a thin UV-curable acrylic layer to improve the surface smoothness and the adhesion between the anode and the substrate. A 200-nm-thick Ag electrode was deposited on the flexible substrate through a shadow mask with an array of 2 mm × 2 mm openings by thermal evaporation. The Ag contact was then modified by a 0.3-nm-thick plasma-polymerized fluorocarbon film (CF_X) to improve the carrier injection property in OLEDs[18]. The modified Ag/CF_X serves as a high conductivity anode and also a mirror to redirect the internal emitting light towards the upper semitransparent cathode to improve the light output. Phenyl-substituted poly(p-phenylenevinylene) (Ph-PPV) films with various thicknesses were then spin-coated as the emissive layer. The specimens were loaded into an evaporation chamber at a base pressure of $\sim 10^{-4}$ Pa for semitransparent cathode deposition. The semitransparent cathode consists of LiF (0.3 nm)/Ca (5 nm)/Ag (15 nm)/tris-(8-ydroxyquinoline) aluminum (Alq_3, 52 nm). The EL is measured with a SPEX750M spectral-photometer, and the J-V-L characteristics are measured with a Keithley 2420 source measure unit and calibrated silicon photodiode.

RESULTS AND DISCUSSIONS

The layered structures of flexible top-emitting OLEDs reported in this paper are schematically depicted in Fig. 1 (a). The device structure is Al-PET /Ag (200nm) /CFx (0.3nm)/Ph-PPV(80-150nm) /semitransparent cathode. An organic microcavity, consisting of an emissive layer (EL) of Ph-PPV sandwiched between the metal anode and the semitransparent cathode, was formed color tuning and efficiency enhancement. Electron and hole injection are enhanced by interface modification at the metal/organic contacts, and color is tuned by varying the thickness of the Ph-PPV layer. The optical thickness of the active EL layer in the microcavity is in the order of few hundred nanometers and its thickness may be comparable to the emission wavelength. A typical flexible top emitting OLED with yellow emission in bending condition is shown in Fig. 1(b). The devices remained the similar emission performance after repeated bending. Such flexible electroluminescent devices enable to be bent or rolled into any shape without affecting the EL performance. This demonstrates that OLED with top emission structure on Al-PET may provide an inexpensive approach for flexible EL displays and thus make possible new product concepts.

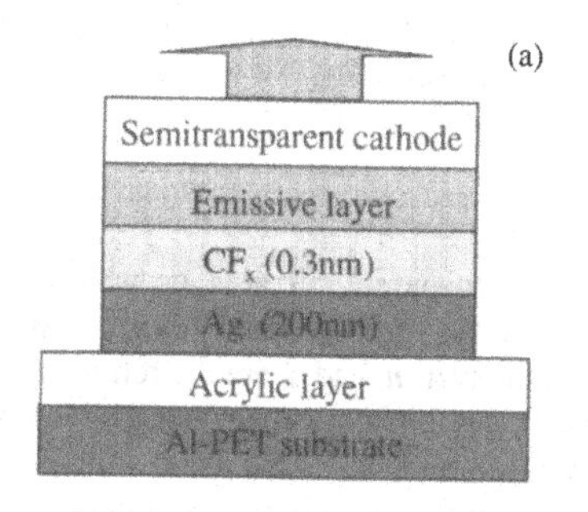

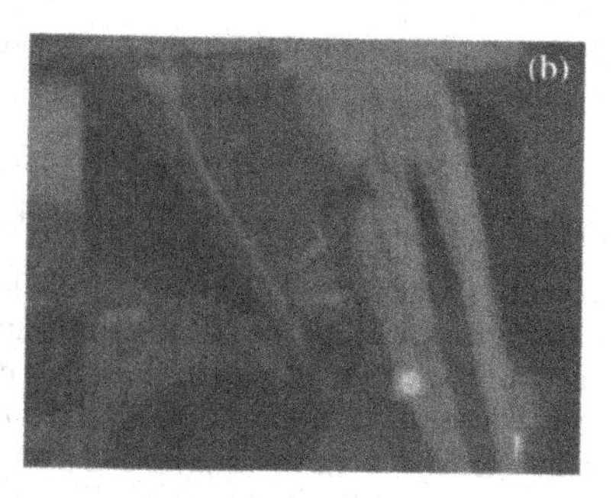

Fig.1 A cross sectional view of a flexible OLED with a top emission architecture on Al-PET foil (a) and a photograph of such a flexible OLED operating in a continuous bending test (b).

The EL spectra measured for a set of flexible microcavity OLEDs with different EL layer thicknesses and a conventional non-cavity top-emitting OLED are shown in Fig.2. The EL peak position of the OLEDs, with a Ph-PPV thickness varied from 80 to 150 nm, exhibits a clear red shift in the wavelength from 530 nm to 610 nm, showing an optical microcavity effect. The photo images taken for microcavity OLEDs and a non-cavity OLED are also illustrated on the top of the corresponding EL curves in Fig.2. The device with such a microcavity structure can be used for color tuning and efficiency enhancement. It is clear, as seen in Fig.2, that the full width at half maximum (FWHM) of EL peak for a non-cavity OLED was 137nm, The values of FWHM obtained for the microcavity OLEDs with emitting layer thickness of 80 nm, 110 nm, and 150 nm were 120nm, 77nm and 25nm, respectively. These observations are attributed to the optical microcavity effect. It is well known that the emission from the Fabry-Perot cavity is determined by the resonance modes of the cavity, and the spectral position of the cavity modes can determined by the optical thickness of the cavity:

$$L = k(\lambda_k / 2) \tag{1}$$

where k=1, 2, 3... is the mode index, L is the optical thickness of the cavity, and λ_k is the mode wavelength of the cavity.

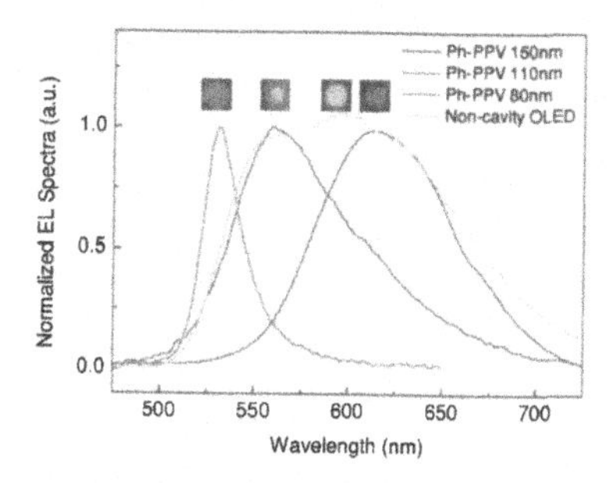

Fig.2 EL spectra measured for a set of structurally identical devices with different emissive layer thicknesses and a conventional non-cavity OLED, the inserted color photos are the corresponding photo images taken for the devices.

In this case, the optical thickness of the cavity can be calculated, taking into account a substantial penetration depth into the semitransparent mirror, by

$$L = \frac{\lambda_v}{2}\left(\frac{n_{eff}}{\Delta n}\right) + \sum_i n_i d_i + \left|\frac{\Phi_m}{4\pi}\lambda_v\right| \quad (2)$$

the first term is the effective penetration depth in the semitransparent mirror layer, where λ_v is the vacuum wavelength, n_{eff} is the effective refractive index of the semitransparent mirror, Δn is the difference between the indices of the materials of the such layer, n_i and d_i are the refractive index and the thickness of organic layer. The last term is the optical thickness contributed by the phase shift at the interface of the metal layer and the Ph-PPV layer, and Φ_m is the phase shift at the interface, depending on the refractive indices of the metal and the Ph-PPV layer at the interfaces:

$$\Phi_m = \arctan\left(\frac{2n_m k_m}{n_s^2 - n_m^2 - k_m^2}\right) \quad (3)$$

where n_s is the refractive index of Ph-PPV in contact with the metal, and n_m, k_m are the real and imaginary parts of the refractive index of the metal.

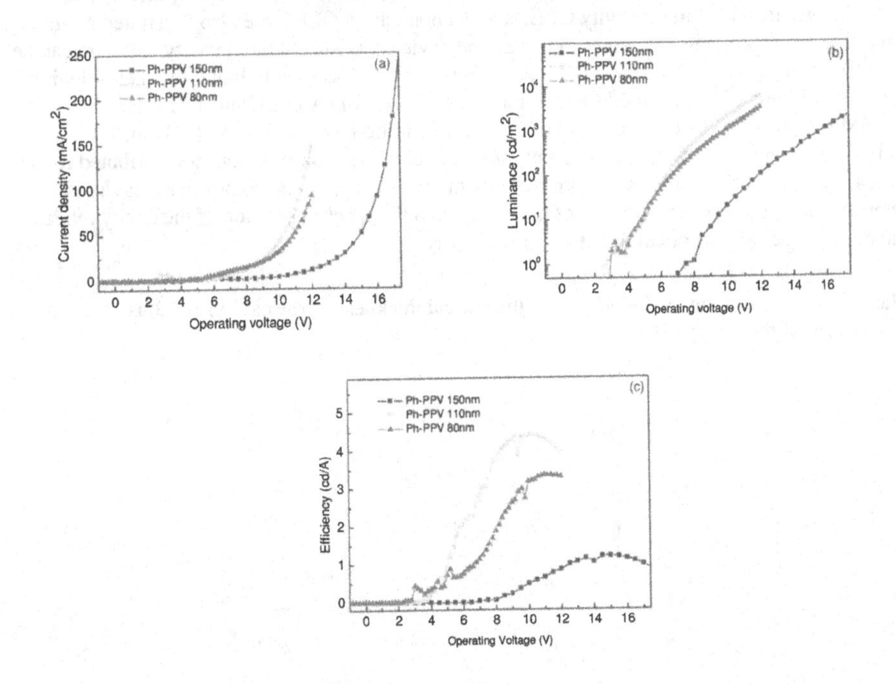

Fig.3 Characteristics of (a) current density vs the operating voltage, (b) luminance vs operating voltage and (c) efficiency vs voltage of the flexible OLEDs with different Ph-PPV thickness.

The current density-voltage, luminance-voltage and efficiency-voltage characteristics of the devices with different Ph-PPV thickness are shown in Figs. 3 (a), (b) and (c), respectively. The turn-on voltage for the devices with Ph-PPV thickness of 80 and 110nm is around 2.5V, it is increased to ~7.5V when a thicker Ph-PPV layer of 150nm was used in the device with the identical configuration. This is because the presence of the thicker polymer makes the whole device more resistive as hence a higher driving voltage is required. The luminance of $6000 cd/m^2$ is obtained at voltage of 12V for the OLED with Ph-PPV thickness of 110nm. It also can be seen from Fig. 3 that the EL efficiency of the devices is varied quite substantially with different Ph-PPV thicknesses used in the devices. The maximum EL efficiency of 4.56cd/A was obtained for OLED with a Ph-PPV layer thickness of 110 nm at the operating voltage of 10V. The EL efficiency measured for the devices with Ph-PPV thickness of 80nm and 150nm is 3.4cd/A and 1.2cd/A, respectively.

Present OLED technologies are focused on rigid substrates but flexible devices are fast gaining attention due to its lightweight, low cost and physical flexibility. To date, the research efforts surrounding flexible OLEDs have been centered on fabricating OLED on transparent flexible plastic substrates. We are currently extending our work to include design and fabrication of a flexible OLED using a top emission OLED architecture. The flexible substrate consists of a plastic layer laminated to or coated with a metal layer. This substrate has the potential to meet permeability standards in excess of the most demanding display requirements. The robustness of this substrate is also very high. This technology may provide a cost-effective approach for mass production, such as roll-to-roll processing, which is a widely used industrial process.

CONCLUSION

In summary, a flexible ITO-free OLED on Al-PET substrate is demonstrated. When a top emitting OLED is formed on a metal surface of a flexible substrate, the metal surface can serve as a part of the anode for the top emitting OLED as well as a barrier to minimize oxygen and moisture permeation. The structure of our flexible OLED consists of 1) an opaque flexible substrate, which can be a metal (Al)-laminated plastic (PET) or a metal film sandwiched between two plastic foils, 2) a planarisation/isolation layer, 3) bilayer Ag/CF_X anode, 4) an active light-emitting layer, and 5) an upper semitransparent cathode. A microcavity structure was formed between a modified Ag anode and a semitransparent cathode. The performance of the flexible OLEDs can be modified by selecting an appropriate emissive Ph-PPV layer thickness.

REFERENCES:

1. C.W. Tang, and S.A. Vanslyke, Appl. Phys. Lett. **51** (12), 913 (1987).
2. L.S. Hung, and C.H. Chen, Mater. Sci. Engi. R **39** 143 (2002).
3. J.H. Burroughes, D.D.C. Bradley, A. R. Brown, R.N. Marks, K. Mackay, R. H. Friend, P.L. Burn, and A.B. Holmes, Nature **347**, 539 (1990).
4. R. H. Friend, R.W.Gymer, A.B. Holmes, J.H. Burroughes, R.N. Marks, C. Taliani, D.D.C. Bradley, D.A. Dos Santos, J.L. Bredas, M. Logdlund, and W.R. Salaneck, Nature **397**, 121 (1999).
5. M.H. Lu, M.S. Weaver, T.X. Zhu, M. Rothman, R.C. Kwong, and J.J. Brown, Appl. Phys. Lett. **81**, 3921 (2002).

6. S.L. Lai, M.Y. Chan, M.K. Fung, C.S. Hung, and S.T. Lee, Chem. Phys. Lett. 366, 128 (2002).
7. P.E. Burrows, G. Gu, S. R. Forrest, E.P. Vicenzi, and T.X. Zhou, J. Appl. Phys. 87, 3080 (2000).
8. A.N. Krasnov, Appl. Phys. Lett. 80, 3853 (2002).
9. A. Plichta, A. Weber, and A. Habeck, Mater. Res. Soc. Symp. Proc. 769, Warrendale, PA, 2003), paper H9.1.
10. K.S. Ong, J.Q. Hu, R. Shrestha, F.R. Zhu, and S.J. Chua, Thin Solid Films, in press (2004).
11. G. Gu, P. E. Burrows, S. Venkatesh, and S. R. Forrest, Opt. Lett. 22, 172 (1997).
12. G. Gustafsson, G. M. Treacy, Y. Cao, F. Klavetter, N. Colaneri and A. J. Heeger, Synth. Met. 57, 4123 (1993).
13. A.B. Chwang, M.R. Rothman, S.Y. Mao, R.H. Hewitt, M.S. Weaver, J.A. Silvermail, K. Rajan, M. Hack, J.J. Brown, X. Chu, L. Moro, T. Krajewski, and N. Rutherford, Appl. Phys. Lett. 83, 413 (2003).
14. A. Sugimoto, H. Ochi, S. Fujimura, A. Yoshida, T. Miyadera, and M. Tsuchida, IEEE J. Sel. Top Quant. 10(1), 107 (2004).
15. Shizuo Tokito, Tetsuo Tsutsui, and Yasunori Taga, J. Appl. Phys. 86(5), 2407(1999).
16. V. Bulovic, V.B. Khalfin, G. Gu, P.E. Burrows, D.Z. Garbuzov, S.R. and Forrest, Phys. Rev. B, 58(7), 3730(1998).
17. A.B. Djurisic, and A.D. Rakic, Appl. Optics. 41(36), 7650 (2002).
18. Y.Q. Li, J.X. Tang, Z.Y. Xie, L.S. Hung, S.S. Lau, Chem. Phys. Lett. 386 (1-3): 128(2004).

Organic Photonic Bandgap Structures

Mater. Res. Soc. Symp. Proc. Vol. 846 © 2005 Materials Research Society DD12.2

Complex, 3D Photonic Crystals Fabricated by Atomic Layer Deposition

J. S. King, D. Gaillot, T. Yamashita, C. Neff, E. Graugnard, and C. J. Summers

School of Materials Science & Engineering, Georgia Institute of Technology
771 Ferst Drive, Atlanta, GA 30332-0245, USA
Phone: (404) 894 – 8414, FAX: (404) 894 – 9140

Abstract

Recently we have demonstrated the potential of Atomic Layer Deposition (ALD) for the fabrication of advanced luminescent photonic crystal (PC) structures based on the inverse opal architecture.[1-3] PC's offer efficiency enhancement, decreased threshold, and other enhancements that improve phosphor performance. 3D PC structures are being extensively modeled, revealing that changes in the structures such as shifting the distribution of dielectric material can significantly improve photonic band gap (PBG) properties. For example, in the inverted "shell" structure, the width of the PBG can be increased from 4.25% to 8.6%.[4] Similarly, the PBG width can also be increased to 9.6% by formation of a non-close-packed structure.[5] Using the FDTD method, we have found that the PBG in a TiO_2 non-closed packed structure can be as high as 5%. The performance of these structures depends critically on precisely and accurately placed high dielectric material. Using ALD, we have demonstrated infiltration of TiO_2 films with extremely smooth surfaces (0.2- 0.4 nm RMS roughness) while maintaining a high level of control over the infiltration coating thickness, enabling formation of composite infiltrated and inverse opals with nano-scale precision.

Here we report progress in fabrication of multi-layered and non-close packed PCs using ALD. Two and three-layer inverse opals were formed by the deposition of thin layers of ZnS:Mn and TiO_2 in stacked configuration, each exhibiting luminescence when excited by UV light. Evidence for modification of the emission characteristics by high order PBGs (gaps other than between the 2nd and 3rd bands) has been observed. In addition, non-close packed inverse opals have been formed by infiltrating heavily sintered silica opals with TiO_2, etching the spheres with hydrofluoric acid, and backfilling the resulting inverse opal. Resulting structures were characterized using specular reflectance and transmission, photoluminescence, and SEM. This work demonstrates the enormous potential that ALD offers for the realization of high performance photonic crystal structures.

Introduction

Two & three-dimensional photonic crystal (PC) structures have the potential for controlling the emission wavelength, luminosity, efficiency, time response and threshold properties of phosphor materials, and are therefore attractive for luminescent structures. PC's based on the infiltration of synthetic opals have been established as promising structures for obtaining the required periodicity.[6-9] For an inverse opal with sufficient refractive index contrast (> 2.8), a complete PBG forms between high order photonic bands.[10] However, the requirements to achieve a PBG for visible wavelengths in 3D

PCs are beyond the limit of current material properties, so it is necessary to employ new material structures for this purpose. Multi-layered and non-close-packed architectures are two routes that have high potential for advanced photonic crystal structures.[5,11,12] Multi-layered structures allow for increased functionality and flexibility in materials selection. For example, luminescent multi-layered inverse opals offer the potential for high index structures while retaining luminescent characteristics. Non-close-packed inverse opals increase photonic band gap widths as well as potentially decrease minimum index requirements for attaining a full PBG.

Recently, we reported successful formation of ZnS:Mn and TiO_2 inverse opals using Atomic Layer Deposition (ALD).[2,3] When doped, ZnS is highly luminescent with a wide range of emission wavelengths. However, its refractive index is not high enough (~2.5) for the formation of a full photonic band gap, but is sufficient to produce a pseudo-photonic band gap. TiO_2 has a higher refractive index, approaching 3.0 in the blue region.[13] In this study, we report results of the formation of ZnS:Mn and TiO_2 multi-layered inverse opal films fabricated using ALD. Evidence for modification of the emission characteristics by high order PBGs (gaps other than between the 2nd and 3rd bands) has been observed. In addition, non-close packed inverse opals have been formed by infiltrating heavily sintered silica opals with TiO_2, etching the spheres with hydrofluoric acid, and backfilling the resulting inverse opal using ALD.

Experimental details

Silica opal films were grown on silicon substrates in a manner similar to that of Park, *et al*, as described elsewhere.[2,14] For multi-layered inverse opals, the interstitial volume of the opal was next filled with ZnS:Mn or TiO_2 by ALD. $ZnCl_2$, H_2S, and $MnCl_2$ precursors were used for ZnS:Mn growth, and $TiCl_4$ and H_2O were used for TiO_2.[1-3] The top surface of the opal was next removed using an ion mill to expose the silica spheres. Etching the infiltrated films in a 2% HF solution resulted in the removal of the silica spheres, and the formation of structurally stable multi-layered inverse opals. The films were characterized using specular reflectivity and Scanning Electron Microscopy (SEM). Photoluminescence (PL) was measured using pulsed N_2 laser excitation (337 nm).

Non-close-packed inverse opals were formed by first sintering the opal films at 1000° C for 3 hours prior to ALD infiltration. This temperature is higher than is usually used for close-packed inverse opals (800 °C). The higher temperature yields an increase in the neck size between the spheres that comprise the opal, and a subsequent reduction in available air volume for infiltration. The increased sphere contact point diameter facilitates significant backfilling of the resulting inverse opal. The backfilling step results in the formation of cylinder-like connectivity between the air spheres, yielding a non-close-packed inverse opal. The resulting structures were characterized using SEM and specular reflectivity.

Results

For structural characterizations, the resulting films were examined in a scanning electron microscope (SEM). Figure 1 shows a SEM image of a cross-section cut through a (111) plane of a TiO_2/ZnS:Mn/TiO_2/air (24/10/10 nm) inverse opal, after ion milling.

The original sphere diameter for this structure was 433 nm. The ZnS:Mn and TiO_2 layers are labeled in the image. The layers are highly conformal and spatially distinct, and a multi-layer inverse opal has been successfully formed.

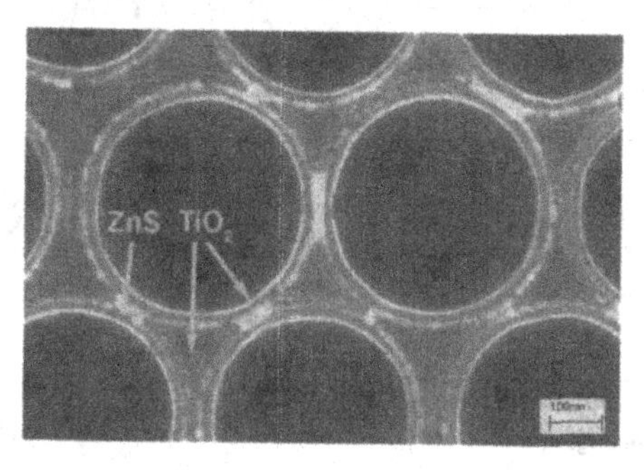

Figure. 1: Ion milled cross section of the (111) plane of a 433 nm (sphere diameter) TiO_2/ZnS:Mn/TiO_2/air (24/10/10 nm) layered inverse opal photonic crystal.

During fabrication, each processing step was followed by examining the shift of the primary Γ-L PPBG peak found in specular reflectance as a function of wavelength. This peak corresponds to the PPBG that forms between the 2nd and 3rd photonic bands. The PPBG should widen as the refractive index contrast increases. Figure 2 is a plot of the reflection spectra measured at 15° from normal incidence during the formation of a three layer inverse opal using a 433 nm SiO_2 opal template. The peaks were centered at 947 nm for the sintered opal (curve a), and shifted to beyond 1100 nm (beyond the range of the spectrometer used) after infiltrating with 15 nm of ZnS:Mn (curve b) and with a second layer consisting of 19 nm of TiO_2 (curve c). After removing the SiO_2 spheres, forming a 2 layer inverse opal, the peak shifted to 1018 nm (curve d), due to the resulting decrease in average refractive index. Backfilling the air voids with 2 nm of TiO_2 shifted the Γ-L PPBG peak to 1031 nm (curve e). An additional 8 nm infiltration yielded a shift in the peak to 1100 nm. The broad reflectivity peaks for the 2 and 3 layer inverse opals (curves d,e, and f) at ~ 570 - 700 nm are attributed to high-order Γ-L PPBGs that form between the 5th and 6th and the 8th and 9th bands in an inverse opal.

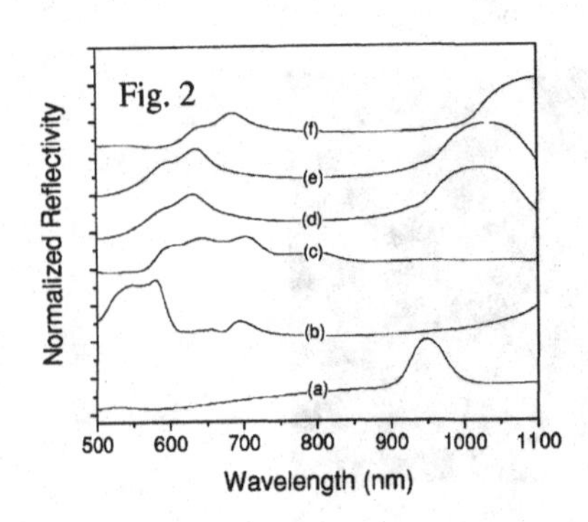

Figure 2. Reflectivity spectra for a 433 nm (a) sintered SiO_2 opal, (b) ZnS:Mn (15 nm) infiltrated opal, (c) ZnS:Mn/TiO_2 (15/19 nm) infiltrated SiO_2 opal, (d) ZnS:Mn/TiO_2 inverse opal, (e) TiO_2/ZnS:Mn/TiO_2 (2/15/19 nm) inverse opal, and (f) TiO_2/ZnS:Mn/TiO_2 (10 nm/15/19 nm) inverse opal. (15° from normal)

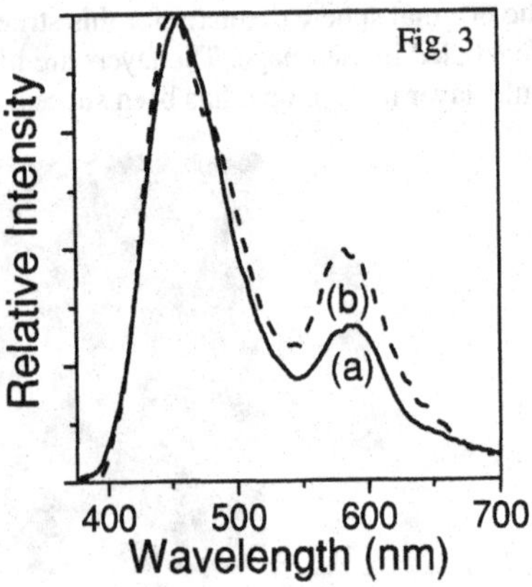

Figure 3. Photoluminescence data for 433 nm (a) TiO_2/ZnS:Mn/TiO_2/Air (2/19/15 nm) inverse opal and (b) TiO_2/ZnS:Mn/TiO_2/Air (19/15/10 nm) inverse opal. (337 nm excitation)

Figure 3 shows PL measured for the three layer inverse opals formed by backfilling with 2 and 10 nm thick TiO_2 layers. Peaks at both 460 and 585 were observed for both, consistent with Cl^- defect and Mn^{2+} emission, respectively. An increase in the 585 nm peak relative intensity was observed after the second backfilling. This change is attributed to the shifting of the high order PPBG (shown in figure 2) off of the 585 nm luminescence peak.

SEM images of the non-close-packed inverse opal are shown in Figure 4. The inverse opal that resulted from infiltration of a heavily sintered 433 nm opal is shown in Figure 4(a). The connecting channels between the air spheres are clearly very large, leaving a large amount of room for TiO_2 backfilling. After backfilling with 120 ALD cycles, the structure in Figure 4(b) resulted. Clearly the air spheres have reduced in diameter, and formed cylinder-like connections between each other, yielding a non-close-packed inverse opal.

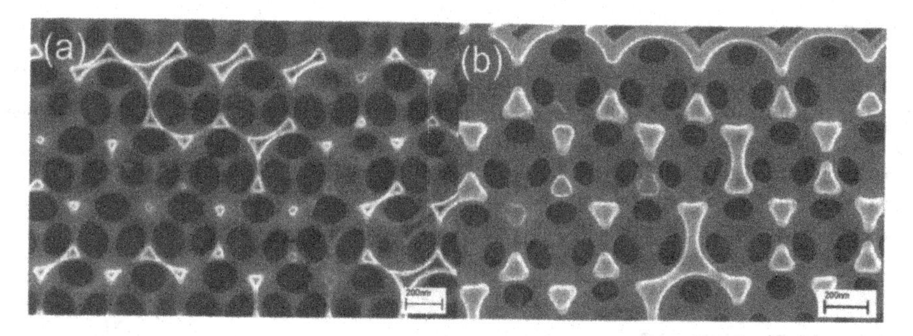

Figure 4. SEM images of ion milled cross sections of the (111) plane of (a) 433 nm sphere diameter TiO_2 inverse opal, formed from a heavily sintered SiO_2 opal and (b) non-close-packed inverse opal formed after 120 TiO_2 ALD backfilling cycles.

Specular reflectivity (15°) was measured during each processing step of the 433 nm non-close-packed inverse opal, as shown in Figure 5. Curve (a) is the reflectivity for the heavily sintered opal, which yielded a PPBG peak at 885 nm. After infiltration, the peak shifted to 955 nm, a much smaller shift than is typically observed for a close packed opal, due to the reduced amount of air space available for infilling. After etching, the peak shifted to 739 nm, a much greater shift than is usually observed in an inverse opal due to the reduced filling fraction of TiO_2. The infiltrated TiO_2 was enough to be structurally stable, even with the reduced filling fraction. After backfilling with 120 ALD cycles, a non-close-packed inverse opal was formed, yielding a shift of the PPBG peak to 850 nm. In addition to the 2/3 PPBG, the non-close packed inverse opal has a high order peak positioned at 500 nm.

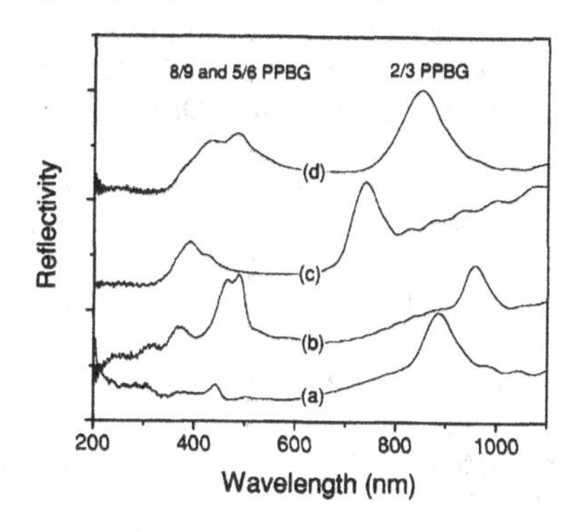

Figure 5. Reflectivity spectra for a 433 nm (a) heavily sintered SiO_2 opal (1000° C 3 hrs), (b) TiO_2 infiltrated SiO_2 opal, (c) inverse TiO_2 opal, and (d) TiO_2 backfilled (120 ALD cycles) inverse opal (non-close-packed inverse opal). (15° from normal)

CONCLUSIONS

We have reported precise tuning of photonic crystal optical properties using ALD of ZnS:Mn and TiO_2. Independent control of the luminescent and photonic band gap properties has been demonstrated through successful realization of multi-layered ZnS:Mn/TiO_2 inverse opals. In addition, non-close packed inverse opals have been formed by infiltrating heavily sintered silica opals with TiO_2, etching the spheres with hydrofluoric acid, and backfilling the resulting inverse opal using ALD. The data presented demonstrates that ALD uniquely enables the formation of complex photonic crystal structures.

ACKNOWLEDGEMENTS

The authors acknowledge support for this project from the U.S. Army Research Office under MURI contract # DAAA19-01-0603. JK acknowledges partial support by the Georgia Institute of Technology Molecular Design Institute, under prime contract N00014-95-1-1116 from the Office of Naval Research. They also thank S. Blomquist, E. Forsyth, and D. Morton of the U.S. Army Research Laboratory for their assistance and use of their facilities.

REFERENCES

1 J. S. King, C. W. Neff, S. Blomquist, E. Forsythe, D. Morton, and C. J. Summers, Phys. Stat. Sol. (b) **241** (3), 763 (2004).
2 J.S. King, C.W. Neff, C.J. Summers, W. Park, S. Blomquist, E. Forsythe, and D. Morton, App. Phys. Lett. **83** (13), 2566 (2003).
3 J.S. King, E. Graugnard, and C.J. Summers, Adv. Mat., Accepted for publication (2004).
4 K. Busch and S. John, Phys. Rev. E **58** (3), 3896 (1998).
5 M. Doosje, B.J. Hoenders, and J. Knoester, J. Opt. Soc. of Am. B **17** (4), 600 (2000).
6 A. Blanco, E. Chomski, S. Grabtchak, M. Ibisate, S. John, S.W. Leonard, C. Lopez, F. Meseguer, H. Miguez, J.P. Mondla, G.A. Ozin, O. Toader, and H.M. van Driel, Nature **405** (6785), 437 (2000).
7 Y.A. Vlasov, X.-Z. Bo, J.C. Sturm, and D.J. Norris, Nature **414** (6861), 289 (2001).
8 H.M. Yates, W.R. Flavell, M.E. Pemble, N.P. Johnson, S.G. Romanov, and C.M. Sotomayor-Torres, J. Cryst. Growth **170** (1-4), 611 (1997).
9 S.G. Romanov, N.P. Johnson, A.V. Fokin, V.Y. Butko, H.M. Yates, M.E. Pemble, and C.M. Sotomayor Torres, App. Phys. Lett. **70** (16), 2091 (1997).
10 H.S. Sozuer, J.W. Haus, and R. Inguva, Phys. Rev. B (Condensed Matter) **45** (24), 13962 (1992).
11 Florencio Garcia-Santamaria, Marta Ibisate, Isabelle Rodriguez, Francisco Meseguer, and Cefe Lopez, Adv. Mat. **15** (10), 788 (2003).
12 Roberto Fenollosa and Francisco Meseguer, Adv. Mat. **15** (15), 1282 (2003).
13 Edward D. Palik, Ghosh, Gorachand., *Handbook of optical constants of solids*. (Academic Press, San Diego, 1998).
14 S.H. Park, D. Qin, and Y. Xia, Adv. Mat. **10** (13), 1028 (1998).

Mater. Res. Soc. Symp. Proc. Vol. 846 © 2005 Materials Research Society DD12.3

Prototyping of three-dimensional photonic crystal structures using electron-beam lithography

G. Subramania, J.M. Rivera
Sandia National Laboratories,
P.O Box 5800, Albuquerque, New Mexico 87185

Abstract

We demonstrate the fabrication of a three-dimensional woodpile photonic crystal in the near-infrared regime using a layer-by-layer approach involving electron-beam lithography and spin-on-glass planarization. Using this approach we have shown that we can make structures with lattice spacings as small as 550 nm with silicon as well as gold thus allowing for fabrication of photonic crystals with omnidirectional gap in the visible and near-IR. As a proof of concept we performed optical reflectivity and transmission measurements on a silicon structure which reveal peaks and valleys expected for a photonic band gap structure. The approach described here can be scaled down to smaller lattice constants (down to ~400 nm) and can also be used with a variety of materials (dielectric and metallic) thus enabling rapid prototyping full three-dimensional photonic bandgap based photonic devices in the visible.

Introduction

Since the publication of the original paper by Eli Yablanovitch [1]and Sajeev John [2] in 1987 tremendous progress has been made in the area of photonic band gap structures. The unique properties of photonic band gap structures or photonic crystals as they are also called, arise from the creation of forbidden light propagation modes which allow for molding and confinement of light over small spatial regimes. These properties have enormous implications in the field of quantum optics and photonics particularly, for applications in integrated optics, optical computing and energy efficient lighting. Theoretical models predict that a three dimensionally periodic structure with a right geometry and material parameters can give rise to an omnidirectional electromagnetic forbidden gap which provides the best case scenario to harness the full potential of the photonic band gap. A large number of applications such as the ones mentioned above, require photonic crystals that operate in the near infrared (e.g. communications and optical computing) and visible (e.g. light emission) part of the electromagnetic spectrum. This implies that structures with submicrometer periodicity are required which makes fabrication of three dimensional structures of these kinds extremely challenging. Early successes in this direction were achieved using techniques such as state-of-the-art submicron lithography in the CMOS process [3, 4] and wafer fusion [5]. Technological demands required by these techniques make fabrication of these structures quite difficult hence the slower progress compared to the two dimensional photonic crystals. Alternative approaches have been taken by various groups to fabricate 3D photonic crystals over large area which include self-assembly[6, 13],three dimensional interference lithography [7], pore diameter modulation [8], autocloning [9], glancing angle deposition(GLAD) [10], microtransfer molding [11] and micromanipulation [12]. An extensive review of various techniques can be found in [13]. While these techniques are an important step forwards that address the crucial problem of large area fabrication of three dimensional photonic crystal fabrication there are still challenges associated with each of the above techniques that need to be resolved.

In this proceedings, we describe a rapid prototyping approach where by we can fabricate 3-D photonic crystal with submicron (<= 0.5 μm) periodicity in a layer by layer fashion with a high degree of accuracy using electron beam lithography. While, this is not a means for very large area fabrication as it could involve prohibitively long direct write times it certainly offers other advantages such as making small area devices (e.g. microcavities, emission sources, filters) whose properties can be experimentally verified. The main advantage of this technique is in the flexibility in the 3D geometries that can be fabricated as well as in the choice of materials.

Fabrication

In this report we decribe the fabrication of " Iowa State woodpile" (a.k.a lincoln-log) structures with lattice spacings as small as 550nm out of silicon as well as gold. A recent publication [16] describes a similar method by which alternating layers of 2D slab photonic crystals have formed to obtain a 3D photonic crystal demonstrating flexibility in the choice of geometry. There are three main challenges to fabricating three dimensional photonic crystals for short wavelength (a) accurately defining the feature sizes (b) alignment of the next layer to the previous with high precision (c) planarization of each layer. Electron beam (e-beam) direct write systems provides the required capabilities to tackle the first two issue. To solve the problem of planarization we use Honeywell's T-12B spin on glass. We used this technique to form a lincoln log structure, a geometry that can give omnidirectional photonic gap [14] using two materials one with silicon the other gold. The motivation to use silicon was to be able to perform optical characterization and compare the results with earlier reports in order to check the proof of concept for this approach.

The procedure for fabricating the structure is a follows. We typically used a 1 inch square substrate of gallium arsenide or silicon. First step is to define a set of alignment marks corresponding to each chip. The alignment mark is typically a 4μm wide cross-hair pattern of ~50nm thick gold (fig 1). The e-beam system (JEOL JBX-5FE) scans the two arms of the cross hair to accurately determine the center which forms the coordinate origin from which patterns are referenced. Therefore it is important that the edges of the mark be very sharp. The next step involves putting down some E-beam resist (polymethylmethacrylate) on the sample substrate followed by a direct write of the first layer patterns referenced to the appropriate chip mark. The next step involves electron beam evaporation of silicon or gold as the case may be.

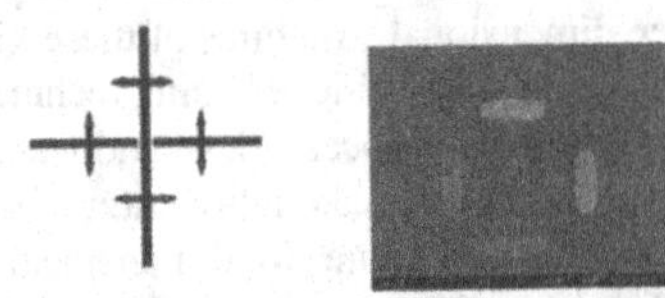

Figure 1. Alignment mark : Left shows schematic of e-beam scan of the arms, right shows an SEM image of gold alignment marks and the actual scan marks.

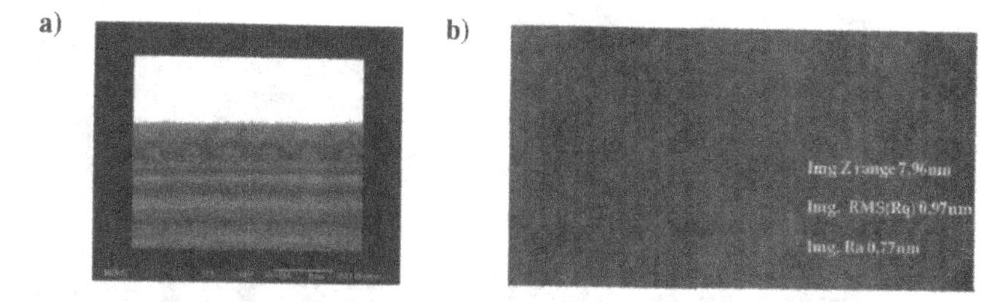

Figure 2. a) SEM image of showing excellent filling of the inter-rod spacing. b) AFM image of the surface over the device region after spin on glass planarization.

Electron beam evaporation does not always give the densest film particularly in the case of silicon and may require some sort of post process annealing. We chose this technique for its simplicity so that we can remove the excess material by a lift off procedure by soaking the sample in acetone. However, a deposition and etch approach can also be used which adds some additional steps. Next is the planarization step. We spin on Honeywell's T-12B spin on glass which excellently fills in the gaps between the silicon/gold lines (figure 2a). An atomic force microscopy measurement reveals a planarization of 1nm root mean square waviness (figure 2b). The excess spin on glass is removed by etching in a CHF3 plasma to expose the top of the first layer. This step has to be carefully timed and stopped to prevent over etching otherwise, this can destroy the planarization achieved earlier. Then we repeat the steps to make the required number of layers. Using the above procedure we have fabricated 5 layers of silicon woodpile photonic structure with lattice spacing as small as 550nm (figure 3a,b).

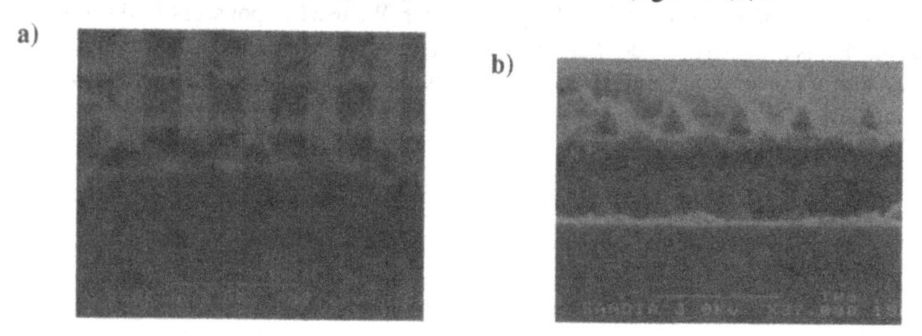

Figure 3. a) Shows angled view 5layers of silicon rod woodpile photonic crystal with lattice spacing of 550nm. **b)** shows cross section image demonstrating a very good alignment.

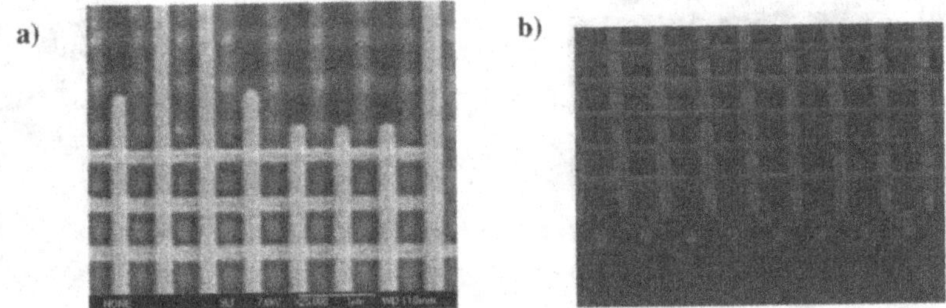

Figure 4 a) SEM image of the top of a gold photonic crystal where the underlying layers are seen quite clearly due to strong electron scatter from gold. The image shows a very good alignment. **b)** sectional view showing the cross-section.

The evaporation followed by a lift-off approach also easily allows us to fabricate woodpile photonic structure with gold (figure 4a,b). Three-dimensionally periodic metallic structures are quite important of wideband filters with a tunable cut-off edge as well as in light emission.

Results

We fabricated silicon "woodpile"structure with three different lattice spacing of 660nm, 600nm and 550nm for optical characterization to compare with results from previous reports fabricated using standard CMOS processing[4]. The rod width was kept around 200nm in all cases and the height was ~200nm. We performed reflectance and transmission spectroscopy on the devices silicon photonic crystal using a microscope accessory connected to a Nicolet FTIR spectrometer. We used a spot size of 60μm X 60μm which is a slightly smaller than the device size of 80 micrometer square. Figure 5 shows optical transmission and reflectance spectrum from a 4 layer , 660nm lattice spacing device.

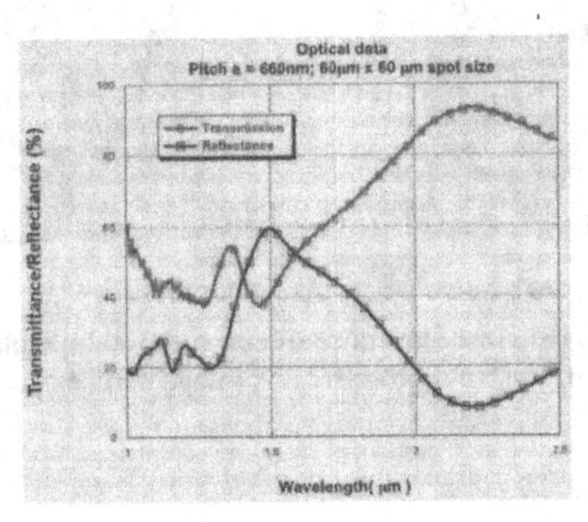

Figure 5. Transmission(circle) and reflectance(square) spectra from a 4 layer 660nm lattice pitch silicon "woodpile" device.

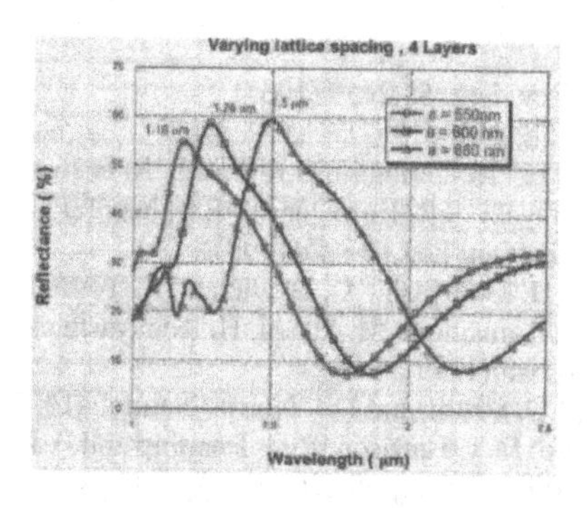

Figure 6. Reflectance spectra from 4 layer device with lattice spacings of 660nm, 600nm and 550nm. The peak position shows a systematic shift towards shorter wavelengths for smaller lattice constant.

Figure 6 shows reflectivity spectra from Si structures with lattice spacing of 550nm, 600nm and 660nm with the peaks moving towards shorter wavelength for smaller lattice constant consistent as expected. The peak position is in close agreement with that reported earlier for similar structures [15]. However the peak magnitude of 60% is somewhat smaller than that for the earlier reported value of ~85%. One possible reason is the density of evaporated silicon being lower than that of CVD silicon. Second is the presence of the spin on glass background that was not removed for these measurements. The third contribution can be from the trapezoidal profile of the silicon rods resulting from the lift-off techniques.

Summary

In summary, we demonstrate a technique of rapid prototyping three dimensionally periodic photonic crystal for near-IR and visible wavelengths using a combination of electron beam lithography and spin on glass planarization in a layer-by-layer fashion. As an example, we have fabricated five layers of silicon photonic crystal ("Iowa State woodpile") with lattice spacing as small as 550nm demonstrating the capabilities of this technique to make 3D photonic crystals for the visible regime. To show the versatility of this approach for other materials we have also demonstrated fabrication of 4 layers of gold photonic crystal. Metallic photonic crystals are good candidates for photonic crystal light emitters and broadband filters. By using this approach we plan to develop prototype photonic crystal devices in the visible and near-IR regime that are based on omni directional photonic band gap.

Acknowledgments

The research at Sandia National Laboratories is supported by U.S. Department of Energy. Sandia is a multiprogram laboratory operated by Sandia Corporation, a Lockheed Martin Company, for the U.S. Department of Energy's National Nuclear Security Administration under contract DE-AC04-94AL 85000. The authors would like to acknowledge Karen Cross for the atomic force microscopy (AFM) measurements.

References

1. E. Yablanovitch, Phys. Rev. Lett. **58**,2059 (1987)
2. S. John Phys. Rev. Lett. **58**, 2486 (1987).
3. S.Y. Lin , J.G. Fleming,D.L. Hetherington,B.K. Smith, R. Biswas,K.M. Ho, M.M. Sigalas, W. Zubrzycki, S.R. Kurtz, and J. Bur, Nature(London) **394**,251(1998)
4. J.G. Fleming and S.Y. Lin, Opt. Lett., **24**,49(1999).
5. S. Noda, N. Yamamoto, H. Kobayashi, M. Okano and K. Tomoda, Appl. Phys. Lett. **75**, 904(1999); S. Noda, N. Yamamoto, M. Imada, H. Kobayashi, M. Okano, J. Lightwave Technology,**17**,1949(1999).
6. Y.A. Vlasov, X-Z. Bo, J.G. Strum, and D.J. Norris, Nature, **414**, 289(2001).
7. M. Campbell, D.N. Sharp, M.T. Harrison, R.G. Denning and A.J. Tuberfeld, Nature(London) **404**, 53(2000); S. Shoji, S. Kawata, Appl.Phys. Lett. **76**, 2668(2000); X. Wang, J.F. Xu, H.M. Su, Z.H. Zang, Y.L. Chen,H.Z. Wang, Y.K. Pang and W.Y. Tam, Appl. Phys. Lett. **82**, 2212(2003); A. Fiegel, M. Veinger, B. Sfez, A. Arsh, M. Klebenov and V. Lyubin, Appl. Phys. Lett. **83**, 4480(2003).
8.J. Schilling, F. Mueller, S. Matthias, R.B. Wehrspohn, U. Goesle, K. Busch, Appl. Phys. Lett. **78**, 1180(2001).
9. T. Kawashima, T. Sato, Y. Ohtera, and S. Kawakami, IEEE J. Quan. Elec.**38**, 899(2002).
10. S.R. Kennedy, M.J. Brett, O. Toader, S. John, Nano Letters **2**, 59(2002).
11. W.Y. Leung, H. Kang, K. Constant, D. Cann, R. Biswas, M.M. Sigalas, K.M. Ho, J. Appl. Phys. **93**, 5866(2003).
12. K. Aoki, H. Miyezaki, H. Hireyama, K. Inoshita, T. Baba, N. Shinya, Y. Aoyagi, Appl. Phys. Lett. **81**, 3122(2002).
13. A.F. Koenderink, P.M. Johnson, J.F.G. Lopez, W.L. Vos, Competus Rendus de l'Academie des Sciences, **3**, 67(2002).
14. K.M. Ho, C.T. Chan, C.M. Soukoulis, R. Biswas and M. Sigalas, Solid State Commmun. **89**, 413(1994).
15. S.Y. Lin and J.G. Fleming, J. Lightwave Technology, **17**, 1944(1999).
16. M. Qi, E. Lidorikis, P.T. Rakich, S.G. Johnson, J.D. Joannopoulos, E.P. Ippen and H. Smith **429**, 538(2004).

Mater. Res. Soc. Symp. Proc. Vol. 846 © 2005 Materials Research Society DD12.9

Two-dimensional Magneto-photonic Crystal Circulators

Zheng Wang[1], Shanhui Fan
Department of Electrical Engineering,
Stanford University, Stanford, CA 94305
[1]Department of Applied Physics,
Stanford University, Stanford, CA 94305-4090

ABSTRACT

Previous research has demonstrated enhanced Faraday rotation in one-dimensional magnetic photonic crystals, where nonreciprocity in magneto-optical cavities is resonantly enhanced to provide optical isolation in optical paths on the scale of a few microns. In this paper, we study the nonreciprocity of two-dimensional magnetic photonic crystal resonators to allow further miniaturization and monolithic in-plane integration with current integrated optical devices. The nonreciprocal magnetic resonators are constructed by alternating the magnetization directions of the ferromagnetic domains in cavities side-coupled to photonic crystal waveguides. We show analytically that the gyrotropic splitting and the strength of the magnetic hybridization of the cavity modes are determined by the overlap integral between the domain magnetization vector and the modal cross product. With a large overlap obtained from optimizing the domain structures, we circularly hybridize two nearly degenerate modes to form a pair of counter-rotating whispering-gallery like modes, oscillating at different frequencies. As a physical realization, we synthesize two singly-degenerate circularly-hybridized modes in a two-dimensional crystal formed of a triangular air hole lattice in bismuth iron garnet with a TE bandgap. We tune the magnetic splitting and the decay constants of the rotating modes to demonstrate numerically a three-port optical circulator with a 30dB extinction bandwidth of 35GHz at 1550nm. An alternative implementation of a four-port circulator is achieved by side-coupling a point defect to two parallel waveguides. Our numerical experiments are performed with finite-difference time-domain simulations and agree well with the analytical coupled-mode theory predictions.

INTRODUCTION

Integrated non-reciprocal optical devices, such as optical isolators and circulators, can play very important roles in densely integrated optical circuits. Besides their usual application in eliminating the detrimental reflections to laser sources, integrated non-reciprocal devices also simplify the design of large scale optical circuits by suppressing multiple reflections between components, and thereby improving tolerance with respect to fabrication imperfections and environmental fluctuations.

Enhanced Faraday rotation in one-dimensional magneto-optical photonic crystal (PC) cavities has attracted much research attention recently [1]. Here we explore nonreciprocal resonators in two-dimensional photonic crystals. Such resonators are of great advantages in on-chip circuits, due to their strong field confinement and fabrication compatibility. We numerically demonstrate a broadband three-port circulator in two-dimensional PCs, where the domain structures of the ferromagnetic material in the resonator are spatially engineered to produce the strong non-reciprocal effects.

THEORY OF MAGNETO-OPTICAL HYBRIDIZATION IN PHOTONIC CRYSTAL

In the presence of gyrotropic medium, the eigenmodes of a photonic crystal defect can show drastically different properties from regular defect states. For simplicity, we start with a non-magnetic cavity supporting nearly-degenerate TE modes in a 2D PC consisting of a triangular array of air holes with a radius of $0.35a$ in dielectric, where a is the lattice constant. The ferromagnetic dielectric bismuth iron garnet (BIG) has a dielectric constant of 6.25 and is magnetized along out-of-plane directions [3]. The modes are classified as an even mode $|e\rangle$ (Figure 1.a) and an odd mode $|o\rangle$ (Figure 1.b), with respect to the mirror plane. The gyrotropic effect of the ferromagnetic material in the vicinity of the cavity is taken into account by treating the off-diagonal elements in the complex permittivity and permeability tensors as perturbations [2]. For TE modes, the gyromagnetic effect vanishes since the magnetic field is a scalar field. The effect of the gyrotropic medium is therefore entirely due to the off-diagonal elements of the dielectric tensor. In the presence of gyrotrophic medium, The strength of the magneto-coupling between the two nearly degenerate TE modes can be derived as

$$V_{ij} = \frac{i}{2}\sqrt{\omega_i\omega_j}\int \hat{\mathbf{z}}\cdot\left(\varepsilon_a\vec{E}_i^*\times\vec{E}_j + \mu_a\vec{H}_i^*\times\vec{H}_j\right)dV = \frac{i}{2}\sqrt{\omega_i\omega_j}\int \hat{\mathbf{z}}\cdot\varepsilon_a\vec{E}_i^*\times\vec{E}_j dV\,, \qquad (1)$$

where ω_i, $\vec{E}_i$ and $\vec{H}_i$ are the frequency, electrical field and magnetic field respectively for the i-th mode. ε_a and μ_a are the imaginary part of the off-diagonal elements in the permittivity and permeability tensors respectively. Their signs are determined by the direction of the local

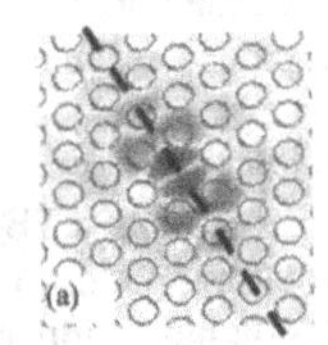

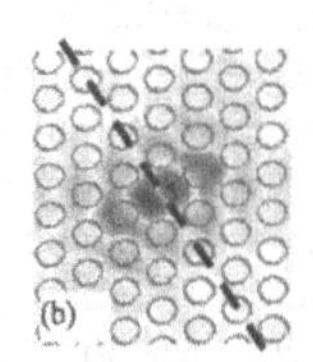

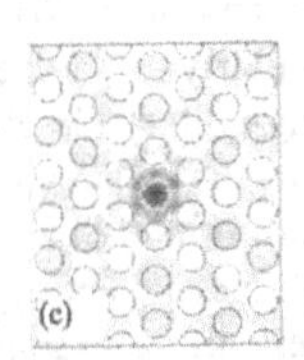

Figure 1 (a) and (b) The out-of-plane H component of two doubly-degenerate cavity modes in a point defect formed of a missing air hole in a photonic crystal with TE bandgap. The mirror planes are marked with dark dash lines. (c) The spatial pattern of E field cross-product of mode a and b. The blue and red colors stand for large positive and negative values.

magnetization. The modal coupling strength is thus governed by the overlap between the cross-product of the E field of the two modes, as shown in Figure 1.c, with the magnetization vector of the material in the cavity. Since the modal cross product changes sign rapidly in the cavity, when a single magnetic domain is used, the coupling strength is almost zero. To accomplish a large magneto-optical coupling, it is necessary to properly align the magnetization of the magnetic domain with the modal cross product (Figure 2.a), to obtain a large coupling between the two modes.

In the strong magneto-coupling limit $|V_{ij}| >> |\omega_e - \omega_o|$, the new eigenmodes in the system become two circularly hybridized modes $|e\rangle \pm i|o\rangle$ at split frequencies $\omega_{oe} \pm V_{ij}$ respectively. These eigenmodes are superposition of the two original with a ±90° phase lag in between, and spin toward opposite directions in time at different resonant frequencies. Hence they exhibit non-reciprocity and broken time-reversal symmetry.

THREE-PORT JUNCTION CIRCULATORS

Using non-reciprocal magneto-PC cavities in place of regular PC defects one can create devices with nonreciprocal transport characteristics. For concreteness, let us consider a Y-junction circulator formed of a magneto-PC cavity described above coupled to three waveguides

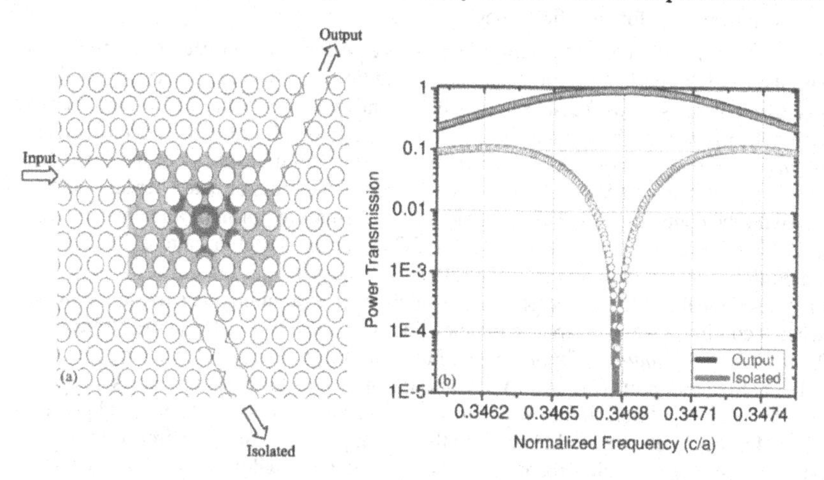

Figure 2 (a) A three-port junction isolator constructed as a point defect side-coupled to three waveguides. Black circles represents the air holes with r=0.35a in bismuth-iron-garnet (ε=6.25). The light and dark gray areas in the PC cavity illustrate the magnetic domains with opposite magnetization direction (both being out-of-plane) with ε_a=0.02 (b) Transmission spectra at the output port and the isolated port. The finite-difference time-domain spectra (dots) agree well with the simple coupled-mode theory analysis (solid curves).

from sides, as illustrated in Figure 2.a. The arrangement of the waveguides is chosen to preserve the three fold rotational symmetry and obtain equal coupling between the two dipole modes to any waveguide in the absence of gyrotropy. The degeneracy of the two modes is maintained in terms of both resonant frequency and decay constant. The three waveguides are constructed by enlarging the radius of a row of air holes to 0.55*a*, where *a* is the lattice constant, to support a single mode at the mid-gap frequencies.

With gyrotropic media in the cavity, the eigenmodes are two counter-rotating modes hybridized from the two dipole modes mentioned before. With temporal coupled-mode theory [4,5], the power transmission coefficient at the output port and the isolated port (Figure 2.a) are found as:

$$T_{output} = \left| \frac{2}{3} \left(\frac{\exp(j2\pi/3)}{1 + j(\omega - \omega_+)\gamma_+} + \frac{\exp(j4\pi/3)}{1 + j(\omega - \omega_-)\gamma_-} \right) \right|^2$$
$$T_{isolated} = \left| \frac{2}{3} \left(\frac{\exp(j4\pi/3)}{1 + j(\omega - \omega_+)\gamma_+} + \frac{\exp(j2\pi/3)}{1 + j(\omega - \omega_-)\gamma_-} \right) \right|^2, \tag{2}$$

where γ_+ and γ_- are the decay constants of mode $|e\rangle + i|o\rangle$ and $|e\rangle - i|o\rangle$ respectively. The ideal circulator transmission, i.e. complete transmission at the output port and zero transmission at the isolated port, can be achieved at a frequency of $\omega_- + \gamma_- / \sqrt{3}$ when the resonance splitting satisfies $|\omega_+ - \omega_-| = (\gamma_+ + \gamma_-)/\sqrt{3}$. This condition can be realized by varying the magnetization strength, i.e. by adjusting ε_a for specific values of $\gamma_{\pm}$.

To confirm the nonreciprocal transport properties, we perform finite-difference time-domain (FDTD) simulations with perfect matched boundary conditions [6]. The imaginary part of the off-diagonal elements ε_a is set as 0.02463 to match with the quality factor of 365.6 for the cavity modes, so that the magneto-hybridized modes oscillates at a frequency of 0.346508 *(c/a)* and 0.347055 *(c/a)* respectively. As shown in Figure 2.b, the transmission spectra of the device through FDTD calculations demonstrate nearly ideal three-port circulator characteristics and agree nicely with the coupled-mode theory prediction. In a small footprint of a few wavelengths squared, this device demonstrates a bandwidth of 115GHz for an extinction ratio of 20dB and 35GHz at a 30dB extinction ratio operated at 1550nm wavelength.

The non-reciprocal transport process is also evident from examining the corresponding field patterns when a continous wave is input into the device. The wave is excited at the frequency 0.346770*(c/a), where maximum isolation occurs.* For a non-magneto resonator, the light is transmitted to both output ports (Figure 3a). In contrast, the magneto-resonator shows complete transmission from the input port to the output port and zero transmission to the isolated ports (Figure 3.b). Moreover, reflected light back into the output port is completely dropped to the isolated port (Figure 3c). By using absorbing materials at the isolated port, the input port is therefore completely free from the back reflections from the output port.

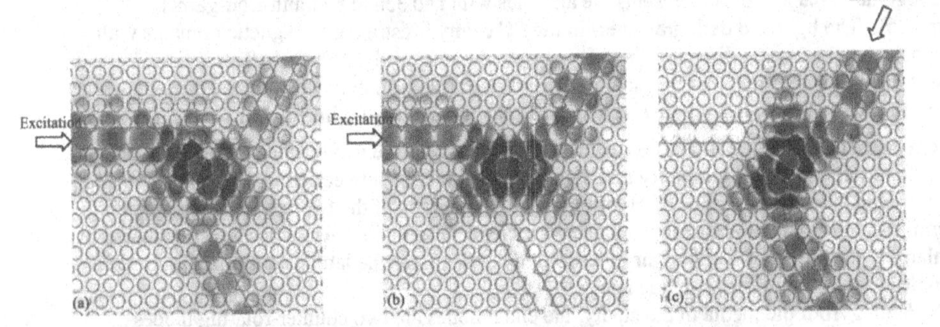

Figure 3 Field patterns of the three-port junction circulator excited at the maximum isolation wavelength under various conditions: (a) non-magneto cavity ε_i=0, excited at the input port; (b) magneto-cavity ε_a=0.02, excited at the input port; (c) magneto-cavity ε_a=0.02, excited at the output port. Black circles represents the air holes in bismuth-iron-garnet and the gray areas are regions with large out-of-plane magnetic fields.

Practically, the proposed structure can be implemented using BIG thin films, which exhibit strong gyrotropy with ε_a up to 0.06 [3,7]. From the coupled-mode theory, the bandwidth of the circulator scales with the magneto-optical splitting between the modes. Hence the maximum theoretical bandwidth of such photonic crystal circulator can be up to 87GHz for a 30dB extinction ratio. In the mean time, the quality factor of the resonator can be reduced to 140. We also note that such devices can be similarly designed in 2d PC slabs, because the radiation loss can be designed to be orders of magnitude lower than the waveguide coupling loss [8,9].

FOUR-PORT JUNCTION CIRCULATORS

Replacing the regular reciprocal resonator in a channel add/drop filter with a gyrotropic cavity, we transform this device into a four-port circulator [10]. As shown in Figure 4, the forward wave from the input port experiences a complete transmission via the mode $|e\rangle + i|o\rangle$ to the output port at resonant frequency ω_+. The reversed transfer occurs through the counter-rotating mode $|e\rangle - i|o\rangle$, at a frequency detuned by $2V_{oe}$. When the frequency separation $2V_{oe}$ is much greater than the resonance linewidths $\gamma_\pm$ in the limit of strong magneto coupling, a large extinction ratio can be produced between the two ports. This can be seen in the transmission coefficients derived from the coupled mode theory as:

$$T_{in\rightarrow out} = \left|1 - i\gamma_+/(\omega - \omega_+ + i\gamma_+)\right|^2,$$
$$T_{out\rightarrow in} = \left|1 - i\gamma_-/(\omega - \omega_- + i\gamma_-)\right|^2. \quad (3)$$

Such four-port circulator characteristics is inverted when the signal frequency is near the lower resonance ω_-. We further performed numerical FDTD calculation to confirm the coupled-mode theory. The point defect is created by replacing an air hole with an air ring with an inner radius of $0.55a$ and an outer radius of $0.9a$. By itself, the point defect supports two degenerate quadrupole modes at a frequency of $0.340(c/a)$. We incorporate magneto-optical material of ε_a=0.06 into the crystal with a domain pattern shown in Figure 4.a, the two quadrupole modes

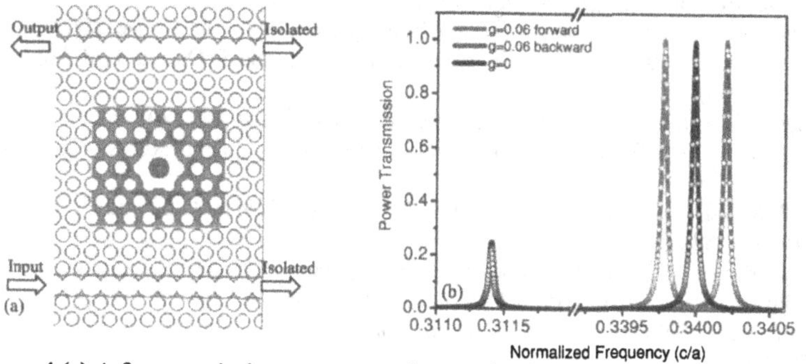

Figure 4 (a) A four-port isolator constructed as a point defect side-coupled to two waveguides. The cyan and magenta colors illustrate the magnetic domains with opposite magnetization direction (both being out-of-plane) with ε_a=0.06 (b) Transmission spectra for a magnetized cavity and a non-magnetized cavity. The finite-difference time-domain spectra (dots) agree well with the simple coupled-mode theory analysis (solid curves).

strongly couple to each other and hybridize into the two counter-spinning modes at a frequency of $0.340208(c/a)$ and $0.339778(c/a)$ respectively. The corresponding forward and backward transmission peaks evidently agree well with the couple-mode theory, as shown in Figure 4.b. They provide maximum isolation of 25.6dB on resonance. On the other hand, the non-magneto-cavity shows a single reciprocal resonance peak transmitting light in both directions.

Practically, in magneto-optical devices, magnetic domains less than 100nm have been artificially prepared[11,12], which represents a scale sufficient for photonic crystals operating at 1.55μm for optical communication purposes.

CONCLUSIONS

We show that the magneto-optical hybridization in photonic crystal defect states can be achieved by aligning the magnetic domains with the modal cross-product. The hybridized modes, when coupled with waveguides, can transport energy and signal uni-directionally. As an example, we demonstrate a three-port optical circulator with a 30dB extinction bandwidth of 35GHz at 1550nm wavelength in a 2D bismuth-iron-garnet crystal. A four-port circulator is also shown with a 25.6 dB maximum extinction ratio.

REFERENCE

[1] I. L. Lyubchanskii, I.L. Lyubchanskii, N.N. Dadoenkova, M.I. Lyubchanskii, E.A. Shapovalov and T.H. Rasing, J. of Phys. D **36**, 277-87 (2003).
[2] A. G. Gurevich and G. A. Melkov, *Magnetization Oscillations and Waves*,(CRC Press,1996, Chapter 4).
[3] T. Tepper and C. A. Ross, J of Cryst. Growth **255**, 324-331(2003).
[4] H. A. Haus, *Waves and fields in optoelectronics*, (Prentice-Hall, 1984)
[5] D. M. Pozar, *Microwave Engineering*, (Wiley, 1998, Chapter 9)
[6] A. Taflove and S. Hagness, *Computational Electrodynamics*, (Artech House, 2000)
[7] N. Adachi, V. P. Denysenkov, S. I. Khartsev, A. M. Grishin and T. Okuda, J. of Appl. Phys. **88**, 2734-2739 (2000).
[8] Y. Akahane, T. Asano, B. S. Song and S. Noda, Nature **425**, 944-947 (2003)
[9] M. Notomi, A. Shinya, S. Mitsugi, E. Kuramochi and H.Y. Ryu, Opt. Express **12**, 1708-1719 (2004)
[10] S. Fan, P. R. Villeneuve, J. D. Joannopoulos, M. J. Khan, C. Manolatou and H. A. Haus Phys. Rev. B **59**, 15882 (1999).
[11] A. K. Zvezdin and V. A. Kotov, *Modern Magnetooptics and Magnetooptical Materials*, (Institute of Physics Pub., Bristol ; Philadelphia, PA, 1997)
[12] A. Hubert and R. Schafer, *Magnetic domains : the Analysis of magnetic microstructures* (Springer, Berlin, 2000).

Mater. Res. Soc. Symp. Proc. Vol. 846 © 2005 Materials Research Society DD12.11

OPTICAL PROPERTIES OF POLYSTYRENE OPALS INFILTRATED WITH CYANINE DYES IN THE FORM OF J-AGGREGATES

F. Marabelli[1], D. Comoretto[2], D. Bajoni[1], M. Galli[1], and L. Fornasari[1]
[1] INFM-Dipartimento di Fisica "A. Volta", Università degli Studi di Pavia, via Bassi 6, 27100-Pavia (Italy)
[2] INFM-Dipartimento di Chimica e Chimica Industriale, Università degli Studi di Genova, via Dodecaneso 31, 16146-Genova (Italy)

ABSTRACT

We report preliminary data on the infiltration of polystyrene opals with cyanine dye water solutions, which are known to form J-aggregates. Different approaches are used to infiltrate cyanines inside opals. The degree of infiltration is evaluated by scanning and transmission electron microscopy as well as micro reflectance spectroscopy. Data show that in spite of the low wettability of polystyrene with water solutions, some infiltration can be achieved. Even though this degree of infiltration is still not enough to observe strong modification on the photonic band structures, further strategies to improve it are in progress.

INTRODUCTION

Photonic crystals are materials having a periodical and regular modulation of the dielectric constant. This modulation can be achieved in one, two or three dimensions giving rise to different structures, which control the propagation of the electromagnetic waves [1]. Among three dimensional structures, artificial direct- and inverse- opals have been extensively studied [2] due to the relatively easy growth techniques, which makes them the prototype model for the investigation of fundamental physical properties of 3D photonic crystals [3]. More recently, the possibility to modulate the photonic properties of such photonic crystals by infiltration with photoactive materials has attracted a deep attention [4]. The modulation of optical properties of silica opals upon infiltration with organic dyes [5, 6] and inorganic materials [7] was recently reported. However, as far as we know, few data are reported on the infiltration of polystyrene opals. As a matter of fact, the infiltration process must satisfy different requirements: the solvent for the photoactive molecules must be a non solvent for polystyrene spheres. Moreover, wettability of opal surface by this liquid may prevent the infiltration process. In this case, specific strategies must be developed. In spite of these problems, since high quality polystyrene opals can be grown, and organic molecules possess high linear and nonlinear polarizability useful to be combined with photonic crystal properties, the investigation of the infiltration with conjugated dyes is appealing since it allows to prepare full organic three dimensional infiltrated photonic crystals. Recently, we have grown high quality polystyrene opals, which allowed the investigation of new physical phenomena related to their photonic band structure [8]. Now, we extend our work to the infiltration process. Aim of this work is to report our preliminary results on the infiltration of polystyrene opals with carbocyanine, which may form, in the suitable conditions, J-aggregates. These fluorescent aggregates exhibit very sharp features on the optical spectra [9], which may interact with the photonic crystal band structure [10].

EXPERIMENTAL

Polystyrene artificial opals with different sphere diameters (222, 260, 300, 340 and 426 nm, refractive index 1.59) are grown from commercial monodisperse sphere suspensions (Duke Scientific). Two different geometrical configurations are employed to grow films and bulk opals. In the first case, two glass or fused silica windows are sandwiched with a spacer and then sphere suspension is inserted from the top allowing the sedimentation of the particles on the vertical walls for several days. After deposition and drying, the two glasses are separated living the opal films on either walls. Bulk opals are obtained by filling with the sphere suspensions a glass/polystyrene tube with a window sealed at the bottom and again by allowing deposition and drying.

1,1'-Diethyl-2-2'-cyanine iodide (PIC) and polyvinylsulphonate (PVS) were purchased from Aldrich and used without any further purification. Deionized water solutions of PIC were mixed at various volumes with water solutions of PVS. These mixtures did not show any evidence of J-aggregate absorption band at about 580 nm. However, drop cast films with these mixtures clearly showed the presence of J-aggregates.

The infiltration process were afforded by different approaches:

- PIC/PVS solution was dropped onto the opal surface;
- opals were dipped in a vial containing PIV/PVS solutions and then air was gently drained away from the vial in order to remove it from the opal interstices. The process was stopped at the beginning of boiling and then repeated several times.
- PIC/PVS solution was dropped onto the opal surface and air were sucked from the back opal surface with a pump in order to promote infiltration;
- PIC/PVS solutions were mixed with monodisperse polystyrene suspension and then opals were grown starting from these new suspensions.

A Leo Stereoscan 440 by LEO Electron Microscopy Ltd was used for Scanning Electron Microscopy (SEM) measurements. Transmission Electron Microscopy (TEM) was performed by a JEOL JEM 2010 (JEOL Ltd) equipped with energy dispersive spectrometry (EDS) Oxford Link Si(Li) probe.

Variable-angle specular reflectance was measured in the 0.4-4 eV spectral range by means of a Fourier-Transform spectrophotometer Bruker IFS66. The light of a Xe arc-lamp was collimated and then focused to a spot of 100 μm diameter on the sample surface. The sample was mounted on a home-made θ-2θ goniometer, that allows incidence angles to be varied between 5 and 70 degrees, with an angular resolution of 1 deg.

RESULTS AND DISCUSSION

The first two infiltration procedures did not provide good results. As a matter of fact, the wettability of polystyrene with water is very poor and it is therefore difficult to achieve opal infiltration. However, when PIC/PVS solution drops cast on the opal dry, they may live solute materials on the surface. This can be observed by the existence of few small red spots on the opal surface. When these opals are observed at the SEM, we notice that the sample surface has been partially disordered by the infiltration process and that the solute coverage is very limited. The solute deposited after solution evaporation on an opal fracture surface shows two different

Fig. 1. SEM images of a 260 nm opal infiltrated by drop casting with PIC/PVS solutions. (a) 50Kx of a solute drop; (b) 25Kx of several rod-like aggregates.

morphologies reported in Fig. 1. In particular, Fig. 1a shows a spot of PIC/PVS resembling a jam covering the surface. Similar kind of coverage was already observed in silica opals infiltrated with polydiacetylenes where the polymer solvent poorly wet silica [11]. A completely different morphology is instead observed in other regions of the same sample. Fig. 1b shows the presence of long rods having about 100 nm diameter and length up to 10 μm. Rods seems to lean on the surface instead of stemming from the bulk opal thus confirming that this infiltration procedure failed. The reduced dimension of the spots covering the surface and of rods prevented the EDS analysis of their constituents. This is particularly puzzling for rods since they can be due both to the formation of giant J-aggregates or to molecular crystals.

More promising seems to be the procedure to grow opals directly from sphere suspensions added with PIC/PVS solution. When observed by naked eye, these opals have an appearance very different with respect to the bare ones. Bare opals grown with 260 nm diameter spheres show bright golden and green reflections depending on the viewing angle. Infiltrated samples retain such features, but for a pink background due to diffuse light scattering. Due to this remarkable color change, some degree of infiltration is expected for these samples. A qualitative estimate of the infiltration can be obtained from reflectance spectra. Fig. 2a shows the comparison of the near normal incidence reflectance spectrum for bare and infiltrated opals. The stop band of bare opal is peaked at about 2.15 eV and shows a relatively high reflectance (50% relative to an aluminum mirror). The reflectance spectrum of infiltrated opals shows a peak at 2.1 eV and possesses a more symmetric lineshape. It is well known that the spectral position of the stop band is bathochromically shifted upon infiltration due to the reduced dielectric contrast between spheres and voids, which are now partially filled and then possess a dielectric constant higher than that of air. In spite of the remarkable change in color observed by naked eye, a minor change is observed in the reflectance spectra of bare and infiltrated opals. Even though the J-aggregate optical transition at 2.14 eV is overlapped to the photonic stop band (Fig. 2b), the expected effect, a dip in the reflectivity peak due to light propagation inside the photonic stop band [5,10], was not observed.

Let's now analyze in more detail the spectral results. The different line shape and intensity of the reflectance spectra can be assigned to the different quality of the surface for the two samples. As a matter of fact, our experience shows that the shape of the reflectance spectra may be affected by disorder [8,11,12]. The spectral shift upon infiltration, related to a change in the dielectric contrast, is very limited, indicating a reduced infiltration of the opals in striking contrast with the color appearance. The pink color of infiltrated opals is due to diffusion of light which is not noticeable in specular reflectance. Similar results have also been detected for opals infiltrated by sucking air from their back surface.

In order to evaluate the degree of infiltration of these samples, we performed a detailed electron microscopy analysis. Fig. 3a shows a SEM image of a 260 nm opal grown with PIC/PVS solution. The opal surface shows more defects than that of bare bulk opals [3,11,12]. In particular, several spheres seem to be removed and displaced from the surface; but what is

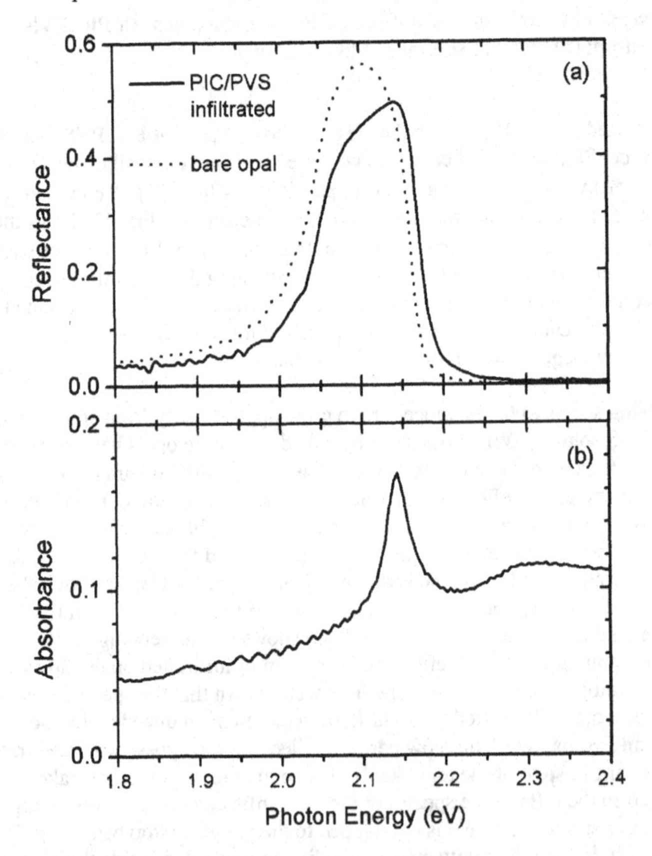

Fig. 2. (a) Near-normal incidence reflectance spectra of PIC/PVS infiltrated (full line) and bare (dashed line) opals. (b) Absorbance spectrum of PIC/PVS J-aggregates cast film.

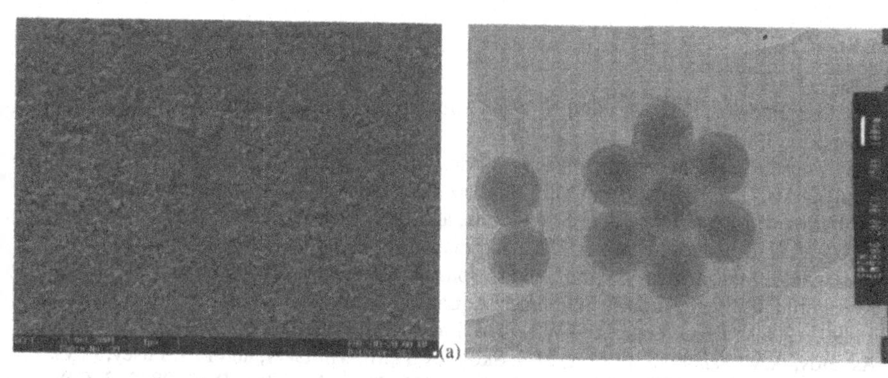

Fig. 3. (a) SEM image (10Kx) of a 260 nm opal infiltrated with PIC/PVS. (b) TEM images of the same film.

puzzling is that there is no evidence of PIC/PVS on the surface. Moreover, when those samples are broken and the crack surface is observed, no evidence of infiltrated material is found too. Since somewhere PIC/PVS molecules have to be found, we performed a TEM analysis of the same sample (Fig. 3b). It is now evident that PIC/PVS molecules infiltrated the sample and partially filled the interstices between spheres.

EDS analysis confirms that the elemental composition of these films correspond to PIC and PVS molecules. Moreover, preliminary analysis seem to indicate that PIC/PVS film is in an ordered state since an electron diffraction patterns corresponding to 0.6 Å spacing can be detected, even though organic film quickly degrades upon electron irradiation thus preventing a detailed analysis of these structures.

These data suggest that more concentrated PIC/PVS solutions would be necessary to achieve a proper degree of infiltration. However, with concentrated solutions the formation of large molecular aggregates may be favored, which may close the surface interstices, and then stop the infiltration process.

CONCLUSIONS

In conclusion, we showed that in spite of the low wettability of polystyrene with water solution, we succeeded to infiltrate polystyrene opals with PIC/PVS by growing opals starting from nanosphere suspensions where molecules were also dissolved. We are currently working to improve the infiltration degree by optimizing the PIC/PVS concentration..

ACKNOWLEDGMENTS

D.C. acknowledges support from the Ministry of the University and Scientific and Technological Research through the PRIN 2004 project. Technical help of M. Michetti and C. Uliana for the electron microscopy images is also acknowledged.

REFERENCES

1 J.D. Joannopoulos, R.D. Meade, J.N. Winn *Photonic Crsytals*, Princeton University Press, Singapore, 1995.
2 See for instance recent papers by Yu. A. Vlasov, V.N. Astratov, A.V. Baryshev, A.A. Kaplyanskii, O.Z. Karimov, and M.F. Limonov, Phys. Rev. E 61, 5784 (2000); H.P. Schriemer, H.M. van Driel, A.F. Koenderink, and W.L. Vos, Phys. Rev. A 63, 011801 (2000); S.G. Romanov, T. Maka, C.M. Sotomayor Torres, M. Muller, R. Zentel, D. Cassagne, J. Manzanares-Martinez, and C. Jouanin, Phys. Rev. E 63, 056603 (2001); A.F. Koenderink, L. Bechger, H.P. Schriemer, A. Lagendijk, and W.L. Vos, Phys. Rev. Lett. 88, 143903 (2002); J.F.G. Lopez and W.L. Vos, Phys. Rev. E 66, 036616 (2002); V.N. Astratov, A.M. Adawi, S. Fricker, M.S. Skolnick, D.M. Whittaker, and P.N. Pusey, Phys. Rev. B 66, 165215 (2002); H.M. van Driel and W.L. Vos, Phys. Rev. B 62, 9872 (2000). W.L. Vos and H.M. van Driel, Phys. Lett. A 272, 101 (2000); H. Miguez, V. Kitaev, and G.A. Ozin, Appl. Phys. Lett. 84, 1239 (2004); J.F. Galisteo-Lopez, E. Palacios-Lidon, E. Castillo-Martinez and C. Lopez, Phys. Rev. B68, 115109 (2003); and references therein.
3 C. Lopez, Adv. Mater. 15, 1679 (2003); D.J. Norris, E.G. Arlinghaus, L. Meng, R. Heiny, and L.E. Scriven, Adv. Mater. 16, 1393 (2004). Y. Xia, B. Gates, and Zhi-Yuan Li, Adv. Mater. 13, 409 (2001).
4 See for instance *Tuning the Optical Response of Photonic Bandgap Structures*, SPIE Proc. 5511 (2004).
5 N. Eradat, A.Y. Sivachenko, M.E. Raikh, Z.V. Vardeny, A.A. Zakhidov, and R.H. Baughman, Appl. Phys. Lett. 80, 3491 (2002).
6 M. Deutsch, Y.A. Vlasov, D.J. Norris, Adv. Mater.12, 1176 (2000).
7 K. Sumioka, H. Nagahama, and T. Tsutsui, Appl. Phys. Lett. 78, 1328 (2001).
8 E. Pavarini, L.C. Andreani, C. Soci, M. Galli, F. Marabelli, and D. Comoretto, Phys. Rev. B, submitted. D. Comoretto, E. Pavarini, M. Galli, C. Soci, F. Marabelli, L.C. Andreani, SPIE Proceedings 5511, 135 (2004).
9 *J-Aggregates*, edited by T. Kobayashi, World Scientific, Singapore, 1996.
10 A.Y. Sivachenko, M.E. Raikh, and V.Z. Vardeny, Phys. Rev A64, 013809 (2001).
11 D. Comoretto, F. Marabelli, C. Soci, M. Galli, M. Patrini, E. Pavarini, and L.C. Andreani, Synth. Met. 139, 633 (2003).
12 D. Comoretto, R. Grassi, F. Marabelli, and L.C. Andreani, Mater. Sci. and Engin. C 23, 61 (2003).

AUTHOR INDEX

Abe, Kenichi, 127
Abruña, Hector D., 301
Adachi, Chihaya, 73
Ågren, H., 3
Akino, Nobuhiko, 27
Aoki, Atsushi, 127
Arkhipov, V.I., 141, 183
Asahi, Tsuyoshi, 263

Bae, Byeong-Soo, 251
Baev, A., 3
Bajoni, D., 333
Barnakov, Y., 231
Berry, Rebecca, 301
Bhatambrekar, Nishant, 115, 121
Blanchet, Graciela B., 159
Boilot, Jean-Pierre, 171
Boothby, Clare E., 165
Borghs, G., 141, 183
Botek, Edith, 13
Bozio, Renato, 39
Brett, Michael J., 281
Brooks, Jason, 73
Brown, Julie J., 73
Buissette, Valérie, 171
Buker, Nicholas, 107
Bukhari, Alma, 301

Canteenwala, Taizoon, 275
Carpenter, Michael A., 269
Champagne, Benoît, 13
Chan, Khai Leok, 165
Chan, Sherina, 301
Chen, Aiqing, 81
Chen, Antao, 115
Cheyns, D., 183
Chiang, Long Y., 275
Cho, Sung Yong, 165
Choi, Kyung M., 207
Clot, Olivier, 115, 121
Collini, Elisabetta, 39
Comoretto, D., 333
Cooper, Thomas M., 47

Dadson, J.B., 231
Dalton, Larry R., 107, 115, 121
Damour, Megan, 301
Deutsch, Miriam, 81

Efstathiadis, Harry, 269
Eliseev, Andrei A., 237
Evans, Nicholas R., 165

Fan, Shanhui, 327
Ferrante, Camilla, 39
Firestone, Kimberly A., 107
Fleitz, Paul A., 47
Flores-Torres, Samuel, 301
Fornasari, L., 333
Friend, Richard H., 165
Friguglietti, Leigh, 301

Gacoin, Thierry, 171
Gaillot, D., 315
Galli, M., 333
Gel'mukhanov, F., 3
Genevich, Joseph, 301
Genoe, J., 141, 183
Goldberg, Velda, 301
Golding, T., 193
Gorman, B., 193
Goushi, Kenichi, 73
Graugnard, E., 315
Grunlan, Jaime C., 289
Gunn, Erica, 301

Hall, Benjamin C., 47
Haller, Marnie, 107
Hammond, Scott R., 115, 121
Hao, X.T., 307
Hasegawa, Keisuke, 81
Hauck, Tanya, 275
Hayer, Anna, 165
Heremans, P., 141, 183
Ho, Karen, 301
Holder, Elisabeth, 67, 295
Holmes, Andrew B., 165
Huang, Heh-Lung, 243
Hunter, D., 231

Janssen, D., 183
Jen, Alex K.-Y., 53, 107, 115, 121

Jhu, Miao-Cai, 243
Johnson, Ashley, 301
Jura, Marek, 59

Kaloyeros, Alain E., 269
Kamada, Kenji, 3, 13
Kaneko, Futao, 99
Kaneto, Keiichi, 151
Kang, Dong Jun, 251
Kaplan, Michael, 301
Karna, Shashi P., 133
Kasai, Hitoshi, 257, 263
Kato, Keizo, 99
Kawamura, Yuichiro, 73
Kim, Jin-Ki, 251
King, J.S., 315
Kirakosyan, Arman S., 93
Kishi, Ryohei, 13
Koentjoro, Olivia F., 59
Köhler, Anna, 165
Koshizaki, Naoto, 223
Kozodaev, Dmitry, 67, 295
Kubo, Takashi, 13
Kynast, Ulrich, 237

Lao, David, 107
Lee, Hee Hyun, 159
Lellig, Christoph, 177
Le Mercier, Thierry, 171
Li, J., 193
Li, Y.Q., 307
Liddell, Chekesha M., 201
Lin, Wenbin, 53
Lin, Yin Yin, 301
Liu, Jia-Ming, 243
Lohmeijer, Bas G.G., 67
Lowenthal, Alisabet, 301
Lukashin, Alexey V., 237

Mak, Chris S.K., 165
Malliaras, George G., 301
Marabelli, F., 333
Marin, Veronica, 67, 295
Masuhara, Hiroshi, 263
Matsui, Jun, 127
Matsune, Hideki, 263
McLean, Daniel G., 47
Meier, Michael A.R., 67, 295
Menard, Etienne, 159
Minkov, I., 3
Mitsuishi, Masaya, 127
Miyashita, Tokuji, 127
Mohanty, S., 231
Moreau, Mélanie, 171
Morkoc, H., 193

Nakagawa, Nozomi, 13
Nakanishi, Hachiro, 257, 263
Nakano, Masayoshi, 13
Nakasuji, Kazuhiro, 13
Neff, C., 315
Neogi, A., 193
Neogi, P.B., 193
Nichols, William T., 223
Nitta, Tomoshige, 13
Niu, Yu-Hua, 53

Ohdaira, Yasuo, 99
Ohta, Koji, 13
Oikawa, Hidetoshi, 257
Ong, K.S., 307
Onodera, Tsunenobu, 257
Osgood Jr., R.M., 87

Padmawar, Prashant A., 275
Panoiu, N.C., 87
Pineda, Andrew C., 133
Polyutov, S., 3
Pradhan, A.K., 231
Pritzker, Kenneth P.H., 275
Purvis, Lafe, 107
Pustovit, Vitaliy N., 213

Raithby, Paul R., 59
Reuter, Heike, 177
Reynaert, J., 141, 183
Rivera, J.M., 321
Robinson, Bruce, 115, 121
Rogers, John A., 159, 207
Rogers, Joy E., 47
Rohde, Charles, 81
Rommel, Harrison L., 115, 121

Sarkar, A., 193
Sasabe, Hiroyuki, 73
Sasaki, Takeshi, 223
Schlabach, Sabine, 177
Schubert, Ulrich S., 67, 295
Sekiguchi, Takashi, 257
Shahbazyan, Tigran V., 93, 213
Sharp, Emma L., 59
Shen, Kou-Hui, 243
Shinbo, Kazunari, 99
Siddique, Rezina, 269
Sinness, Jessica, 115, 121
Sirinakis, George, 269
Sit, Jeremy C., 281
Slagle, Jonathan E., 47
Slinker, Jason D., 301
Snoeberger, Robert, 107
Soltzberg, Leonard, 301
Song, Yanning, 201
Sorge, Jason B., 281
Subramania, G., 321
Summers, C.J., 315
Sun, Richard, 269
Suth, Seiyan, 301
Szabó, Dorothée V., 177

Tai, Oliver Y.-H., 33
Takashima, Wataru, 151
Tan, L.W., 307
Tassi, Nancy G., 159
Tekin, Emine, 67
To, Regina, 301
Toyoshima, Susumu, 99
Tretyakov, Yuri D., 237
Tseng, Mei-Rurng, 243

Ujimoto, Mitsuru, 151

van Popta, Andy C., 281
Verma, Sarika, 275
Vollath, Dieter, 177

Wang, C.H., 33
Wang, Yuxiao, 33
Wang, Zheng, 327
Watkins, Scott E., 165
Williams, Charlotte K., 165
Wilson, Paul J., 59

Yamada, Satoru, 13
Yamaguchi, Kizashi, 13
Yamashita, T., 315
Yoon, Jong-Won, 223
Yopak, Regina, 301

Zempo, Yasunari, 27
Zhang, Kai, 231
Zhang, Lin, 53
Zhu, F.R., 307
Zhuravleva, Natalia G., 237

SUBJECT INDEX

acrylamide polymer, 127
all optical poling, 33
atomic layer deposition, 315

bismuth iron garnet, 327

C_{60} dyad conjugate, 275
ceramics, 231
chiral, 53
chromophore, 115
    synthesis, 121
conjugated polymer, 27, 165
contact-limited, 141
crystal, 141

degradation, 301
dendrimers, 121
depletion layer, 151
discotic, 115
di-yne, 59

electrochromic, 289
electron-beam lithography, 321
electro-optic(s), 107, 115, 121
emission light, 99
erbium ion doped glasses, 207

flexible substrate, 307

gallium nitride, 133
GaN quantum dots, 193
glancing angle deposition, 281
gold nanoparticles, 269

holographic interference, 251
hybrid semiconductors, guanosine, 193

infiltrated opals, 333
inkjet printing, 67
inverse opal, 315
iridium, 243
    complex, 73
iridium(III) complexes, 67

J aggregates, 39

Langmuir-Blodgett film, 127
large area, 159
laser amplifier materials, 207
layer-by-layer, 289
LDH, 237
LED(s), 53, 165
light-emitting, 183
    iridium(III) complex, 295
luminescence, 171, 177, 237
luminescent nanoparticles, 231

material, 243
memory effect, 151
metal, 81
microcavity, 307
microcontact printing, 159
micro-optcal device, 251
molecular luminescence, 99
morphology, 295

nanocomposite(s), 177, 269
    electrodes, 223
    materials, 171
nanocrystals, 171
nanoparticle(s), 93, 213
nanoshell, 81
nitro-stilbene derivative, 3
NLO, 107
nonlinear
    optical properties, 133
    optics, 13, 107
non-spherical, 201

OLED, 243, 301
open-shell diradical, 13
optical
    circulator, 327
    nanowires, 87
    nonlinearities, 87
    response, 27
    spectroscopy, 59
optics, 81
organic, 141
    light emitting diode, 73
    nanoparticle, 263
oxide/anthracene/PMMA, 177

phosphorescence, 165
photocurrent, 127
photo-HYBRIMER, 251
photoluminescence, 73
photonic crystal(s), 193, 315, 321, 327, 333
plasmon modes, 87
plastic electronics, 159
platinum, 59
  poly-ynes, 47
polyalkylthiophene, 151
polydiacetylene, 257
polyethyleneglycolated C_{60}, 275
polymer film, 33
polymeric terpyridine-PEG macroligand, 295
polystyrene opals, 333
porphyrin, 39
p-quinodimethane, 13
property screening, 67
Pt/TiO_2, 223
pulsed laser ablation and sputtering, 223

Raman scattering, 213
rare earth
  doped glasses, 207
  picolinate, 237

scanning electron microscopy, 257
second harmonic generation, 33
self-assembled fullerene, 275
silicon, 321
silver, 257
single particle spectroscopy, 263
size effect, 263
spectroscopic ellipsometry, 269, 281
square, 53
structure-property relationship, 47
supernatant, 201
surface plasmon, 213
  excitation, 99

TDDFT, 27
tetracene, 183
thin films, 231
time dependent Hartree-Fock, 133
titanium dioxide, 281
top-emitting OLED, 307
transistor, 183
transition metal complexes, 301
triplet excited state absorption, 47
tungstate, 289
two-photon absorption, 3, 39

ultrafast spectroscopy, 93

vibrational modes, 93
vibrations, 3

zinc sulfide, 201

Printed in the United States
By Bookmasters